OPEN SOURCE GIS

A GRASS GIS Approach
Third Edition

OPEN SOURCE GIS
A GRASS GIS Approach
Third Edition

by

Markus Neteler
FBK-irst & CEA, Trento, Italy

Helena Mitasova
North Carolina State University, USA

 Springer

Markus Neteler
FBK-irst, Istituto per la Ricerca
  Scientifica e Tecnologica
Via Sommarive,18
38050 Trento, ITALY

CEA, Centre for Alpine Ecology
38100 Viote del Monte Bondone
Trento, Italy
neteler@osgeo.org

Helena Mitasova
North Carolina State University
Department of Marine, Earth
  and Atmospheric Sciences
Raleigh NC 27695-8208, USA
hmitaso@unity.ncsu.edu

ISBN-13: 978-1-4419-4206-7   e-ISBN-13: 978-0-387-68574-8

Printed on acid-free paper.

9 8 7 6 5 4 3 2 1

springer.com

to our friends and to all GRASS developers, present and past

Foreword

GRASS GIS software was developed in response to the need for improved analysis of landscape "trade offs" in managing government lands and the emerging potential of computer-based land analysis tools. During the last decades of the 20th century, government land managers in the U.S. (and across the world) faced increasing requirements from legislation and stakeholder groups to examine and evaluate alternative actions. To fulfill these new requirements, land managers needed new tools.

During this same era, computational capabilities wondrously improved. Tasks requiring days and months with paper and acetate overlays could be accomplished with this newly emerging geographic information technology within minutes. But even in the mid-1980s, GIS technology involved significant capital investment. Managers wanted to see results before they spent their limited funds on new technologies.

The U.S. Army Construction Engineering Research Laboratory (CERL) in Champaign, Illinois has the mission of developing and infusing new technologies for managing U.S. Department of Defense installations. These installations include millions of acres of lands needed for military training and testing. Other uses included wildlife management, hunting and fishing and forestry, grazing and agricultural production. Other priorities were added through legislation – such as protecting endangered species and habitats, protecting cultural sites, and limiting the on and off-post impacts of noise, ordnance, contaminants and sediments.

Military land managers were unable to cope with the challenge of examining proposed new actions (such as new weapon firing ranges or new vehicle training routes) without improved methods to gather, integrate and visualize their data and to examine alternative courses of action. Acquiring emerging proprietary technologies and digital data wasn't even a consideration – the cost was too high and the expertise required to learn, operate and manage the technology was beyond their resources.

Given this need, a group of then young researchers at CERL elected to develop their own set of initial landscape analysis tools. Initially, this in-house software development effort was designed to "bridge the gap" as commercial proprietary technology developed. The other costs involved in implementing GIS (acquiring data and hardware, learning GIS skills and computer maintenance skills) were so high; CERL decided that no-fee software could reduce the technology hurdle involved in implementing GIS. This proved to be true – and U.S. military installations were some of the first government managers to become active users of this new technology.

Once our efforts began, software development took on a life of its own. The Open Source code and Internet accessible software soon sparked the creative energies of numerous other organizations and individuals, and many began to use GRASS and contribute capabilities. At CERL, a small-scale skunk works project became the biggest and hottest program in the lab. Dozens of persons were employed developing new tools, building digital databases, assisting with complex applications and fielding the technology across the Department of Defense.

The needs we addressed drove the design criteria for GRASS. Because of the requirement to analyze alternative actions and to evaluate impacts of actions on continuous surfaces of differing elevations and vegetation and soil types, GRASS development was focused on raster analysis tools. Also, because of the need for digital and "real time" data, GRASS also incorporated remotely sensed image integration and analysis tools. At the time, this focus set GRASS apart from marketplace capabilities, which were primarily based on vector data and tools and did not include image analysis.

To nurture a "growing" GRASS community, CERL and other organizations established forums for sharing and contributing software. For several years, the lab (and lab partners) also offered newsletters, developed formal interagency partnerships (primarily with the U.S. Department of Agriculture and National Park Service) and held annual software user meetings. During the early 1990s, this GRASS community helped to initiate the Open GIS Foundation (now the Open GIS Consortium) as an international organization focused on advancing openness and interoperability for geospatial technologies.

But by the mid-1990s, many of the original military installation GIS users were switching to proprietary marketplace GIS technologies. In the intervening years, marketplace GIS vendors had added raster analysis tools, much like those in GRASS. Installation managers had become dependent on GIS, and were now willing to buy from the marketplace. Generally, the government is expected to buy off the marketplace, unless there are no comparable marketplace options. Plus, installation managers wanted GIS software just like the systems that were showing up in the offices of supporting contractors and local and state government offices across-their-fence lines. As a result, CERL managers decided they had achieved their purpose of "bridging the gap" in introducing this new technology. CERL entered into agreements with GIS vendors, and helped installations transition their data to proprietary systems. Army research programs were directed to new challenges.

Fortunately, in the years since CERL stopped active development and support of GRASS, the Universities of Hannover (Germany), Baylor, Texas (U.S.A.), and recently the ITC-irst – Centro per la Ricerca Scientifica e Tecnologica (Italy) have continued to coordinate the development of GRASS GIS, performed by a team of developers from all over the world. Thanks to their efforts, GRASS GIS keeps getting better, and valuable and reliable Open Source GIS capabilities are still available through the Internet.

Those of us at CERL are grateful for these academic efforts. GRASS remains an unique capability that continues to play an important role in education and in the advancement of scientific understanding and resource management. The analysis tools within GRASS and the access to source code provide important benefits in our ability to understand and model geospatial phenomena. Plus, developers of this Open Source GIS continue to pioneer and advance capabilities that later emerge in the proprietary geospatial marketplace.

Thanks to the authors, this book should help sustain these important roles for GRASS GIS for years to come.

USA CERL   *William D. Goran*

Preface third edition

Geographical Resources Analysis Support System (GRASS) is one of the largest Free Software Geographical Information System (GIS) projects released under the GNU General Public License (GPL). It combines powerful raster, vector, and geospatial processing engines into a single integrated software suite and includes tools for spatial analysis, modeling, image processing and sophisticated visualization.

With this third edition of *Open Source GIS: A GRASS GIS Approach*, we enter the new era of GRASS 6, the first release that includes substantial new code developed by the International GRASS Development Team. It comes at a time when dramatic growth in acceptance of the Open Source concept fuels further development of Free and Open Source Software for Geoinformatics (FOSS4G) and brings interoperability to a new level of efficiency. The major FOSS4G projects, including GRASS, have become part of the OSGeo foundation – an organization established in 2006 to "support and promote the collaborative development of open geospatial technologies and data." Following the spirit of the foundation, GRASS is tightly integrated with the latest GDAL/OGR and PROJ libraries supporting range of raster and vector formats, as well as projections. GRASS toolkits for Quantum GIS (QGIS) and R Project for Statistical Computing have been developed thanks to strong links with these projects.

The third edition of *Open Source GIS: A GRASS GIS Approach* reflects these new developments. The first chapter includes information about the OSGeo foundation. Chapter three that introduces GRASS and the new sample data set, has added information about the new graphical user interfaces that can be used with GRASS 6. The properties of GRASS raster and vector data are described in chapter four, which also includes extensive material on importing data in various formats, and an introduction to new geocoding tool. The raster chapter has been enhanced with new examples, more comprehensive topographic analysis and modeling, and introduction to voxel data processing. The chapter on vector data has been completely rewritten to reflect introduction of a new vector data format and attribute support through database management system (DBMS) in GRASS 6. This chapter now includes new sections on attribute database management and SQL support, vector networks analysis, linear reference systems, and lidar data applications. The site data chapter of earlier book editions was integrated within the chapter six as vector point data processing section. The visualization chapter reflects the changes in 2D display, nviz, and use of Paraview. Image processing was reduced and updated, orthophoto chapter was eliminated to make space for more new material. Application chapter was merged with raster analysis. Equations and SQLite-ODBC connection guide were added into Appendix. All chapters were enhanced with numerous practical examples using the first release of a free, comprehensive, state-of-the-art geospatial data set. The examples are based on the GRASS 6.3 version from July 2007.

Finally, we briefly recall history of GRASS and this book: GRASS was developed in 1982-1995 by the U.S. Army Corps of Engineers Construction Engineering Research Laboratory (CERL) in Champaign, Illinois to support land management at military installations. After CERL withdrew from further GRASS development in 1995, the GRASS 4.2.1 release, published in 1998, was coordinated by this book's author at the Institute of Physical Geography and Landscape Ecology, University of Hannover. The development of the GRASS 5.0 release started in 1999 when GRASS was released under GPL. Since 2001, the "GRASS Development Team" has its headquarters at FBKITC-irst (Centro per la Ricerca Scientifica e Tecnologica), Trento, Italy. GRASS 5.0.0 was officially released in 2002, accompanied by the first FOSS4G – GRASS users conference held in September 2002 in Trento, Italy, and by the publication of the first edition of this book.

The book has its own history. It started as "GRASS Recipes" written in 1995 for students at the Institute of Landscape Architecture, University of Hannover. In 1996, the first continuous German text was written and later published in "Geosynthesis" series at the Geographical Institute, University of Hannover. The first english edition of the book, published in June 2002, was the result of collaborative work of a number of translators and a new coauthor. It was written for the GRASS 5.0pre3 release. The second edition, published in 2004, was based on the GRASS 5.3 release and included updates reflecting the system enhancements and the feedback from our readers. This third edition is based on GRASS 6 and represents a fundamental update and enhancement of the material.

The GRASS project's Web site, providing access to the GRASS software and documentation, can be reached at "GRASS Headquarters" at http://grass.itc.it and a number of mirror sites. The material related to this book can be accessed at http://grassbook.org.

Trento, Italy   *Markus Neteler*
Raleigh, USA   *Helena Mitasova*
                                                                              August 2007

Acknowledgments

First and foremost, we would like to thank the large number of developers who designed, implemented, and enhanced GRASS over the 25 years of its existence. We especially appreciate the help from the members of the current GRASS Development Team who answered our numerous questions and implemented the bug fixes and improvements that were needed to run the examples included in the book.

This 3rd edition has been substantially rewritten using a new, modern data set that was prepared thanks to agencies providing public access to geospatial data. We are especially grateful to the North Carolina (NC) Center for Geographic Information and Analysis, Wake County GIS, NC State Climate Office, NC Department of Transportation, and USGS for making their data available. Advice and assistance with the data set by Julia Harrell, Silvia Terziotti, Robert Austin, Adeola Dokun, Jeff Essic, and Doug Newcomb, and computer system assistance by Micah Colon are greatly appreciated.

We are grateful to Martin Landa, Trento (Italy), for endless testing, bugfixing and implementation of new vector functionality needed for practical examples included in this book. Roger Bivand, Norwegian School of Economics and Business Administration, provided valuable assistance with "spgrass6" classes in R, and Maurizio Napolitano, Trento (Italy), helped to set up OpenLayers with GRASS and UMN/MapServer. Jaro Hofierka, Presov (Slovakia), helped with the raster and vector chapters as well as Appendix equations. Roberto Antolin, Milano (Italy), gave us advice on applications of the lidar tools in GRASS. We are especially grateful to Aldo Clerici, Parma University (Italy), and Jachym Cepicky (Czech Republic) for their excellent technical comments. Our thanks also goes to Ondrej and Lubos Mitas for their continuous help and advice that contributed to the clarity and technical accuracy of the book.

We greatly appreciate the support of our research work related to this book by Cesare Furlanello and FBKITC-irst – Centro per la Ricerca Scientifica e Tecnologica, Italy, as well as by Russell Harmon from the Army Research Office and by the National Research Council.

Previous support for the GRASS software development by William D. Goran and USA CERL, Douglas Johnston and the University of Illinois Geographic Modeling Systems Laboratory, as well as the University of Hannover, Institute of Physical Geography, is also acknowledged. We are grateful to James Westervelt, Michael Shapiro, David Gerdes and William Brown for major code design of GRASS.

Finally, we would like to express our thanks for patience, encouragement, and assistance to Susan Lagerstrom-Fife, Sharon Palleschi and Springer Author Support.

Contents

1

Open Source software and GIS

Over the past decade, Geographical Information Systems (GIS) have evolved from a highly specialized niche to a technology that affects nearly every aspect of our lives, from finding driving directions to managing natural disasters. While just a few years ago the use of GIS was restricted to a group of researchers, planners and government workers, now almost everybody can create customized maps or overlay GIS data. On the other hand, many complex problems related to urban and regional planning, environmental protection, or business management, require sophisticated tools and special expertise. Therefore the current GIS technology spans a wide range of applications from viewing maps and images on the web to spatial analysis, modeling and simulations.

GIS is often described as integration of data, hardware, and software designed for management, processing, analysis and visualization of georeferenced data. The software component has a major impact on the capabilities to effectively solve a wide range of problems using geospatial data. To ensure the continuous innovation and improvement of the GIS software, existence of diverse approaches to GIS software development is crucial. Besides the widely used proprietary systems, an Open Source GIS plays an important role in adaptation of GIS technology by stimulating new experimental approaches and by providing access to GIS for the users who cannot or do not want to use proprietary products.

1.1 Open Source concept

The idea of Open Source software has been around for almost as long as software has been developed. The results of research and development at the universities and government laboratories have been often made available in the form of Public Domain software packages. Richard M. Stallman first defined the concept of Free Software in form of four freedoms:

0. freedom: The freedom to run the program, for any purpose.
1. freedom: The freedom to study how the program works, and adapt it to your needs.
2. freedom: The freedom to redistribute copies.
3. freedom: The freedom to improve the program, and release your improvements to the public, so that the whole community benefits.

Software following these four principles is called "Free Software". In 1984, Richard M. Stallman started to work on the GNU-Project and in 1985 he created the "Free Software Foundation" to support the Free Software concept. The license of the GNU-Project, the GNU General Public License not only grants the four freedoms described above, but it also protects them. The user of the software is also protected since these freedoms are guaranteed. Because of this protection, the GPL has been the most widely used license for Free Software. The basic idea behind free software is based on the assumption that by allowing the programmers to read, redistribute, and modify the source code, the software evolves: it gets improved, bugs are fixed and capabilities expanded. The ubiquitous availability of the source code and the continuous, often instantaneous peer-review of the code contribute significantly to this process. You can learn more about the ideas behind the Open Source at the Open Source[1] and Free Software[2] Web sites.

Full access to the source code is particularly important for GIS because the underlying algorithms can be complex and can greatly influence the results of spatial analysis and modeling. To fully understand system's functionality, which is not as obvious as it may be, for example, for a word processing software, it is important to be able to review and verify the implementation of a particular function. While an average user may not be able to trace bugs within a complex source code, there is a number of specialists willing to test, analyze and fix the code. The different backgrounds and expertise of these developers and users contribute to the synergethic effects leading to faster and more cost effective software development of a stable and robust product.

The Open Source Geospatial Foundation   Over the past few years a growing number of Open Source GIS, Web mapping, and GPS projects has been established with different goals. Most of them are listed at the "FreeGIS portal" Web site[3]. Smaller projects are usually based on individual developer's initiative, when the lack of available software for a specific application is solved by his own development and the result is then made available to the public on

---

[1] Open Source Web Site, http://www.opensource.org

[2] Free Software pages,
http://www.gnu.org/philosophy/free-software-for-freedom.html

[3] FreeGIS Web Portal, http://www.freegis.org

the Internet. Depending on the level of required expertise, other programmers may join the project and further develop, improve and extend these tools. Some projects are finished quickly, others evolve over time. In general, the Open Source development is very dynamic. The Open Source licenses and the free access through the Internet enable the new contributors take over an abandoned project and continue the development. The overall idea differs significantly from the strategies used in the proprietary GIS development industries.

In February 2006, the Open Source Geospatial Foundation (OSGeo[4]) has been created to support and promote worldwide use and collaborative development of Open Source geospatial technologies and data. It includes GRASS as one of its founding projects. Mature open source geospatial software that undergoes rigorous review of its code and development structure becomes an official OSGeo project. Web mapping systems, desktop applications, geospatial libraries and a metadata catalog are represented. The foundation supports outreach and advocacy activities that promote Open Source concepts and provides financial, organizational and legal help to the broader Open Source geospatial community. It also builds shared infrastructure for improving cross-project collaboration. OSGeo has been a stimulating force for cooperative developments of sister projects, leveraging each other efforts by developing shared architecture components and expanding interoperability.

1.2 GRASS as an Open Source GIS

GRASS (Geographical Resources Analysis Support System) is a raster/vector GIS combined with integrated image processing and data visualization subsystems. It includes more than 350 modules for management, processing, analysis and visualization of georeferenced data. As we have mentioned in the Preface, the key development in the recent GRASS history was the adoption of GNU GPL (General Public License, see http://www.gnu.org) in 1999. By this, GRASS embraces the Open Source philosophy, well known from the GNU/Linux development model, which stimulated its wide acceptance (Raymond, 1987 and Raymond, 1999, for a discussion see also Wheeler, 2003). This license protects the GRASS developers against misuse of their code contribution within proprietary projects which do not allow free access to their source code. The GPL ensures that all code based on GPL'ed code must be published again under GPL. The benefits of using other developers' code further increases the motivation to participate. For the GRASS users, the license offers various advantages besides full access to the source code, especially the low cost, access to the new features and capabilities developed between the releases and possibility to provide releases more often than it is common for proprietary products. Finally, full access to the source code is also an investment protection for the future. In case that the project is withdrawn by the

---

[4] OSGeo Web Site, http://osgeo.org

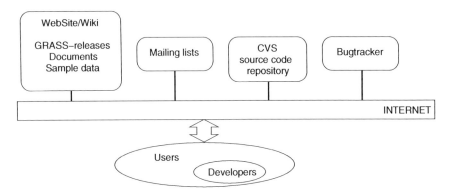

Fig. 1.1. GRASS Development Model: Developers' and users' interaction with semi-automated development tools over Internet

current developers, others may take over the development, while keeping free access to the source code.

Unlike most proprietary GIS, GRASS provides complete access to its internal structure and algorithms. Advanced users who want to write their own GIS modules may therefore learn from existing modules as well as by reading the "GRASS Programmer's Manual" (GRASS Development Team, 2006). The documented GRASS GIS libraries with the Application Programming Interface (API) make the new module development more efficient and allow to integrate new functionality into GRASS. Applications can be also written with shell or Python scripts to automate the GIS workflow (see Section 9).

The GRASS Development Model is similar to other Open Source projects (Figure 1.1). The backbone of the project is the Internet which supports the software distribution, user support, centralized management of the GRASS development through CVS (Concurrent Versioning System, the source code repository server), as well as a bugtracking system, several mailing lists, and a Wiki collaborative help system. The GRASS Development Team is coordinated from FBK-irst (formerly known as ITC-irst) – Centro per la Ricerca Scientifica e Tecnologica, Trento (Italy) and includes developers from all over the world. The dynamic and very open team continuously improves and extends the GRASS capabilities. Communication with other like-minded GIS projects is achieved through the Open Source Geospatial Foundation.

GRASS is available via Internet and on CD-ROM as precompiled binary versions for different UNIX, MacOS X and MS-Windows platforms along with the complete C-source code. While GRASS is Free Software with protection of the authors' and users' rights through the GPL, commercial services related to GRASS can be offered and are welcome by both the developers and users community.

1.3 The North Carolina sample data set

A new, modern sample GIS data set has been prepared for this edition. It is available from the book related Web site.[5] This data set is a comprehensive collection of raster, vector and imagery data covering parts of North Carolina (NC), USA. The data were prepared from public data sources provided by the North Carolina state and local government agencies and Global Land Cover Facility (GLCF). Data are provided at three hierarchical levels (Figure 1.2): entire NC with raster data at 500m resolution (boundary in geographic coordinates: 37N-33N,75W-85W); Southwest Wake county with raster data at resolutions 30m-10m (boundary coordinates 35:48:34.6N-35:41:15.0N, 78:46:28.6W-78:36:29.9W), and a small watershed in rural area with data resolutions of 1m-3m. The data set includes section of the NC capital city Raleigh and its surroundings. The coordinate system of the ready-to-use GRASS data set is NC State Plane (Lambert Conformal Conic projection), metric units and NAD83 geodetic datum. Additional data are provided in geographic coordinates and NC State Plane, english units (feet) in various external formats. More complete data description can be found at the GRASS book site.

Vector data include administrative boundaries, census data, zipcodes, firestations, hospitals, roads and railroads, public schools and colleges, bus routes, points of interest, precipitation, hydrography maps, geodetic points, soils and geological maps. Raster data include elevation (NED 3arc-sec, SRTM-V1 30m, lidar derived DEMs at 1m and 6m), slope, aspect, watershed basins, geology, and landuse. The resolution of raster maps is 500m, 30m, 10m, and 1m. Imagery data include 1m resolution orthophoto, several LANDSAT-TM5/7 scenes and a MODIS daily Land Surface Temperature (LST) time series. Also multiple-return lidar data are included. The examples throughout this book are based on this data set. Furthermore, new derivative maps are generated and explained.

1.4 How to read this book

This book focuses on the basic principles and functionality of GRASS. After a brief introduction to GIS concepts, map projections and coordinate systems are explained. GRASS is introduced in the third chapter using the North Carolina sample data set provided on the related Web site. The fourth chapter describes the properties of GRASS raster and vector data and provides extensive information on import and export of a wide range of data formats. Management, display, analysis and modeling using raster and vector data is covered in the next two chapters, again using hands-on examples based on the sample data set. Interactive visualization and map creation is covered in chapter seven at a basic level needed to communicate the results of a GIS project

---

[5] North Carolina sample data set download site,
http://www.grassbook.org, Section "Data 3rd Edition"

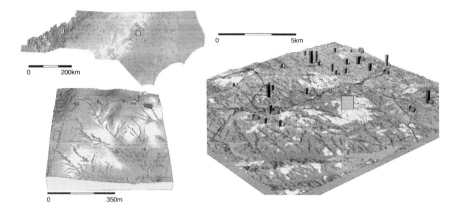

Fig. 1.2. North Carolina sample data set: state, county and field level of detail

effectively. An extensive chapter is devoted to the satellite image processing and analysis as a special case of raster data application. The ninth chapter provides an introduction to GRASS scripting and programming. Chapter ten demonstrates the use of GRASS with other Open Source software. The Appendix provides equations used in some of the modules. References to literature provide access to detailed information about the given topic.

We use the following conventions throughout the book. Commands which you can type in are written in typewriter font, for example: r.mapcalc. Terminology related to GRASS is written in capital letters, such as LOCATION, MAPSET, DATABASE, and GRID RESOLUTION. Wherever [...] appears within the description of GRASS workflow, we have omitted some less important screen output. Lines starting with # symbol indicate comments that are not executed by the shell. Long lines representing UNIX or GRASS commands are broken with <\> which means that the command continues on the next line. This character is usually not necessary when typing, we often used it here for formatting reasons. If you use <\>, be sure not to have blank space after the <\> character. Otherwise the subsequent line(s) are ignored. Text from the graphical user interface menus is written using a different font, for example: Display. Because GRASS is updated fairly frequently, there may be some differences between the command options and parameters in this book and the latest release. It is therefore useful to verify the most recent command usage in the related manual page.

You can download ready-to-use databases which we use throughout the book as well as updates to this book from the related Web site at
http://www.grassbook.org

GIS concepts

To use GIS effectively, it is important to understand the basic GIS terminology and functionality. While each GIS software has slightly different naming conventions, there are certain principles common to all systems. At first, we briefly describe the GIS basics in general (for in depth information read Longley et al., 2005, Clarke, 2002, or Burrough and McDonnell, 1998) and then we explain the principles of map projections and coordinate systems that are used to georeference the data.

2.1 General GIS principles

Data in a GIS database provide a simplified, digital representation of Earth features for a given region. Georeferenced data can be organized within GIS using different criteria, for example, as thematic layers or spatial objects. Each thematic layer can be stored using an appropriate data model depending on the source of data and their potential use.

2.1.1 Geospatial data models

Georeferenced data include a *spatial* (geometrical or graphical) component describing the location or spatial distribution of geographic phenomenon and an *attribute* component used to describe its properties. The spatial component can be represented using one of the two basic approaches (Figure 2.1):

- *field* representation, where each regularly distributed point or an area element (pixel) in the space has an assigned value (a number or no-data), leading to the *raster data model*;
- *geometrical objects* representation, where geographic features are defined as lines, points, and areas given by their coordinates, leading to the *vector data model*.

Raster data

Vector data

Attribute data

cat	soilname	area
1	Ro	243017.6
2	Wo	13426100.1
3	Au	433044.4
4	GrC	466433.7
5	PkC	119344.9

Fig. 2.1. Data models in GIS – raster and vector data with attribute table:
Raster data: rows and columns of values representing spatial phenomenon;
Vector data: representation by points, lines and areas;
Attributes: descriptive data stored in a database table

Depending on scale, representation of a geographic feature can change; for example, a river can be handled as a line at small scale or as a continuous 3D field (body of water) at large scale. Similarly, a city can be represented as a point or as an area. Note that we use the terms small and large scale in the cartographic sense, for example, 1:1million is small scale, 1:1000 is large scale.

To effectively use GIS, it is useful to understand the basic properties and applications of each data model (in older GIS literature, the raster and vector data models have been often referred to as raster and vector data formats).

Raster data model Raster is a regular matrix of values (Figure 2.1). If the values are assigned to *grid points*, the raster usually represents a continuous field (elevation, temperature, chemical concentration) and is sometimes called *lattice*. If the values are assigned to *grid cells* (area units, pixels), it represents an image (satellite image, scanned map, converted vector map). If the cell values represent category numbers, one or more attributes can be assigned to that cell using a database. For example, a soil type with category number 3 can have attributes describing its texture, acidity, color and other properties. The grid cells are organized and accessed by *rows* and *columns*. The area represented by a square grid cell is computed from the length of its side, called *resolution*. Resolution controls the level of spatial detail captured by the raster data. Most data are represented by a 2D raster, with the grid cell (unit area) called a *pixel*; volume data can be stored as a 3D raster with a unit volume called a *voxel* (volume pixel). General d-dimensional raster formats are used for spatio-temporal or multispectral data (e.g. HDF format[1]).

The raster data model is often used for physical and biological subsystems of the geosphere such as elevation, temperature, water flow, or vegetation. However, it can also be used for data usually represented by lines and polygons such as roads or soil properties, especially for scanned maps. The raster data model was designed with a focus on analysis, modeling and image processing.

---

[1] HDF format and tools, http://www.hdfgroup.org/

Its main advantage is its simplicity, both in terms of data management as well as the algorithms for analysis and modeling, including map algebra. This data model is not particularly efficient for networks and other types of data heavily dependent on lines, such as property boundaries. GRASS has extensive support for the raster data model.

Vector data model   Vector data model is used to represent areas, lines and points (Figure 2.1). We describe the vector data model using GRASS terminology; in other systems, the definitions may be slightly different.

The vector data model is based on *arc-node* representation, consisting of non-intersecting lines called *arcs*. An arc is stored as a series of points given by (x, y) or (x, y, z) coordinate pairs or triplets (with height). The two endpoints of an arc are called *nodes*. Points along a line are called *vertices*. Two consecutive (x, y) or (x, y, z) pairs define an *arc segment*. The arcs form higher level map features: *lines* (e.g., roads or streams) or *areas* (e.g., farms or forest stands). Arcs that outline areas (polygons) are called *area edges* or *boundaries*. A complete area description includes a *centroid*. In GRASS, 3D polygons are called *faces* (they do not need a centroid but can be visualized). A 3D volume is a closed set of *faces* including a 3D centroid (*kernel*). Not all GIS software packages support 3D vector data types. Linear features or polygon boundaries are drawn by straight lines connecting the points defining the *arc segments*. To reduce the number of points needed to store complex curves, some GIS include mathematically defined *curve sections* or *splines* that are used to compute the points with the required density at the time of drawing.

In addition to the coordinate information, the vector data model often includes information about the data *topology* which describes the relative position of objects to each other (see Section 6.3.1 for more details on vector data topology).

Each map feature is assigned a category number which is used to link the geometric data with descriptive, attribute data (such as category labels or multiple attributes stored in a database). For example, in a vector map "roads", a line can be assigned category number 2 with a text attribute "gravel road" and a numerical attribute representing its width in map units.

Point features (e.g., a city or a bridge) or point samples of continuous fields (e.g., elevation, precipitation), are represented as independent points given by their coordinates. A value or a set of attributes (numerical or text) is assigned to each point.

Vector data are most efficient for discrete features which can be described by lines with simple geometry, such as roads, utility networks, property boundaries, building footprints, etc. Continuous spatial data can be represented by vector data model using isolines, point clouds or various types of irregular meshes; however, such representations usually lead to more complex algorithms for analysis and modeling than the raster data model. GRASS 6 provides support for both the 2D and 3D multi-attribute vector model.

point   line   area   surface   volume

Fig. 2.2. Data dimensions in a Geographical Information System (after Rase, 1998:19)

Attributes – GIS and databases   Attributes are descriptive data providing information associated with the geometrical data. Attributes are usually managed in external or internal GIS database management systems (DBMS). The databases use the corresponding coordinates or identification numbers to link the attributes to the geometrical data. Some systems, such as PostGIS[2] or, with some limitations, MySQL also allow the user to store geometrical data into the database.

For raster data, GRASS supports only a single attribute for each cell category. For vector data, GRASS offers a generic SQL-DBMS abstraction layer with two choices for internal databases (limited DBF file driver, and SQLite driver) and several full featured interfaces to external databases (PostgreSQL, MySQL, and ODBC interface to various DBMS). Multiple attributes can be stored and managed for each vector object. One or several attribute tables can be linked to a vector map.

Data model transformations   The same phenomenon or feature can be represented by different data models. GIS usually includes tools for transformation between the vector and raster data model. For example, elevation can be measured as vector point data, then interpolated into a raster map which is then used to derive contour lines as vector data. Note that transformations between different data models are usually not lossless (there can be a loss or distortion of information or spatial displacement due to the transformation).

Dimensions of geospatial data   In general, Earth and its features are located and evolve in 3D space and time. However, for most applications a projection of geospatial data to a flat plane is sufficient; therefore two-dimensional representation of geographical features (with data georeferenced by their horizontal coordinates) is the most common. Elevation as a third dimension is usually stored as a separate raster map representing a surface within three-dimensional space (often referred to, not quite correctly, as a 2.5-dimensional representation, Figure 2.2). Elevation can be also added as a z-coordinate or as an attribute to vector data. If there is more than a single

---

[2] PostGIS DBMS, `http://postgis.refractions.net`

z-value associated with a given horizontal location, the data represent a volume and are three-dimensional (e.g., chemical concentrations in groundwater, or air temperature). Three-dimensional data can change in time, adding the fourth dimension. GIS provides the most comprehensive support for 2D data. GRASS 6 includes a 3D raster model for volume data and a 3D vector model for multi-attribute vector data (see Brandon et al., 1999; Neteler, 2001; Blazek et al., 2002); however, only a limited number of modules is available for true volume data processing and analysis.

2.1.2 Organization of GIS data and system functionality

GIS can be implemented as a comprehensive, multipurpose system (e.g. GRASS, ArcGIS), as a specialized, application oriented tool (e.g. GeoServer, MapQuest), or as a subsystem of a larger software package supporting handling of geospatial data needed in its applications (e.g., hydrologic modeling system, geostatistical analysis software, or a real estate services Web site). The multipurpose systems are often built from smaller components or modules which can be used independently in application oriented systems.

The multipurpose GIS usually stores the georeferenced data as thematic *maps*. Each geographic feature or theme, such as streams, roads, vegetation, or cities is stored in a separate map using the vector or raster data model. The maps can then be combined to create different types of new maps as well as perform analysis of spatial relations. GRASS and most of the proprietary GIS products are based on this data organization.

A large volume of geospatial data is nowadays distributed through *Internet based GIS* and *Web Services*. The data sets are stored on central server(s) and users access the data as well as the display and analysis tools through the Internet. Examples are the browser based interactive maps and virtual globes (Google Earth, NASA WorldWind etc.), National Map of the U.S.[3], UMN/MapServer Gallery[4]. Almost every multipurpose GIS software includes tools supporting development of Web-based applications. GRASS can be used with UMN/MapServer, an Open Source project for developing Web-based GIS applications which supports a variety of spatial requests like making maps, scale-bars, and point, area and feature queries (see Chapter 10). Creation of interactive maps, including MapServer, OpenEV, GDAL/OGR, and PostGIS on the Internet, is described in Mitchell (2005) and Erle et al. (2005). The availability of public programming interfaces by many Web mapping providers inspires implementation of "mashups" that aggregate different (Web based) services into new value-added applications.

Other projects such as JGrass/uDig[5] are using JAVA to implement a client/server model. A new approach is the implementation of OGC Web Processing Service (WPS) in Python, the PyWPS software (see Section 9.3.2).

---

[3] National Map of the U.S., http://nationalmap.gov

[4] UMN/MapServer Gallery, http://mapserver.gis.umn.edu

[5] JGrass/uDig (JAVA GRASS Client-Server) Web site, http://www.jgrass.org

Internet GIS can be enhanced by interactive 3D viewing capabilities using GeoVRML[6] as well as by multimedia features adding photographs, video, animations or sound to the georeferenced data.

While creating digital and hardcopy maps has been the core GIS function over the past decade, the emphasis is shifting towards Web Services, spatial analysis and modeling. GIS functionality is rapidly evolving and currently covers a wide range of areas, for example:

- integration of geospatial data from various sources: projections and coordinate transformations, format conversions, spatial interpolation, transformations between data models;
- visualization and communication of digital georeferenced data in form of digital and paper maps, animations, virtual reality (computer cartography);
- spatial analysis: spatial query, spatial overlay (combination of spatial data to find locations with given properties), neighborhood operations, geostatistics and spatial statistics;
- image processing: satellite and airborne image processing, remote sensing applications;
- network analysis and optimization;
- simulation of spatial processes: socioeconomic such as transportation, urban growth, population migration as well as physical and biological, such as water and pollutant flow, ecosystem evolution, etc.

The most rapid and innovative development in geospatial technologies is currently linked to integration of geospatial information within various aspects of Web capabilities and services such as:

- Geospatial Web and Semantic Web (content can be read and used by software agents);
- Service Oriented Architecture (SOA) and Web Services – for example, PyWPS, GeoServer, UMN/MapServer, deegree;
- Geotagging and GeoRSS: addition of geographical identification to various media to support mapping and location-based search;
- Sensor Web: processing and serving real time georeferenced data acquired by multiple sensors;
- Map tiling for projection on virtual globes – or example, OSGeo tiling project;
- building communities that share geospatial data and develop geospatial applications and mashups; using geospatial concepts within Web 2.0.

OSGeo foundation plays a major role in the development of these new technologies.

GIS functionality is used to solve spatial problems in almost every area of our lives. Here are a few examples. In the area of socioeconomic applications,

---

[6] GeoVRML Web site, http://www.geovrml.org

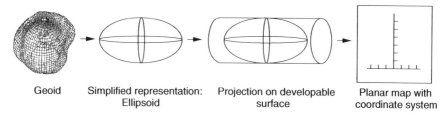

Geoid   Simplified representation: Projection on developable   Planar map with
                    Ellipsoid   surface   coordinate system

Fig. 2.3. Earth's surface representation in map projections and coordinate systems

GIS can be used to find directions, locate a hospital within a given distance from a school, find optimal locations for a new manufacturing facility, design voter districts with given composition and number of voters, identify crime hot spots in a city, select optimal evacuation routes, manage urban growth. GIS plays an important role in conservation of natural resources, agriculture, and management of natural disasters, such as identification and prevention of soil erosion risk, forest resource management, ecosystem analysis and modeling, planning of conservation measures, flood prediction and management, pollutant modeling, and more.

2.2 Map projections and coordinate systems

The basic property of GIS, as opposed to other types of information systems, is that the stored data are georeferenced. That means that the data have defined location on Earth using coordinates within a georeferenced coordinate system. The fact that Earth is an irregular, approximately spherical object makes the definition of an appropriate coordinate system rather complex. The coordinate system either has to be defined on a sphere or ellipsoid, leading to a system of geographic coordinates or the sphere has to be projected on a surface that can be developed into a plane where we can define the cartesian system of coordinates (*easting, northing and elevation*; see Sections 2.2.2).

2.2.1 Map projection principles

When working with GRASS, the projection and coordinate system must be defined whenever a new project (LOCATION in GRASS terminology) is created. The map projection definition is stored in an internal file within the given LOCATION. It is used whenever the data need to be projected into a different projection or when calculations requiring information about the Earth's curvature are performed. Different parameters are needed to define different projections and coordinate systems; therefore, it is important to understand the map projection terminology.

Shape of Earth   Shape of Earth is usually approximated by a mathematical model represented by an *ellipsoid* (also called a spheroid). A variety of
cartographic ellipsoids have been designed to provide the best-fit properties
for certain portions of the Earth's surface, for example, Clarke 1866 for North
America, Bessel 1841 for several European countries, or the current WGS
1984 used worldwide. While the ellipsoid describes the shape of Earth by a
relatively simple mathematical function, the *geoid*, an equipotential surface of
the Earth's gravity, undulates due to the spatially variable distribution of the
Earth's mass, see Figure 2.3. For map projections, the ellipsoids are usually
sufficient for horizontal positioning; however, the geoid has to be used for high
accuracy elevation calculations.

Geodetic or map datum   A set of constants specifying the coordinate system used for calculating the coordinates of points on Earth is called a *geodetic
datum*. Horizontal datums define the origin and orientation of a coordinate
system used to calculate the horizontal coordinates (usually northing and
easting). Vertical datums define the coordinate system origin for calculating
the elevation coordinate, such as mean sea level. For maps to match, their
coordinates must be computed using the same datum. Different datums mean
a shift in the origin of the coordinate system, and that means a shift of the
entire map.

Map projection   To transform the curved Earth surface into a plane (flat
sheet of paper or a computer screen), a *map projection* is used. Direct projection of a spherical object to a plane cannot be performed without distortion.
The most common approach is to project the spheroid onto a *developable
surface*, such as a cylinder or a cone that can be developed into a plane without deformation (tearing or stretching), see Figure 2.3. A large number of
different projections have been designed with the aim to minimize the distortion and preserve certain properties (for a mathematical description refer
to Bugayevskiy and Snyder, 2000:20-22). The *conformal* projection preserves
angles (shapes for small areas) and is often used for navigation and national
grid systems. The *equidistant* projection preserves certain relative distances
and is used for measurement of length. The *equivalent* projection preserves
area and is used for measurement of areas. Each of the properties (angle, distance, area) is preserved at the expense of the others. The map projection
is usually selected depending on the application because there is no perfect
solution to the projection problem. Most coordinate systems used for land
surface mapping use conformal projections.

    The developable surfaces can either touch the spheroid (tangent case) or
intersect it (secant case). The most commonly used surfaces are a cylinder
(*cylidrical* projection), a cone (conic projection), and a plane (*azimuthal* projection). The points or lines where the developable surface touches or intersects
the spheroid are called *standard points* and *standard lines* with zero distortion
(e.g. standard parallel for a tangent cone or two standard parallels for a secant

cone). That means that the projected maps do not have uniform scale for the entire area, and that the true map scale is preserved only along the standard lines. To minimize distortions, some projections reduce the scale along the standard parallel(s) or central meridian(s). This is expressed as a *scale factor* smaller than 1.0 in the definition of such a projection.

Transverse projections use developable surfaces rotated by 90° so that the standard (tangent) line is a meridian called *central meridian* instead of a standard parallel. *Oblique projections* may use any rotation defined by azimuth where *azimuth* is an angle between a map's central line of projection and the meridian it intersects, measured clockwise from north. Snyder (1987) provides an excellent manual on map projections with map examples for many important projections.

Coordinate system   To accurately identify a location on Earth, a *coordinate system* is required. It is defined by its origin (e.g., prime meridian, datum), coordinate axes (e.g. x, y, z), and units (angle: degree, gon, radiant; length: meter, feet). The following general coordinate systems are commonly used in GIS:

- *geographic* (global) coordinate system (latitude-longitude);
- *planar (cartesian) georeferenced* coordinate system (easting, northing, elevation) which includes projection from an ellipsoid to a plane, with origin and axes tied to the Earth surface;
- *planar non-georeferenced* coordinate system, such as image coordinate system with origin and axes defined arbitrarily (e.g. image corner) without defining its position on Earth.

Note that for planar georeferenced systems *false easting* and *false northing* may be used. These are selected offset constants added to coordinates to ensure that all values in the given area are positive.

For mapping purposes, each country has one or more *national grid systems*. Information about national grid systems can be obtained from the national cartographic institutes or from the ASPRS Web site[7]. A national grid system is defined by a set of parameters such as ellipsoid, datum, projection, coordinate system origin and axes, etc. Examples of worldwide and national grid systems are UTM (Universal Transverse Mercator), Gauss-Krüger, Gauss-Boaga, or State Plane. Information about the grid system used to georeference digital geospatial data is a crucial component of the metadata and allows the user to integrate and combine data obtained from different sources.

---

[7] Information about national grid systems:
  - ASPRS – Grids & Datums, http://www.asprs.org/resources/grids/
  - European coordinate systems, http://www.mapref.org
  - A comprehensive, general list of projection transformations is available at
  http://www.remotesensing.org/geotiff/proj_list/

2.2.2 Common coordinate systems and datums

Geographic coordinate system: latitude-longitude The most common coordinate system used for global data is the spherical coordinate system which determines the location of a point on the globe using latitude and longitude. It is based on a grid of meridians and parallels, where *meridians* are the longitude lines connecting the north and south poles and *parallels* are the latitude lines which form circles around the Earth parallel with the *equator*. The longitude of a point is then defined as an angle between its meridian and the *prime meridian* (0°, passing through the Royal Observatory in Greenwich, near London, UK). The latitude of a point is defined as an angle between the normal to the spheroid passing through the given point and the equator plane. The longitude is measured 0-180° east from prime meridian and 0-180° west, where 180° longitude is the international date line. Latitude is measured 0-90° north and 0-90° south from equator.

Geographic coordinates can be expressed in decimal degrees or sexagesimal degrees. Decimal values of west (W) and south (S) are expressed as negative numbers, north (N) and east (E) as positive numbers (e.g. Murcia, Spain: -1.1333°, 37.9833°). Values given in sexagesimal system always use positive numbers together with N, S, E, W (Murcia, Spain: 1:07:59.88W, 37:58:59.88N). It is not difficult to convert between these notations, see the GRASS Wiki[8].

Universal Transverse Mercator Grid System The Universal Transverse Mercator (UTM) Grid System is used by many national mapping agencies for topographic and thematic mapping, georeferencing of satellite imagery and in numerous geographical data servers. It applies to almost the entire globe (area between 84° N and 80° S). The pole areas are covered by the Universal Polar Stereographic (UPS) Grid System, please refer to Robinson et al. (1995).

UTM is based on a Transverse Mercator (conformal, cylindrical) projection with strips (zones) running north-south rather than east-west as in the standard Mercator projection. UTM divides the globe into 60 zones with a width of 6° longitude, numbered 1 to 60, starting at 180° longitude (west). Each of these zones will then form the basis of a separate map projection to avoid unacceptable distortions and scale variations. Each zone is further divided into strips of 8° latitude with letters assigned to from C to X northwards, omitting the letters I and O, beginning at 80° south (Robinson et al., 1995:101). For example, Trento (Italy; 11.133E, 46.067N) belongs to UTM zone 32, strip T. A conversion script from latitude-longitude to UTM zone/strip is available from the GRASS Wiki[9].

---

[8] GRASS Wiki, Converting degree notations,
   http://grass.gdf-hannover.de/wiki/Convert_degree
[9] Download section in the GRASS AddOns Web site,
   http://grass.gdf-hannover.de/wiki/GRASS_AddOns

The origin of each zone (central meridian) is assigned an easting of 500,000m (false easting, Maling, 1992:358). For the northern hemisphere the equator has northing set to zero, while for the southern hemisphere it has northing 10,000,000m (false northing). To minimize the distortion in each zone, the scale along the central meridian is 0.9996, leading to a secant case of the Transverse Mercator projection with two parallel lines of zero distortion. Note that UTM is used with different ellipsoids, depending on the country and time of mapping.

For GIS applications, it is important to realize that each UTM zone is a different projection using a different system of coordinates. Combining maps from different UTM zones into a single map using only one UTM zone (which can be done relatively easily using GIS map projection modules) will result in significant distortion in the location, distances and shapes of the objects that originated in a different zone map and are outside the area of the given zone. To overcome the problem, a different coordinate system should be used and the data re-projected. For a quick reference, you can find the UTM zone numbers in the Unit 013 "Coordinate System Overview" of the NCGIA Core Curriculum in GIS.[10]

Lambert Conformal Conic Projection based systems The Lambert Conformal Conic (LCC) projection is one of the most common projections for middle latitudes. It uses a single secant cone, cutting the Earth along two standard parallels or a tangent cone with a single standard parallel. When working with LCC based coordinate systems, the following parameters have to be provided: the standard parallel(s) (one or two), the longitude of the central meridian, the latitude of projection origin (central parallel), false easting and, sometimes, false northing (you may recall that false easting and northing are shifts of the origin of the coordinate system from the central meridian and parallel).

State Plane Coordinate System The State Plane Coordinate System used by state mapping agencies in the USA is based on different projections depending on the individual state shape and location, usually LCC or a Transverse Mercator with parameters optimized for each state. Various combinations of datums (NAD27, NAD83) and units (feet, meters) have been used, so it is important to obtain all relevant coordinate system information (usually stored in the metadata file) when working with the data georeferenced in the State Plane Coordinate System. GIS projection modules often allow to define the State Plane system by providing the name of the state and the county, however, the parameters should always be checked, especially when working with older data.

---

[10] Unit 013 Coordinate System Overview in the NCGIA Core Curriculum in GIS, http://www.ncgia.ucsb.edu/education/curricula/ giscc/units/u013/u013.html

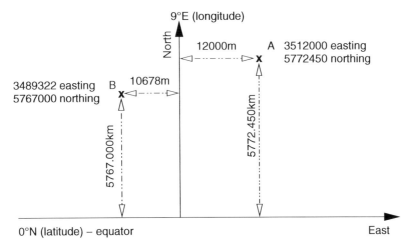

Fig. 2.4. Example for the Gauss-Krüger Grid System with two points A and B

Gauss-Krüger Grid System The Gauss-Krüger Grid System is used in several European and other countries. It is based on the Transverse Mercator Projection and the Bessel ellipsoid. The zones are $3°$ wide, leading to 120 strips. The zone number is divided by 3 according to longitude of central meridian. Adjacent zones have a small overlapping area. The scale along the central meridian (scale factor) is 1.0.

Figure 2.4 illustrates the coordinate system, the northing values are positive north from the equator, the easting values are measured from the central meridian. To avoid negative values, a false easting of 500,000m is defined in addition to the third of the longitude of the central meridian. For example, the false easting for the $9°$ E central meridian is 3,500,000m ($9/3 = 3$, value composed with 500,000m to 3,500,000m).

North American and European Datums In general, a large number of georeferencing datums exists, here we focus on three examples. The North American Datum 1983 (NAD83) is a geodetic reference system which uses as its origin the Earth's center of mass, whereas the old North American Datum 1927 (NAD27) had a different origin, making it useful only in North America. GPS receivers which are mostly based on the WGS84 datum (other local datums can be selected in the GPS receiver's menu) also use the Earth's center of mass as their system's origin.

When using maps based on different datums, a datum transformation to a common datum is required. For example, a change from NAD27 to NAD83 system leads to a shift for the entire map. Overlapping maps based on different datums of the same region would not co-register properly without datum transformation. In the continental United States, a few common assignments

between datums and ellipsoids are in use: NAD27 datum with Clarke 1866 ellipsoid, NAD83 datum with GRS80 ellipsoid, and WGS84 datum with WGS84
ellipsoid.

It is important to know that the NAD27 and NAD83 datums are 2D horizontal datums used for horizontal coordinates (easting and northing). Separate
vertical datums used with these systems are NGVD29 and NAVD88. GRASS
does not handle yet separate vertical datums, so these transformations need
to be done outside GRASS. WGS84 is a 3D datum (x, y and height).

3

Getting started with GRASS

In this chapter, we begin working with GRASS. First, we explain the GRASS software installation and the structure of its database. Then we use a sample data set to perform basic GIS tasks. We also include a number of examples illustrating how to start a GRASS project using different coordinate systems.

3.1 First steps

GRASS, as a multipurpose GIS, with data organized as raster and vector maps, provides a wide range of tools to support most of the GIS functionality outlined in the previous sections. An overview is given in Table 3.1. Detailed explanation of each module, often with a usage example, is given in the GRASS users manual (see your GRASS installation or Web site[1]). The latest GRASS 6.2 release has been designed as a 3D GIS with support for 3D raster and 3D vector data (see Blazek et al., 2002; Neteler, 2001).

3.1.1 Download and install GRASS

GRASS software can be downloaded freely from the main GRASS Web site:
    http://grass.itc.it

The main GRASS site is mirrored in several countries for faster access including the GRASS USA mirror at http://grass.ibiblio.org.

You can find there the source code (portable version for all operating systems) as well as the latest ready-to-install binaries for GNU/Linux, MacOS X and MS-Windows (optionally with the Cygwin tools). For some GNU/Linux distributions, easy to install binary packages are also provided. MS-Windows users may be interested in the Quantum GIS (QGIS[2]) package that includes

---

[1] GRASS users manuals, http://grass.itc.it/gdp/manuals.php
[2] QGIS Web site, http://qgis.org

functionality class	*functionality*
geospatial data integration	import and export of data in various formats
	coordinate systems transformations and projections
	transformations between raster and vector data
	2D/3D spatial interpolation and approximation
2D/3D raster data processing	2D and 3D map algebra
	surface and volume geometry analysis
	topographic parameters and landforms
	flow routing and watershed analysis
	line of sight, insolation
	cost surfaces, shortest path, buffers
	landscape ecology measures
	correlation, covariant analysis
	expert system (Bayes logic)
2D/3D vector data processing	multi-attribute vector data management
	digitizing
	overlay, point and line buffers
	vector network analysis
	spatial autocorrelation
	summary statistics
	multivariate spatial interpolation and approximation
	Voronoi polygons, triangulation
image processing	processing and analysis of multispectral satellite data
	image rectification and orthophoto generation
	principal and canonical component analysis
	reclassification and edge detection
	radiometric correction
visualization	2D display of raster and vector data with zoom and pan
	3D visualization of surfaces and volumes with vector data
	2D and 3D animations
	hardcopy postscript maps
modeling and simulations	hydrologic, erosion and pollutant transport, fire spread
temporal data support	time stamp for raster and vector data
	raster time series analysis
links to Open Source tools	QGIS, R-stats, gstat, UMN/MapServer, Paraview
	GPS tools, GDAL/OGR, PostgreSQL, MySQL

Table 3.1. GRASS functionality

a GRASS installation. GRASS is also available on CD-ROM from various providers.[3] There may be a fee for packaging the CD-ROM and for the customized installation software.

The Web site also includes the "GDP – GRASS Documentation Project"[4], where you can find manual pages, tutorials, information about the externally developed GRASS-modules and various articles. Support for developers and

---

[3] List of GRASS CDs/DVDs, http://grass.itc.it/download/cdrom.php

[4] GDP Web site, http://grass.itc.it/gdp/

users is provided by several mailing lists, you can join them through the Web interface (see the relevant links under "Support section" at the GRASS Web site). Besides international mailing lists in English language, there are also localized lists currently in Czech/Slovak, German, Italian, Japanese, Polish and Spanish. The GRASS project has its Wiki based collaborative help system[5], where you can find (and submit) GRASS add-ons: scripts and modules contributed by the community. You can participate in discussions about future GRASS development and contribute your own tips and tricks for other GRASS users. In addition to the GRASS Wiki site, you can download the add-ons from a source code repository using the SVN client software:

```
svn co \
   https://grasssvn.itc.it/svn/grassaddons/trunk/grassaddons \
   grassaddons
```

You can add a relevant subdirectory to the above URL if you want to download only a selected module. Please see the GRASS Wiki site for detailed instructions.

GRASS binary installation   The GRASS binaries are available for several platforms (RPM, Debian and Gentoo packages as well as installation files for MS-Windows and MacOS X) that can be downloaded from the GRASS server. For GNU/Linux, download the install script `grass-VERSION-install.sh` and the GRASS package `grass-VERSION.tar.gz` (the name depends on the platform). The installation itself should be done as user "root". It requires only one step (check online for appropriate file names):

```
sh grass-VERSION-install.sh grass-VERSION.tar.gz
```

After successful installation, the package file `grass-VERSION.tar.gz` may be deleted. Refer to the Chapter 9 for information on the GRASS source code, the centralized source code server and code compilation.

3.1.2 Database and command structure

GRASS data are stored in a directory referred to as GISDBASE. This directory, often called `grassdata/`, must be created before you start working with GRASS. You can create it using the `mkdir` command or a filemanager either in your home directory or in a shared network directory (e.g. in a network file system NSF) to make it accessible to colleagues. Within this directory, the GRASS GIS data are organized by projects stored in subdirectories called LOCATIONs (Figure 3.1). Each LOCATION is defined by its coordinate system, map projection and geographical boundaries. The subdirectories and files defining a LOCATION are created automatically when GRASS is started for the first time with a new LOCATION (see Section 3.2 for more details). Each

---

[5] GRASS Wiki site, `http://grass.gdf-hannover.de/wiki/`

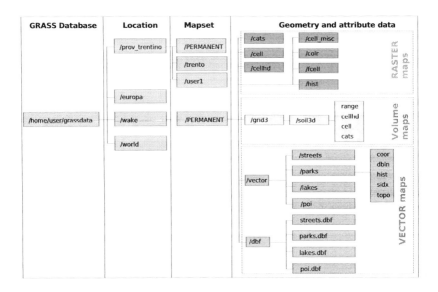

Fig. 3.1. Organization of GRASS data directory, LOCATIONs, MAPSETs, vector and raster maps

LOCATION can have several MAPSETs (subdirectories of the LOCATION, see Figure 3.1) that are used to subdivide the project into different topics, sub-regions, or as workspaces for individual team members. Besides access to his own MAPSET, each user can also read maps in other users' MAPSETs, but he can modify or remove only the maps in his own MAPSET. All MAPSETs include a WIND file that stores the current boundary coordinate values and the currently selected raster resolution.

When creating a new LOCATION, GRASS automatically creates a special MAPSET called PERMANENT designed to store the core data for the project, its default spatial extent in the DEFAULT_WIND file and coordinate system definitions. Only the owner of the PERMANENT MAPSET can add, modify or remove its data; however, these data can be accessed, analyzed, and copied by other users into their own MAPSETs. The PERMANENT MAPSET is therefore useful for providing other users working on the same project with baseline geospatial data such as elevation, roads or streams while keeping them write-protected. To import data into PERMANENT, just start GRASS with the relevant LOCATION and the PERMANENT MAPSET.

The internal organization and management of LOCATION, MAPSETs and maps should be left to GRASS. Operations such as renaming or copying raster and vector maps involve several internal files and should always be done through GRASS commands (we discuss this in detail in Section 3.1.6). Non-GRASS interventions are acceptable only in exceptional situations and when one has a good understanding of GRASS' internal structure.

GRASS modules are organized by name, based on their function class (display, general, imagery, raster, vector or database, etc.). The first letter refers to the function class, followed by a dot and one or two other words, again separated by dots, describing the specific task performed by the module. Table 3.2 lists the most important function classes. For example, v.in.ogr is a vector command for importing vector data in various formats, r.buffer calculates a buffer zone along raster lines and around raster areas, d.rast displays a raster map, i.ortho.photo creates an orthophoto from a scanned/digital aerial image.

Using GRASS on the command line   The general syntax of a GRASS command which is called to run a module is similar to UNIX commands:

```
module [-flag1[flag2...]] parameter1=map1[,map2,...]\
                        [parameter2=number...] [--o] [--q] [--v]
```

where module is the name of the command (see Table 3.2), optional flags enable specific features, and parameter is a name of an input or output file, a constant, name of a method, symbol etc. Note that there must be no space when listing comma-separated names. By default, GRASS does not allow users to overwrite maps with the same name to protect them, but overwriting can be enabled by adding --o to the command line. The level of verbosity can be changed by adding --q or --v. To learn the usage of a command (syntax, flags and parameters), run the command with the help option. We show the next commands only for illustration of the concept, a sample session follows later in this chapter:

```
# 'help' as parameter
d.rast help
```

The next command opens the first map display, called GRASS monitor "x0":

```
# monitor name as parameter
d.mon x0
```

Flags can be specified at any position and combined with parameters:

```
d.vect -c map=soils
```

Do not use white space before or after the = character because the various flags and parameters are white space separated. The first parameter generally does not require the parameter name to be specified, so the previous command can be simplified to:

```
d.vect -c soils
```

prefix	function class	type of command
d.*	display	graphical output
db.*	database	database management
g.*	general	general file operations
i.*	imagery	image processing
m.*	misc	miscellaneous commands
ps.*	postscript	map creation in Postscript format
r.*	raster	2D raster data processing
r3.*	3D raster	3D raster data processing
v.*	vector	2D and 3D vector data processing

Table 3.2. GRASS module function classes

For additional parameter names, it is sufficient to type the initial(s) of the parameter, just to distinguish it/them from other parameters of the command:

```
# full names of parameters
v.to.rast input=roadsmajor output=roadsmajor use=val
# abbreviated names of parameter
v.to.rast roadsmajor out=roadsmajor use=val
```

If you abbreviate too much and parameter names can no longer be distinguished, GRASS will issue an error message.

To read the module-related manual pages, run:

```
# look at help index
g.manual index
# look at a specific command
g.manual d.rast
```

This opens the selected manual page in the default HTML browser of the system. Adding the -m flag, text manual page will be shown in the terminal (but page links won't work here; browse with <space>, leave with <q>):

```
g.manual -m d.rast
```

You can also find these manual pages and tutorials at the GRASS Web site.

3.1.3 Graphical User Interfaces for GRASS 6: QGIS and gis.m

One of the currently easiest approaches to learn GRASS is through Quantum GIS (QGIS).[6] QGIS is a user friendly geographic data viewer (Figure 3.2) with a set of analytical capabilities that runs on GNU/Linux, Unix, MacOS X, and MS-Windows. It supports vector, raster, and database formats and OGC Web Services. QGIS is licensed under the GNU General Public License. It reads all common GIS data formats through the GDAL/OGR libraries, includes a digitizer as well as a GPS track import and export. Projection on the fly can be enabled for easy integration of different data sources. GRASS plugin is

---

[6] Quantum GIS Web site, http://qgis.org

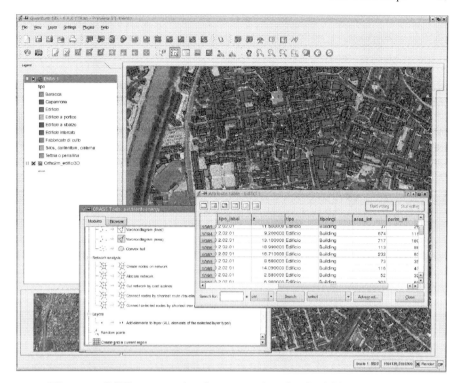

Fig. 3.2. QGIS as a graphical user interface for GRASS with toolbox

provided for reading GRASS raster and vector data directly and a toolbox
enables the user to run QGIS as a graphical user interface for GRASS as it
supports all important GRASS commands. The GRASS shell is also accessible.
QGIS binary packages usually include the latest GRASS version as the GIS
backbone, therefore no extra installation is needed.

GRASS also has its native "GIS manager" graphical user interface called
gis.m (Figure 3.3), we will refer to it where appropriate but we do not explain
its use in detail as it will be replaced in near future with a wxPython based
GUI. It is also possible to use a JAVA-based interface JGrass/uDig designed
for water resources and geomorphology applications.

3.1.4 Starting GRASS with the North Carolina data set

In the following sample session, we assume that you have a working knowledge of running commands, creating directories, and managing files on your
computer system. You need to install the North Carolina (NC) sample data

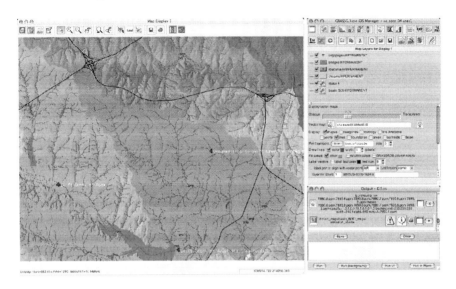

Fig. 3.3. GRASS built-in GIS manager `gis.m`

set that you can download from the GRASS Web site or get on CD-ROM.[7]
It is a comprehensive set of raster, vector and external data covering central
NC region, including a section of the NC capital city Raleigh. The coordinate
system is State Plane NAD83 with metric units. Data are provided at three
hierarchical levels: entire NC with raster data at 500m resolution; Southwest
Wake county with boundary coordinates 215000N-228500N, 630000E-645000E
and resolutions of raster maps ranging from 10m to 30m; and a small watershed in a rural area with data resolutions between 1m-3m; see Section 1.3 for
a more detailed description.

To start, you need to create your GIS data directory, for example,
in your home directory. For shared access you can also store it under
`/usr/local/share/` or in another shared network directory. Depending on
your system set-up you may have to do the latter as a user "root" (or ask your
administrator to do it) and change the access permissions so that you can
read, write and execute within this directory. Since creating it in your home
directory is straightforward, we explain the more complex shared access – as
"root" user, run:

```
# find your group ID (gid)
id
# become root (needs password)
su
```

---

[7] NC data set download,
   `http://www.grassbook.org` (section "Data 3rd Edition")

```
# create directory
mkdir -p /usr/local/share/grassdata
# change its ownership to login name and your group ID
chown yourname /usr/local/share/grassdata
chgrp yourgroup /usr/local/share/grassdata
# make it read-/write-/executable for you and the group
chmod ug+rwx /usr/local/share/grassdata
```

We run mkdir with the flag -p to create all the non-existing parent directories and the desired subdirectory. Then we set the permissions to "read", "write" and "execute" for the user and the group (use your login name for yourname and yourgroup (gid) in the example above).

Then move the downloaded file nc_spm_grassdata-VERSION.tar.gz into this directory, change into it and unpack the file:

```
# replace VERSION with current package version number
mv nc_spm_grassdata_VERSION.tar.gz /usr/local/share/grassdata
cd /usr/local/share/grassdata
tar -xvzf nc_spm_grassdata-VERSION.tar.gz
```

The resulting list of files shows that the data are extracted into a new subdirectory nc_spm, which is the name of your LOCATION. After unpacking, the downloaded "tar.gz" package file can be deleted.

Starting GRASS   You can now call GRASS in the terminal (or from the menu system):

```
grass63
```

and you will see the start menu for selecting a LOCATION and a MAPSET in your GIS data directory (Figure 3.4). Your home directory will be automatically entered as a GIS data directory. Replace it with /home/user/grassdata/ or /usr/local/share/grassdata/ as appropriate. For LOCATION select nc_spm; for MAPSET select user1. Then click on Enter GRASS, and you will see the welcome message and the command line prompt in your window (the version may vary):

```
GRASS 6.3.cvs (nc_spm):~ >
```

Now you are in GRASS and you can call GRASS modules as well as UNIX programs. You can also use the gis.m GUI which opened when you entered GRASS (Figure 3.3). If you don't see it, just type:

```
gis.m
```

Most of the GRASS commands are integrated within this interface and you can find a command for a specific task using the function menus. The interface includes a brief description of the parameters and it also displays the command line version of the module.

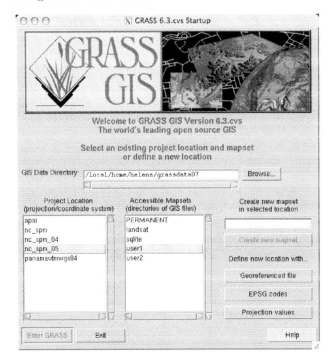

Fig. 3.4. Graphical startup of GRASS

To list the available raster and vector maps, type:

```
g.list rast
g.list vect
```

You can learn more about each raster or vector map in terms of its minimum and maximum coordinates, resolution, and number of classes using the *.info commands, for example:

```
r.info elevation
v.info streams
v.info -c streams
```

The last command with -c flag prints types and names of attribute table columns for a vector map.

3.1.5 GRASS data display and 3D visualization

To view raster and vector maps, we use the standard GRASS monitor "x0" throughout the book. Alternatively, you can also use the built-in graphical user interface gis.m, QGIS with GRASS plugin, JGrass or other software. You can learn how to work with the most recent version of GUI using the on-line help or any of the relevant on-line tutorials. We prefer to use the command

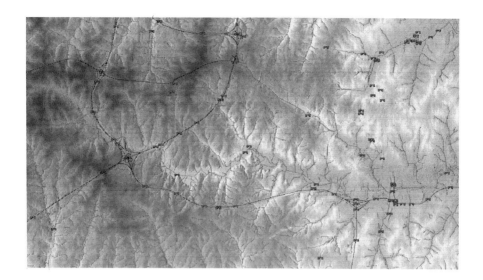

Fig. 3.5. Shaded elevation raster map with overlayed vector streams, major roads and overpasses

line interface (CLI) in the book. Besides the advantage of speed and independence from changing graphical user interfaces, the GRASS shell maintains the command history per mapset which lets you scroll back. Additionally, it is auto-documenting the work that has been done.

To display a map on a unix-like system (GNU/Linux, MacOS X, etc.), first open a GRASS monitor:

```
d.mon x0
```

The default size of the graphics monitor is relatively small, so you may want to resize it to a bigger window using the mouse. MS-Windows users either have to install a X-Server to use the GRASS monitors or they may use gio.m to display maps.

To view the raster elevation map together with vector streams (drawn as blue lines), major roads (drawn as black lines) and overpasses (drawn as symbols) type (Figure 3.5):

```
# set the region to the raster map to be displayed
g.region rast=elevation -p
d.erase

# display elevation map as 2D color map and a shaded color map
d.rast elevation
```

```
d.his h=elevation i=elevation_shade

# display vector line and point maps
d.vect streams col=blue
d.vect roadsmajor
d.vect overpasses icon=extra/bridge size=15 fcol=blue
```

To simultaneously view more than a single raster map (for example, to compare patterns), it is possible to open up to seven GRASS monitors named x0 through x6 using the command d.mon. Use the parameter select to choose in which monitor the map will be displayed, for example (nc_spm LOCATION):

```
g.region rast=aspect -p
d.mon x0
d.rast aspect
d.mon x1
d.rast geology_30m
d.barscale at=0,0
d.mon select=x0
d.vect roadsmajor col=yellow
```

This will display the aspect map in monitor x0, the geological map in monitor x1, and overlay the roads map again in monitor x0. If you are planning to use multiple monitors regularly, it is worth trying the module gis.m which provides a graphical user interface for easily managing multiple monitors and browse through multiple raster and vector maps interactively. To get a list of the maps currently displayed in the GRASS monitor, use d.frame -l.

Zooming You can interactively zoom into a selected location within a map displayed in the GRASS monitor using the command:

```
d.zoom
 Buttons:
 Left:   1. corner
 Middle: Unzoom
 Right:  Quit
```

Use the left mouse button to define the first corner point, then move around the mouse to open the zoom box and use the middle mouse button to set the second zoom box corner. If the zoom level is acceptable, confirm it with right mouse-click. The related mouse button menus are explained in the terminal window. To zoom out with d.zoom, use the middle mouse button. The command requires an open and selected GRASS monitor with at least one map displayed in it, otherwise the graphical user interface will appear and ask for the map name.

Fig. 3.6. Elevation surface and color raster map with overlayed vector streams, major roads and overpasses shown in `nviz`

Region changes and redraw   If you change the region coordinates or raster resolution, run `d.redraw` to tell the GRASS monitor about it. It will fetch the updated region settings and redraw the displayed map(s) accordingly.

Barscale and legends   While we have already used the barscale above, barscale and legends are explained in greater detail in the beginning of the raster and the vector chapters. More sophisticated display methods and output to other formats (PNG, PS etc.) are covered in Chapter 7.

Map queries   Map queries are described in Section 5.1.3 for raster data and Section 6.4.1 for vector data.

Viewing GRASS data in 3D space   We can also load the maps into the NVIZ viewer:

```
nviz elevation vect=streams,roadsmajor point=overpasses
```

Using the green puck, you can modify the view perspective (Figure 3.6). Most sliders should be self-explanatory and are covered in greater detail in the visualization chapter (refer to Section 7.3). In the Visualize ⤳ Raster Surfaces menu you can adjust the raster surface resolution (set fine to a low number or 1 to render the surface at the highest resolution). To drape a new color map over the surface select a new raster map from the Surface Attribute ⤳ color option, for example, `landclass96` from the PERMANENT MAPSET and use DRAW to render your elevation surface with the new color map. You can select colors for the Visualize ⤳ Vector Lines and a different (3D) symbol for the Visualize ⤳ Vector Points (select red color, icon size "200" and icon type "diamond"). You can leave NVIZ through File ⤳ Quit.

In the following chapters, you will see numerous examples of geospatial data processing and analysis performed with the North Carolina sample data set; therefore, at this point, we will just show how to properly end the GRASS session. Exit gis.m by clicking on File ⤳ Exit. If there are still open monitors, close them using the mouse, then exit GRASS by typing:

```
exit
```

It is recommended to close GRASS correctly to ensure that temporary files are not left over. If you exit GRASS but forgot to close gis.m, you can do it at any time later by simply closing the relevant windows using the menu or mouse.

3.1.6 Project data management

When working with GRASS, it is important to understand that each raster or vector map consists of several files which include the data, categories, header, and other information. The general map management commands simplify listing, copying, renaming and deleting maps: use them to maintain the consistency in the GRASS GIS directory. It is not recommended to directly modify the files in the LOCATION or MAPSET directories, unless you are experienced with the system. The map management modules are also applicable to other GRASS related files such as region definitions and imagery groups.

Although it is possible to use all combinations of characters for the GRASS map names if the map name or expression is enclosed within quotes, it is safer to follow the name conventions described below. First, it is important to avoid spaces and special characters, such as a comma, dash, or exclamation mark in GRASS map names. Since GRASS 6, dot is not allowed in the vector map names to maintain compliance with the SQL standard. It is also useful to include at least one letter in raster map names to avoid confusion with numbers being treated as values, especially when using the map algebra module r.mapcalc.

We have already shown that we can use the command g.list to list available raster and vector maps. To list the maps with their titles, use -f flag. If you have many MAPSETs and you want to see the maps stored only in a selected one, use the mapset parameter, for example:

```
# start GRASS with NC data set
grass63

g.list -f vect
g.list -f vect mapset=PERMANENT
```

Remember, when a list exceeds the terminal screen, continue with <SPACE>, go back with <b> and leave with <q>. In case you have many maps available, you may want to list only their subset. You can use wildcards to invoke automated character or name replacement or, optionally, regular expressions. In our example, we want to see all vector maps with the names starting with "s":

```
g.mlist vect pattern="s*"
```

To create a full copy of a map, use the **g.copy** module. You have to specify the map type and add an old and a new map name, separated by comma (no spaces are allowed between the names). As an example, you can copy the map **railroads** from the PERMANENT MAPSET into your own MAPSET:

```
g.copy vect=railroads@PERMANENT,myrailroads
```

To rename a map, you can use **g.rename** and list the old name and the new name, separated by comma:

```
g.rename vect=myrailroads,railnetwork
```

You should also use a GRASS command to remove maps. For example, to remove one of the recently created map copies, type:

```
g.remove vect=railnetwork
```

Multiple maps can be removed by listing them separated by comma. If you need to delete a series of maps, you may carefully (!) use the **g.mremove** module. It allows the use of wildcards or regular expressions similar to **g.mlist**. For example, you can generate several map copies and then delete them in one step:

```
g.copy vect=railroads@PERMANENT,myrailroads1
g.copy vect=railroads@PERMANENT,myrailroads2
g.copy vect=railroads@PERMANENT,myrailroads3
g.list vect

# this only lists the delete candidates
g.mremove vect="myrail*"
# this really removes them
g.mremove vect="myrail*" -f
```

The module will collect the list of map names and ask for confirmation to delete. You should double check the list for any map that you want to keep. You won't be able to undelete it.

Initially, you have access only to the MAPSET PERMANENT (read only, unless you work in it) and your own MAPSET, in our case **user1** (read and write). If several MAPSETs exist for a given LOCATION, for example, when working within a team, you have to add these other MAPSETs to the MAPSET search path to be able to access them. You will have only read access to MAPSETs belonging to other users. To list all MAPSETs available for a given LOCATION and find out which are currently accessible, type:

```
g.mapsets -lp
```

You can add MAPSET **user2** to the search path or re-arrange the search order of MAPSETS for listing their maps as follows:

```
g.mapsets addmapset=user2
g.mapsets mapset=PERMANENT,user1,user2
g.list rast
```

To modify data from another user's MAPSET, copy them to your MAPSET using g.copy. You can restrict others' access to your own MAPSET using the command g.access. MAPSETs to which access is restricted can still be listed in another's MAPSET search path; however, access to the data within these MAPSETs will remain restricted.

To get information about the current LOCATION coordinate system parameters and units, the current MAPSET spatial extent (region) and GRASS environmental variables type:

```
g.proj -p
g.proj -w
g.region -p
g.gisenv
```

You will learn more about these commands in the next section when we describe how to create a new LOCATION, and in the next chapter, Section 4.1.2 that illustrates various options available for region extent and resolution. GRASS environmental variables are often used in scripts as you will learn in Chapter 9.

LOCATION and MAPSET management   The most efficient way to copy a LOCATION or a complete GRASS GIS data directory is to package the directories and then extract them in the destination directory. Care must be taken, if vector attribute data are kept in an external SQL database. For example, to package the nc_spm LOCATION, enter:

```
cd /usr/local/share/grassdata
tar -cvzf /tmp/mync_spm_location.tar.gz nc_spm/
mv mync_spm_location.tar.gz  /some/target/directory/
cd /some/target/directory/
tar -xvzf /tmp/mync_spm_location.tar.gz
```

If the target directory is located on a different computer, you can transfer the file mync_smp_location.tar.gz on an external memory drive (USB, DVD) or through network to the destination machine and extract it there.

You can remove a LOCATION from the GIS data directory as follows:

```
cd /usr/local/share/grassdata
rm -r nc_spm
```

This will remove the entire nc_spm directory. If you want to avoid the delete confirmation prompts for every file/subdirectory, add the flag -f to the rm command. Of course you can also use a file manager. For MAPSETs, the same applies – they can be simply removed or renamed using command line tools or a file manager. Note that this may affect MAPSET search paths (managed with g.mapsets and g.access).

3.2 Starting GRASS with a new project

When starting a new project, you need to define a new LOCATION and its projection and coordinate system. GRASS provides three options through its start-up panel (Figure 3.4):

- use available georeferenced file,
- use European Petroleum Survey Group (EPSG[8]) code, or
- define projection and coordinate system parameters.

New project from a georeferenced file: North Carolina State Plane NAD83 with units meters We will use the North Carolina (NC) geodetic point data provided as a SHAPE file gdc.shp to illustrate the first, easiest case. You can find the data within the geod_pts_spm.zip file in the subdirectory ncexternal/ of the NC sample data set provided for this book. The data include a gdc.prj text file with the projection and datum information: it is State Plane NAD83 with units meters, the same as we use for our sample data set. To create a new LOCATION, start

```
grass63 -gui
```

and select Georeferenced file under Define new location with in the start-up panel. This opens a new panel where you provide:

- Name of new location, for example, mync_spm,
- Path to new location, the path to your GIS data directory,
- Path to georeferenced file, use the Browse button to define this path.

Then click on Define location ⤳ OK. After creating this new LOCATION, it will be selected automatically along with the PERMANENT mapset. To create an additional mapset, type a new name, for example, mymapset into the field under Create new mapset, select it in the list, click Enter GRASS and you should be in your newly created project LOCATION. You can check your projection using g.proj -p (or g.proj -w); it should be Lambert Conformal Conic with NAD83 datum, units meters, that are used for the official State Plane coordinate system in NC. Note that the default region (spatial extent) has not been set and default values (all 1) are used. If you know the coordinates of your project area you can use g.region to define the coordinates, for example, for the LOCATION mync_spm we can define:

```
# use flag -s to save it as default region
g.region n=228500 s=215000 w=630000 e=645000 res=10 -p -s
```

[8] Now called "OGP Surveying & Positioning Committee", http://www.epsg.org/

If you do not know the coordinates, the easiest approach is to import a file (for example, the geodetic points that we used to create the LOCATION mync_spm) and set the region from this file as will be explained in the next chapter.

After having created the new projected LOCATION, it is highly recommended to run g.region -b. The command reprojects the current MAPSET boundary values to latitude-longitude which helps you to verify that the projection definitions are correct. Note that the -b flag reprojects to the WGS84 ellipsoid.

New project from EPSG code: North Carolina State Plane NAD83 with units feet   If you don't have any standard (GDAL/OGR readable, see Section 3.3.3) georeferenced file you can use the European Petroleum Survey Group (EPSG) code, aimed at standardization of common projection definitions. Projections and coordinate systems including geodetic datum can either be manually defined or selected via EPSG code from predefined list entries. The EPSG codes menu window provides a browse and search option for EPSG codes. If you open the EPSG code browser in GRASS, you can search for available North Carolina projections: you will find more than five (including deprecated codes), so you will need to make a decision which one to use. Our sample data set uses "NAD83(HARN) / North Carolina", EPSG code 3358; however, some of our data come in the same projection, but with units feet (EPSG code 2264) and we will use the latter as an example. Again start grass63 -gui, and select EPSG codes in the start-up panel (Figure 3.4). This opens a new panel (Fig. 3.7) where you provide

- Name of new location, for example, change to mync_spf,
- Path to the EPSG-codes file, should appear automatically,
- EPSG code number of projection, the code of the projection.

Then type in the EPSG code into the last entry line, in our case 2264 (click on Browse to see/search the list of all EPSG codes, see Fig. 3.7). Then click on Define location and OK. You should see the new LOCATION in the list and you continue as in the previous example by creating your mapset.

For both of the above described options the spatial extent of the study area – region – needs to be defined. As in the above example, we use g.region to define the boundary coordinates for the LOCATION mync_spf:

```
g.region n=749680 s=705370 e=2116150 w=2066920 res=10 -p

# verify if resulting latitude-longitude coordinates match
g.region -b
```

New project from interactively defined values   The third option is used if Projection values are selected in the start-up panel (Figure 3.4). We will illustrate the procedure in the next section by creating a new LOCATION with geographic (lat/long) coordinate system.

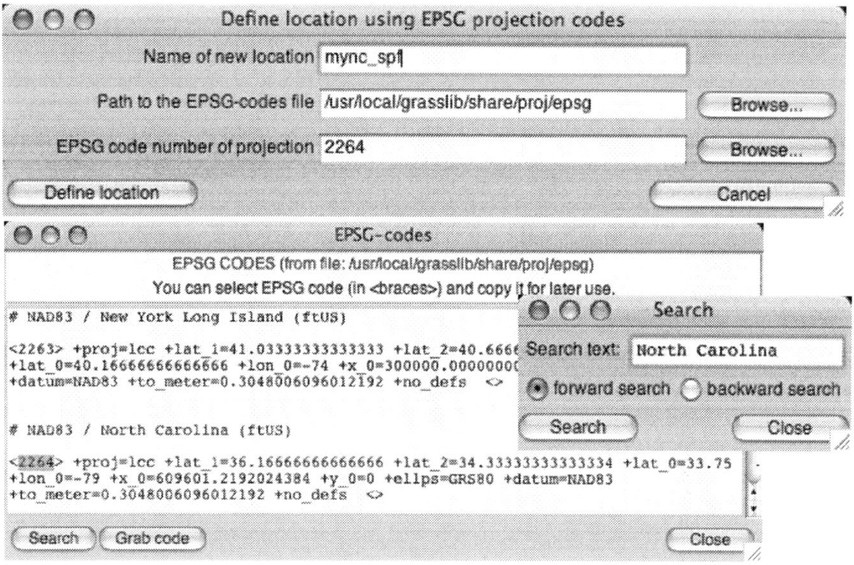

Fig. 3.7. Definition of NC State Plane/Feet from EPSG code

```
------------------------------------------------------------------
                            GRASS 6.3.cvs
DATABASE:  A directory (folder) on disk to contain all GRASS maps
           and data.
LOCATION:  This is the name of a geographic location. It is defined
           by a co-ordinate system and a rectangular boundary.
MAPSET:    Each GRASS session runs under a particular MAPSET. This
           consists of a rectangular REGION and a set of maps.
           Every LOCATION contains at least a MAPSET called
           PERMANENT, which is readable by all sessions.
The REGION defaults to the entire area of the chosen LOCATION.
           You may change it later with the command: g.region
- - - - - - - - - - - - - - -- - - - - - - - - - - - - - - - - -
LOCATION:   mync_ll  (enter list for a list of locations)
MAPSET:     user1_____  (or mapsets within a location)
DATABASE:   /usr/local/share/grassdata_____

      AFTER COMPLETING ALL ANSWERS, HIT <ESC><ENTER> TO CONTINUE
                     (OR <Ctrl-C> TO CANCEL)
------------------------------------------------------------------
```

Fig. 3.8. GRASS text-based startup screen for selection of LOCATION, MAPSET and DATABASE

3.2.1 Defining the coordinate system for a new project

To create a new LOCATION from projection values, first type **grass63 -gui**
and select Projection values in the start-up panel (Figure 3.4). This brings
up the old, text-based GRASS start-up interface   screen (see Figure 3.8).
Alternatively, **grass63 -text** gets you directly to the text-based interface.

For LOCATION enter the name for your new project (in our case **mync_ll**),
for MAPSET you can enter **user1**, and for DATABASE (GIS data directory)
you should have **/usr/local/share/grassdata** (if it is not there, type it in).
Note that this is an old fashioned interface and when you want to change
something, you need to type over it (BACKSPACE will not erase it). Once
you have entered the new LOCATION, MAPSET, and DATABASE, you can
continue with <ESC><ENTER>. Because your LOCATION does not exist yet,
the following menu appears:

```
LOCATION <mync_ll> - doesn't exist
Available locations:
--------------------
mync_spm mync_spf nc_spm
--------------------
Would you like to create location <mync_ll> ? (y/n)
```

Type <y> and you will get the following message:

```
To create a new LOCATION,you will need the following information:
1. The coordinate system for the database
        x,y (for imagery and other unreferenced data)
        Latitude-Longitude
        UTM
        Other Projection
2. The zone for the UTM database
   and all the necessary parameters for projections other than
   Latitude-Longitude, x,y, and UTM
3. The coordinates of the area to become the default region
   and the grid resolution of this region
4. A short, one-line description or title for the location
Do you have all this information for location <mync_ll>? y
```

From the previous sections, you should understand what latitude-longitude
or UTM means and you should know, based on the data that you want to
work with (or from your supervisor, customer or instructor), what coordinate
system you are going to use (see Figure 3.9 for a general idea). You can type
again <y> and you will be asked to specify the new coordinate system:

```
A   x,y
B   Latitude-Longitude
C   UTM
D   Other Projection
```

Type the appropriate letter, in our example, it will be <B> for Latitude-
Longitude. We accept and continue with:

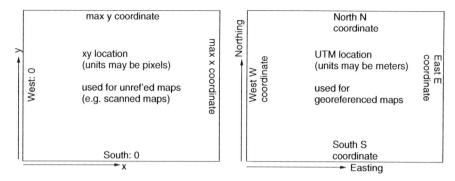

Fig. 3.9. Definition of a xy LOCATION and a projected LOCATION

```
Please enter a one line description for location <mync_ll>
> North Carolina Latitude-Longitude WGS84
ok? (y/n) [y] y
```

The next step is the selection of the geodetic datum:

```
Do you wish to specify a geodetic datum for this location? y
Please specify datum name
Enter 'list' for the list of available datums
or 'custom' if you wish to enter custom parameters
Hit RETURN to cancel request
>
```

To see the list of available datums, enter **list** at this prompt, then select the datum by typing its name:

```
>list
Short Name      Long Name / Description
---
agd66    Australian_Geodetic_Datum_1966
                      (australian ellipsoid)
[...]

wgs84    World_Geodetic_System_1984
                      (wgs84 ellipsoid)
---
Please specify datum name
Enter 'list' for the list of available datums
or 'custom' if you wish to enter custom parameters
Hit RETURN to cancel request
>wgs84
```

For most geodetic datums you have to select the datum transformation parameters:

```
Now select Datum Transformation Parameters
Enter 'list' to see the list of available Parameter sets
Enter the corresponding number, or <RETURN> to cancel request
>list
Number  Details
---
1       Used in Default wgs84 region
        (PROJ.4 Params towgs84=0.000,0.000,0.000)
        Default 3-Parameter Transformation
---
Now select Datum Transformation Parameters
Enter 'list' to see the list of available Parameter sets
Enter the corresponding number, or <RETURN> to cancel request
>1
```

These are all required parameters for latitude-longitude. Next you will be prompted to define your default region by defining the boundary coordinates of the project area and the default raster resolution (here we use sexagesimal degree notation and the spatial extent will cover entire state of NC):

```
             DEFINE THE DEFAULT REGION


             ====== DEFAULT REGION =======
             | NORTH EDGE: 37:00:00N_    |
             |                           |
  WEST EDGE  |                           |EAST EDGE
   85:00:00W_|                           |75:00:00W_
             | SOUTH EDGE: 33:00:00N_    |
             =============================

 PROJECTION: 3 (Latitude-Longitude)        ZONE: 0

                 GRID RESOLUTION
                   East-West:     0:00:30_____
                   North-South:   0:00:30_____

 AFTER COMPLETING ALL ANSWERS, HIT <ESC><ENTER> TO CONTINUE
              (OR <Ctrl-C> TO CANCEL)
```

The default raster resolution (GRID RESOLUTION) is arbitrary, because you can change it later based on the needs of your application. For latitude-longitude LOCATIONs, you have to define the resolution in degree:minutes:seconds (DMS) as well. You can leave this screen with <ESC><ENTER> and then check the list of parameters that appears:

```
   projection:   3 (Latitude-Longitude)
   zone:         0
   north:        37:00:00N
   south:        33:00:00N
   east:         75:00:00W
   west:         85:00:00W
```

```
e-w res:      0:00:30
n-s res:      0:00:30

total rows:            480
total cols:           1200
total cells:       576,000
```

```
Do you accept this region? (y/n) [y] >  y
LOCATION <mync_ll> created!
Hit RETURN -->
```

If everything is correct, type <y> and <ENTER> and you will get back to the startup screen. Type <ESC><ENTER> again and you will get the message that your MAPSET does not exist yet (note that the MAPSET PERMANENT was created automatically):

```
Mapset <<user1>> is not available

Mapsets in location <>
----------------------
(+) PERMANENT

note: you only have access to mapsets marked with (+)
----------------------
Would you like to create < user1 > as a new mapset? (y/n) y
```

Type <y> and your new LOCATION with your MAPSET are created and the GRASS prompt appears. You are now working in GRASS. You can check the definition of your LOCATION by running:

```
g.proj -p
g.region -pm
```

The second command prints the region information with the geodesic resolution reported in meters. Now the LOCATION is ready, and you can start importing data.

The dialog used for the LOCATION definitions will vary depending on the coordinate system. For example, when defining a new LOCATION in the UTM coordinate system (option <C> in the dialog) you will be asked for the Datum name (usually "nad83" in USA), Datum transformation parameters (you can select them based on the state, default is used for NC), and Zone (based on state, NC is in zones 17 and 18).

Most states in USA use State Plane Coordinate System (you will find it in the list for option <D>), and the dialog allows you to create its projection parameters based on the state and county Federal Information Processing Standards (FIPS) codes that are listed for each state and county in the dialog. If you need to work in feet, make sure that you select US Survey Foot for a project in USA.

```
---------------------------------------------------------------
|                  North: rows (y)                          |
|                        (from image)                       |
|                                                           |
| West: 0                               East: cols (x)      |
|                                           (from image)    |
|                                                           |
|                  South: 0                                 |
---------------------------------------------------------------
   Resolution: East-West:    1
               North-South: 1
```

Fig. 3.10. Definition of a region for xy LOCATION suitable for importing an image or scanned map. Units are pixels

3.2.2 Non-georeferenced xy coordinate system

You can define a LOCATION in a general, non-georeferenced coordinate system xy if you need to work with non-georeferenced data, or you do not know the parameters of your coordinate system, or your coordinate system is not supported by GRASS.

To define a new xy LOCATION, type `grass63 -gui` and select Projection values in the start-up panel (Figure 3.4). Type in the new names for LOCATION and MAPSET; for example, `my_xy` and `user1` in the old, text-based GRASS start-up interface screen (see Figure 3.8). Similarly to the procedure described in Section 3.2 proceed to the question Please specify the coordinate system for location my_xy. The coordinate system we need here is <A> "x,y". After entering a one line description, you reach the LOCATION region definition screen. Now define the region size in x and y direction (columns and rows). It should cover at least the size of the image or map that you want to import. The xy LOCATION can be defined larger than needed because the actual memory used depends only on the size of your imported file. When working with imagery data, set the west and south values to 0 (zero) and the north and east values to the number of rows and columns of the image (or more, compare Figure 3.10). The GRID RESOLUTION can be set to 1, because the units are pixels. After leaving this menu and accepting the definition, the new LOCATION is created. You can return to the GRASS startup screen and leave it again to create the MAPSET and to enter GRASS.

3.3 Coordinate system transformations

Geospatial data for a given study area are often provided in different coordinate systems (for example, a combination of data in the UTM, State Plane and geographic coordinates is quite common in USA). It is therefore important to have the capability to transform data between different projections and coordinate systems.

GRASS and its projection support through PROJ4   The projection library needed for GRASS 6 is PROJ 4.5.0 or later which was originally developed by USGS (Evenden, 1995). This library is now maintained by volunteers[9] and contains a stand-alone program cs2cs for the reprojection of coordinate lists. PROJ4 supports datum transformations from version 4.4.5 onwards. The general procedure for a transformation between two projections is internally always performed through geographical coordinates:

Projection1 ⤳ latitude-longitude ⤳ Projection 2

The projection information for each GRASS LOCATION is stored in the PERMANENT MAPSET in files PROJ_INFO and PROJ_UNITS. Depending on the actual projection the following parameters may be included: proj (projection type), name (projection name), ellps (ellipsoid), a (ellipsoid: equatorial radius), es (ellipsoid: eccentricity squared), zone (zone for the area), unfact (conversion factor from meters to other units, e.g. feet), lat_0 (standard parallel), lon_0 (central meridian), k (scale factor), x_0 (false easting) and y_0 (false northing).

To simplify the definition of a projection, PROJ4 provides support for the EPSG codes, aimed at standardization of common projection definitions. Projections and coordinate systems including geodetic datum can either be constructed or selected via EPSG code from predefined list of entries. As mentioned in the previous section, the list of EPSG codes is usually installed at /usr/local/share/proj/epsg.

Depending on the type of data, coordinate transformations can be done in two ways:

- ASCII file coordinate lists can be transformed between any of the more than 120 supported projections by running the external command cs2cs provided by PROJ4 (or m.proj);
- raster and vector maps can be transformed between two existing LOCATIONS with different coordinate systems using the commands r.proj and v.proj.

As an alternative, the external commands gdalwarp (for raster data) and ogr2ogr (for vector data) are available. For map reprojections, see Section 3.3.3.

3.3.1 Coordinate lists

PROJ4 provides the command cs2cs for reprojection of point lists given by coordinate pairs, such as map corners or points from a GPS survey. The

---

[9] PROJ4 Web site, http://proj.maptools.org/

GRASS command m.proj can be used as frontend to cs2cs. You have to define the source and the target projections, then the coordinate pairs are queried or read from an ASCII file, transformed, and written either to the screen or redirected to an output file. You can also list the supported projections using cs2cs -lp, an extended list with flag -P, ellipsoids cs2cs -le, prime meridians cs2cs -lm, datums cs2cs -ld and units cs2cs -lu.

For example, to transform the corner points of the mync_ll LOCATION from latitude-longitude/WGS84, to UTM/WGS84 zone 17 you can use (enter in one line, you can find the required parameters in the epsg file):

```
cs2cs -v +proj=latlong +to +proj=utm +zone=17 +ellps=WGS84 \
      +datum=WGS84 +units=m
```

This command prints out the projection parameters (due to the -v flag) and then waits for input. Now you can type in the longitude and latitude coordinate pairs, optionally with the related elevation; for example, the southwest map corner of the mync_ll LOCATION delivers:

```
# type input, use CTRL-D to leave
78d00'00"W 35d00'00"N
# resulting UTM Zone 17 coordinates
773798.10        3877156.69 0.00
```

which represents the corresponding UTM coordinates in meters.

As a different example, you can transform the elevation map corner coordinates from NC State Plane/NAD83 (nc_spm LOCATION) to LAEA/Sphere/no datum and store the resulting values in a file. To get the elevation map coordinates, start GRASS with the nc_spm LOCATION, and run:

```
r.info elevation
# only output map boundaries
r.info -g elevation
```

and store the reported values into a file nc_spm_NAD83.txt. The input file must be written in plain ASCII format containing row-wise easting and northing:

```
630000 228500
645000 228500
645000 215000
630000 215000
```

Now you can convert the coordinates in the file to the standard raster map projection of the National Atlas of the U.S.[10], which is Lambert Azimuthal Equal Area (LAEA) on a Sphere (see EPSG code 3358):

---

[10] National Atlas of the U.S. download area,
    http://nationalatlas.gov/atlasftp.html

```
# alternative
cat nc_spm_NAD83.txt | cs2cs -v +init=epsg:3358 \
    +to +proj=laea +lat_0=35 +lon_0=-80 +x_0=0 +y_0=0    \
    +ellps=sphere +units=m > ncLAEA.txt
```

The command line will reproject the coordinate pairs stored in file nc_spm_NAD83.txt to coordinates in Lambert Azimuthal Equal Area on a Sphere without geodetic datum and write the result to file ncLAEA.txt where you will find information about the coordinate systems used and the resulting coordinates:

```
110780.21        70423.01
125780.56        70573.67
125917.38        57072.47
110916.94        56921.96
```

Note that you can find the parameters for many projections already formatted for the use with cs2cs command in the epsg file.

3.3.2 Projection of raster and vector maps

The projection of raster and vector maps between two different coordinate systems requires two LOCATIONs: one LOCATION holding the source map and input coordinate system information, and another LOCATION for reading the target coordinate system information and storing the projected map.

While vector maps can be reprojected directly, raster maps require to set region and resolution of the target location appropriately. The easiest way is to generate a vector "box" map of the region of interest in the source location using v.in.region. This vector area map is then reprojected into the target location with v.proj and used to set the region in the target location. Additionally, the target raster resolution is defined. The subsequent run of r.proj reprojects the desired raster map.

To illustrate the procedure, we will project raster and vector maps downloaded from the USGS in the geographic coordinate system to the State Plane system that we will use for most of our work. Our sample data set includes LOCATION nc_ll with a 1/3 arcsecond (approx. 10m) resolution elevation map. We reproject this elevation map from LOCATION nc_ll into the nc_stm LOCATION defined with State Plane coordinate system in Section 3.2.1. For this, we start GRASS with LOCATION nc_ll, set the current region to the elevation map and generate the "box" vector map:

```
# in latitude-longitude LOCATION
g.region rast=elev_ned_03arcsec -p
v.in.region elev_ned_03arcsec_box
```

Then we restart GRASS (or start a parallel session) with nc_stm LOCATION and "pull" the 1/3 arcsecond resolution elevation map from the source LOCATION nc_ll into your current LOCATION at 10m resolution as follows:

```
# in State Plane metric LOCATION
v.proj elev_ned_03arcsec_box location=nc_ll mapset=PERMANENT
# set region to box vector map and update raster resolution
g.region vect=elev_ned_03arcsec_box res=10 -pa
r.proj in=elev_ned_03arcsec out=elev_ned_10m location=nc_ll \
       mapset=PERMANENT method=cubic
d.mon x0
d.rast elev_ned_10m
# box vector map is no longer needed
g.remove vect=elev_ned_03arcsec_box
```

The projected map is saved with the same name as the input map if no output name is defined. The resolution and region (map extent) of the projected map is given by the current region settings in the target LOCATION `nc_stm`. You can verify the settings with `g.region -p` and use this command to limit transformations to subregions with desired resolution. Please refer to the manual page of `r.proj` for the available interpolation methods used during the transformation. Projections can introduce significant errors if the resolution and method are not selected carefully, especially if there is a significant difference between the resolution of the original and projected map.

Similarly, you can project vector maps:

```
# in State Plane metric LOCATION
v.proj in=roads location=nc_ll mapset=PERMANENT
d.vect roads
```

Creation of subregions is not supported for the vector data, therefore entire map is always projected or it has to be clipped beforehand in the source LOCATION.

3.3.3 Reprojecting with GDAL/OGR tools

Sometimes it is more convenient to reproject maps to a desired projection (or coordinate system, ellipsoid, geodetic datum) before importing them into a GRASS LOCATION. The free GDAL/OGR libraries[11] provide a set of tools to perform such map preprocessing. GDAL is a translator library for raster geospatial data formats while OGR supports vector data. For GNU/Linux and MS-Windows systems these libraries and tools are included in the "FW-Tools"[12] which are easy to install. However, the GDAL/OGR tools are usually available once GRASS is installed since the GDAL/OGR libraries are a requirement.

---

[11] GDAL library, http://www.gdal.org/
 OGR library, http://www.gdal.org/ogr/
[12] FWTools Web site, http://fwtools.maptools.org/

GDAL and raster data   GDAL provides several tools for working with raster data, such as `gdalinfo` for reports about the properties of the GDAL-supported raster data set, `gdal_translate` for conversion between different raster formats, including subsetting and resampling; and `gdalwarp` for image reprojection. The general command structure is:

```
# list all supported formats
gdalinfo --formats

# show details of a selected format
gdalinfo --format gtiff

# show raster map information (metadata)
gdalinfo [flags] rastmap

# do raster map format conversion or projection assignment
gdal_translate [flags] [parameters] inrastmap outrastmap

# do raster map reprojection
gdalwarp [flags] [parameters] inrastmap outrastmap
```

The following example shows how to convert a free LANDSAT-TM7 scene for the central North Carolina region, available from GLCF Maryland[13], from UTM/WGS84 to LCC/NAD83 to match the `nc_spm` LOCATION projection. To reproject the image we run `gdalwarp` with the flags `-tr xres yres` to maintain original resolution and `-te W N E S` to cut out the region of interest. We can make use of shell capabilities and launch the command within a GRASS session (`nc_spm` LOCATION):

```
# get information about the image
gdalinfo p016r035_7t20020524_z17_nn30.tif

# set to region of interest
g.region n=228500 s=215000 w=630000 e=645000 res=14.25 -pa

# transfer region coordinates into shell as defined variables
eval `g.region -g`
echo "$w $s $c $n"

# check current location projection
g.proj -wf

# reproject to current location projection set on the fly
# as -t_srs (target spatial reference system)
gdalwarp -t_srs "`g.proj -wf`" -te $w $s $e $n \
        p016r035_7t20020524_z17_nn30.tif
        p016r035_7t20020524_nc_spm_wake_nn30.tif
```

---

[13] GLCF Maryland LANDSAT data for NC region,
   `ftp://ftp.glcf.umiacs.umd.edu/glcf/Landsat/WRS2/p016/r035/`

```
# verification
gdalinfo p016r035_7t20020524_nc_spm_wake_nn30.tif
```

The preprocessed LANDSAT-7 band is now ready for import into the `nc_spm` LOCATION with `r.in.gdal`. Note that the available LANDSAT scenes from GLCF Maryland are already included in the `nc_spm` LOCATION (MAPSETs PERMANENT and landsat). If you only need to cut out the image without reprojection or if you want to change the raster format, then use `gdal_translate` instead.

OGR and vector data   Vector maps can be reprojected and preprocessed using the OGR library, with the SHAPE format as default output. The program `ogrinfo` can be used to report vector map metadata and `ogr2ogr` to perform vector map reprojections. The general command structure is:

```
# list all supported formats
ogrinfo --formats

# show vector map information (metadata)
ogrinfo [flags] [parameters] vectmap [layer [layer ...]]

# do vector map format conversion, projection assignment
# or reprojection
ogr2ogr [flags] [parameters] outvectmap invectmap [parameters]
```

In our OGR example, we use the North Carolina county boundaries map from USGS provided in the SHAPE format in latitude-longitude coordinate system (you will find it as 9.shp in the `boundaries_nc_ll.tar` file stored `ncexternal/` directory). We prepare it for import into the `nc_spm` LOCATION which requires reprojection and spatial subsetting. To display the vector map information run:

```
ogrinfo -summary 9.shp
```

The provided data include projection information in the `9.prj` file; in case this file was missing, you can add it using the `ogr2ogr` command with `-a_srs` flag as we illustrate by the following example:

```
# create a copy 'testll.shp' of map '9.shp'
# and add projection file
ogr2ogr -a_srs '+proj=latlong +ellps=wgs84 +datum=wgs84' \
        testll.shp 9.shp

# verification
ogrinfo -summary testll.shp
more 9.prj
more testll.prj
```

When the SHAPE file set includes the projection information (file with `.prj` extension) we are ready to reproject to LCC/NAD83 coordinate system as required for the `nc_spm` LOCATION. We can simultaneously cut out the region of interest, as an example we will create a subset with county boundaries in the Central NC. For convenience, we use the EPSG code to select the output projection parameters. Note that it is necessary to have the NAD datum shift files installed (see the PROJ4 web page, section "Frequently Asked Questions" (FAQ) for details or the GRASS-Wiki). The EPSG codes are available in an ASCII table:

- GNU/Linux and MacOS X: `/usr/share/proj/epsg` or `/usr/local/share/proj/epsg`,
- MS-Windows: depends on the installation of PROJ4.

Alternatively, the graphical GRASS startup screen provides an EPSG code browser. We have already identified the EPSG code for the `nc_spm` LOCATION as 3358 in Section 3.2 so we can perform the projection and subsetting as follows:

```
# project from LatLong/Sphere to LCC/NAD83
# cut out region of interest (boundary coords. W S E N)
ogr2ogr -t_srs '+init=epsg:3358' \
        -spat -79.55 35.00 -77.70 36.30 \
        boundary_county_ctr 9.shp


# verification
ogrinfo -summary 9.shp boundary_county_ctr
```

The command will cut out the boundaries of three counties located within the defined region, project the coordinates and store the SHAPE file set in a subdirectory called **boundary_county_ctr**. It can then be directly imported into the `nc_spm` LOCATION.

4

GRASS data models and data exchange

GRASS stores the georeferenced data as raster and vector maps. In this chapter, we explain the basic properties of GRASS data models and their management. You will also learn how to import and export data in various raster and vector formats; this task has been greatly simplified by a growing list of formats supported by the GDAL/OGR libraries which make GRASS an interoperable GIS which are used by the import and export modules of GRASS. An overview of data import into GRASS is given in Figure 4.1, export is shown in Figure 4.2.

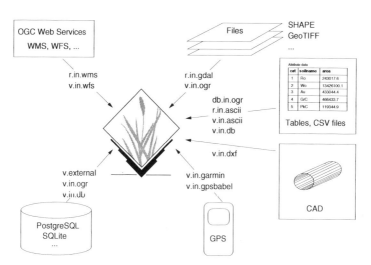

Fig. 4.1. Data import into GRASS

4.1 Raster data

Raster data, stored in GRASS as a matrix of values, represent either a continuous field (surface), an image, or geometric objects (points, lines, areas) corresponding to discrete fields (Figure 4.3). For surfaces, the values in the matrix are assigned to the center points of grid cells. They represent actual measured or computed values, such as elevation, slope, or temperature. For discrete fields, the values are assigned to the entire cell area and represent category numbers.

4.1.1 GRASS 2D and 3D raster data models

A raster map is stored in GRASS as a set of files organized as follows:

- map header which includes a projection code, coordinates representing the spatial extent of the raster map, number of rows and columns, resolution, and information about compression;
- generic matrix of values in a compressed, portable format which depends on the raster data type (integer, floating point or 3D grid);
- optional category file which contains text or numeric labels assigned to the raster map categories;
- optional color table;
- optional timestamp, range of values, quantization rules (for floating point maps), and null (no-data) files;
- history file which contains metadata such as the data source, the command that was used to generate the raster map, or other information provided by the user;
- a raster map that is a reclassification of another map includes a reclassification table instead of a full matrix of values.

All this information is stored in the related subdirectories in the LOCATION_NAME/MAPSET directory. In the following sections, we describe how these components are managed and queried.

Raster data can be stored in GRASS as a 2D integer, 2D floating point (single or double precision), or as a 3D floating point matrix of values (single or double precision). The internal raster format is architecture independent and portable between 32bit and 64bit machines. As a result, a GRASS data directory can be accessed in a heterogeneous network file system (NFS) without compatibility problems. Internally, the integer format is called CELL, single precision floating point is called FCELL, double precision floating point is DCELL, and 3D raster is called GRID3D. The choice of the integer or floating point data depends on the application. Their use can be described in general as follows:

Integer raster maps (CELL type) are used for rasterized geometric objects (points, lines, areas) represented as discrete fields and for some image data. Each raster cell is assigned an integer value called *category number*.

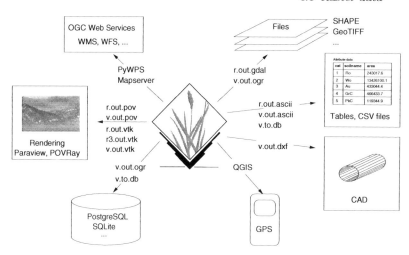

Fig. 4.2. Data export from GRASS

Each of the categories may have a label (usually a character string but a number can be used as well) describing the meaning or properties of these categories, for example, a vegetation type.

Floating point raster maps (FCELL and DCELL types) are used for continuous fields such as elevation or temperature surfaces. It is possible to label these data by defining ranges of values which can be interpreted as categories and assigning each range a label (text or number).

3D floating point raster maps are used for raster volumes stored as a voxel (volume pixel) data model (FCELL and DCELL type) designed to support representation of trivariate continuous fields.

Note that continuous field data can be represented in integer format, good examples are some older digital elevation models. This is a limitation of the data quality and such data should be treated as continuous field representations. We will point out the related, application specific issues later in this chapter and in Chapter 5.

GRASS also allows you to create a new raster map by re-defining the categories in the original raster map as described in Section 5.1.7. The reclassified map internally does not contain raster data: it only provides a table with reclassification rules and serves as a reference to the original raster map. Although it behaves like a regular raster map from the user's point of view, a few GRASS modules may not work with reclassified maps; if that is the case, the module will report an error and suggest that the user generates a true copy of such a raster map (see Section 5.1.5).

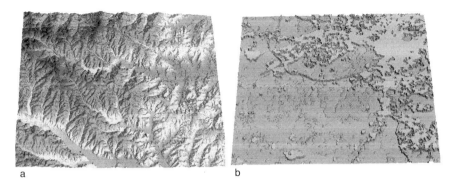

a   b

Fig. 4.3. Types of raster data: a) continuous field, b) discrete areas

4.1.2 Managing regions, raster map resolution and boundaries

GRASS differs from other GIS in the way it handles region (map extent) and resolution. While each raster map has its own spatial extent and resolution defined in its header, the operations with raster data are performed using the "working" (or current) region and the resolution set by the command g.region. If the current region is smaller than the spatial extent of the raster that is being processed, the operation is applied only to the subset of the raster file defined by the current region. If the resolution is different, the raster is automatically resampled (see Section 5.3.3). This approach makes raster analysis, modeling and export very convenient and efficient. Note that the GRID RESOLUTION defined when setting up a LOCATION is the default region resolution and will be used only if the current region is set to the default region.

We will use our sample data set to illustrate the handling of a region and resolution. Start GRASS with LOCATION nc_spm, MAPSET user1 and you can check the current region spatial extent and resolution as follows:

```
g.region -pec
# provides the following output
projection: 99 (Lambert Conformal Conic)
zone:       0
datum:      nad83
ellipsoid:  a=6378137 es=0.006694380022900787
north:      228500
south:      215000
west:       630000
east:       645000
nsres:      10
ewres:      10
rows:       1350
cols:       1500
cells:      2025000
```

```
region north-south extent: 13500.000000
region east-west extent: 15000.000000
region center northing: 221750.000000
region center easting:  637500.000000
```

The reported values are in our case in meters, you can check the units using g.proj -p. Running g.region with the flag -b allows you to print the current region in geographic coordinates (longitude and latitude, referred to WGS84). You can also use the command to change the resolution, for example to 15m, and then save the current region settings as a region file myregion_15m for future use as follows:

```
g.region res=15 save=myregion_15m -p
```

This is sometimes useful when working on different subregions within the given LOCATION. The region can also be defined from an existing raster or vector map, for example:

```
g.region rast=elevation -p
```

will adjust the current region according to the raster map elevation. There are numerous additional options and their combinations for setting the region spatial extent based on coordinate values, here we provide just two examples:

```
g.region n=228000 s=215500 w=632000 e=640000 -p
g.region n=s+500 e=w+500 -p
```

The first command sets the region by defining the coordinates for its northern, southern, eastern and western edges, while the second example creates a small 500m × 500m subregion in the south-west corner of the current region.

If you want to reset to the default region (spatial extent coordinates and raster resolution) of your LOCATION use the -d flag:

```
g.region -dp
```

You can change the definition of your default region to the current region using g.region -s, but you need to run this command from the PERMANENT MAPSET.

It may happen, that region boundaries lead to a modified, non-integer raster resolution. If this is not desired, the -a flag can be used to align the region to resolution. It adjusts all four boundaries to be even multiples of the resolution by slightly enlarging the current region. To see the effect, compare:

```
g.region res=15 -dp
g.region res=15 -adp
```

To print the spatial extent of the 3D region use g.region -p3, you will see the default settings representing a single layer that is 1 unit deep. To change the 3D extent to more realistic values (the vertical extent range from 0 to 200m, with vertical resolution 1m and horizontal resolution 30m) type:

```
g.region b=0 t=200 tbres=1 res3=30 -p3
```

Note that the output will show two horizontal resolutions and row/column numbers (e.g., nsres: 10, nsres3: 30), the first resolution is used when working with 2D rasters, the second (nres3) is for volumes. This allows you to set different horizontal resolutions when working with both volumes and 2D raster maps and to reduce the computational and memory requirements for volume data. To print the current bounding box in latitude-longitude/WGS84 datum, run:

```
g.region -bg
```

This is useful to verify the current projection definition.

4.1.3 Import of georeferenced raster data

When importing raster data, we need to distinguish three general raster format types:

- ASCII raster formats, which can have integer or floating point values, both negative and positive (e.g., ASCII-GRID, GRASS-ASCII etc.);
- Binary image formats, with positive integer values (e.g., JPG, PPM, PNG etc.);
- Binary raster formats: with integer and floating point, negative and positive values; single and multiple bands, single and multiple resolutions (such as ERDAS IMG, HDF, GeoTIFF etc.).

Note that not all formats handle negative and floating point values. When obtaining data, make sure to get information about the coordinate system (projection, datum, etc.). Many formats include this information in support files that are used by the importing modules. All the external data that we use in this chapter to illustrate data import are available in the subdirectory ncexternal/ of the sample data set provided for this book at the GRASS book web site (http://www.grassbook.org).

More than 40 different raster formats can be imported with r.in.gdal command. It uses the GDAL library (see Section 3.3.3) which is required to run GRASS and is included in the GRASS binary releases. The formats are detected automatically and the coordinate system information, if available, is compared with the current LOCATION. The supported formats include the most common georeferenced formats such as GeoTIFF, ArcGRID, ERDAS, USGS SDTS DEM, as well as common image formats that require manual georeferencing, for example, PNG, GIF, JPEG.

If the file is for a region much larger than the region of interest, you can first create a subset of the data using GDAL tools and the current region information (as explained in Section 3.3.3) and then import the smaller, easier to manage subset. As an example, we will cut out a subset and import the NC 1996 land cover raster map (full size is 400MB), provided in the same coordinate system as we use in our `nc_spm` LOCATION, so no reprojection is needed:

```
# set the resolution to match the data
# and get the coordinates for the current region
g.region swwake_30m res=28.5 -ap

# cut out an image subset for the current region
# with boundary coordinates W     N      E      S
gdal_translate -projwin 630000 228500 645000 215000 \
               lc96ras.img lc96ras_cut.img

# import the image subset and display the raster map
r.in.gdal lc96ras_cut.img out=landuse96_28m
d.mon x0
g.region rast=landuse96_28m -p
d.rast landuse96_28m
```

Note that only the raster map with category numbers is imported; you will have to add labels describing the land use types that are provided in a separate text file `lc96catlables.txt` to the categories using `r.support` (see Section 5.1.9 for details).

If your data set does not include coordinate system information (projection, datum, and units) and you are sure that it matches the coordinate system of your LOCATION, the -o flag allows you to use the LOCATION projection information for the imported map and override the projection match check in `r.in.gdal`. This is common with data in the TIFF/TFW format that usually consist of two files: `map.tif` (matrix of values) and `map.tfw` (header with coordinates but without projection information). Make sure to get both files when obtaining data. We have used this option to import a 1m resolution orthophoto, acquired in the year 2001 for the NC Floodplain mapping program, and provided in TIFF/TFW format for a small subarea of our test region:

```
g.region res=1
r.in.gdal -o IMG3720079200P20040216.tif out=ortho_2001_t792_1m
g.region rast=ortho_2001_t792_1m -p

# erase the monitor to set it to the current region and resol.
d.erase
d.rast ortho_2001_t792_1m
```

If the imported TIFF image consists of several bands, they are extracted as separate raster maps into the current MAPSET. A typical example is aerial color images delivered in RGB (red, green, blue) channels. GeoTIFF format is easier to import as it includes projection information and one or several raster maps in a single file. You can automatically extend the LOCATION default region given in the `DEFAULT_WIND` file based on an imported data set that covers a larger area using `r.in.gdal -e`. Import preparation of multispectral satellite data is explained in Section 8.1.2.

Generating a new LOCATION from an external raster map   The
module r.in.gdal provides an additional, very useful functionality by automatically generating a LOCATION from an external raster data set. For this
purpose, it has to be run within another LOCATION (this LOCATION can be
completely unrelated to the imported data and its setting won't be affected
by r.in.gdal execution). For example, you can import a new 30m Shuttle
Radar Topographic Mapping (SRTM) Digital Surface Model (DSM) for our
area, provided as ArcGRID coverage 88123760 in a SRTM_30m_ll.zip file[1] in
geographic coordinate system (WGS84 datum), and at the same time create a
new LOCATION mync_ll2 that is defined with the parameter location. The
projection information is taken from the input data set, in our case stored in
the file 88123760/prj.adf. To import the SRTM Digital Surface Model, run
the r.in.gdal command from the current nc_spm LOCATION:

```
unzip SRTM_30m_ll.zip
cd 88123760/
r.in.gdal 88123760 location=mync_ll2 output=srtm_30m_ll
```

To display the imported SRTM DSM, start GRASS with the new mync_ll2
LOCATION, MAPSET PERMANENT. You should see your imported raster
map srtm_30m_ll when you run g.list rast and you can check the coordinate system using g.proj -p. The region is automatically set to the imported
raster map, so you can open your GRASS monitor using d.mon x0 and display
the map using d.rast srtm_30m_ll.

Generally, if no projection information is present, the new LOCATION will
be set up without the coordinate system definitions. The module g.setproj
can then be used within the new LOCATION to define the projection information. Be careful to use g.setproj only in a new LOCATION because it
will overwrite your existing projection definition and units files! Note that
the module does not perform any coordinate transformation of data (see Section 3.3 to learn how to do that).

Import of ASCII raster files   You can import the GRASS ASCII raster
format using r.in.ascii. The input file must include a header followed by a
matrix of values with the first line starting at the NW corner and the last line
ending at the SE corner. You can try it out within the nc_spm LOCATION
as follows:

```
# create ASCII raster file by exporting the
# elevation map at 30m resolution
g.region res=30 -p
r.out.ascii elevation > elevation.asc
```

---

[1] NC 1 arcsec (30m) LatLong/WGS84 SRTM DEM V1 ArcGRID coverage file,
section "Data 3rd Edition", http://www.grassbook.org

```
# check the header and import the file
head elevation.asc
r.in.ascii elevation.asc out=myelev_30m
```

You can define null values, type (integer, float) and multiplier in the ASCII raster file header (see the manual page for more details). The `r.in.ascii` module can be also used to import SURFER ASCII grid file, by running it with the -s flag. The ARC/INFO ASCII GRID format can be imported by `r.in.gdal`. Data in this format sometimes have an associated `.prj` file with projection information. If not available, you can use the -o flag to override the projection check and use the projection definition from the current LO-CATION.

Raster maps can be directly created from dense point data produced, for example, by lidar or sidescan sonar and provided in the ASCII format as a set of (x,y,z) coordinates (see more about lidar in Section 6.9). You can use the module `r.in.xyz` to create raster maps from point data by computing various statistics using the points located within each cell, a procedure often called binning. You can compute raster maps that represent the number of points located in each cell, range of values in each cell, or compute a DEM using a mean elevation value, as shown in our example for the bare ground lidar elevation data for a small subarea in our sample data set (the input ASCII file produced by the NC Flood Mapping Program BE3720079200WC20020829m.txt is available in the subdirectory **ncexternal/** of the data set provided at the GRASS book site):

```
# set the region extent using the related airphoto but set to
# lower resolution to have several points per grid cell
g.region rast=ortho_2001_t792_1m res=10 -p

# compute a raster map representing number of points per cell
r.in.xyz BE3720079200WC20020829m.txt out=lidar_792_binn10m \
         meth=n
d.rast lidar_792_binn10m

# compute a raster map representing mean elevation for each cell
r.in.xyz BE3720079200WC20020829m.txt out=lidar_792_binmean10m \
         meth=mean
d.rast lidar_792_binmean10m
```

You will learn more about the point cloud data analysis using the `r.in.xyz` command in Section 6.9.

Raster map representing areas, lines or points can be created from given coordinates by `r.in.poly`. The module accepts text files containing coordinate pairs with labels. Either raster area ("A") or raster line ("L") type can be specified. As an example, we use a single area represented by its boundary coordinates stored in a text file **newfacility.txt**:

```
A
638656.00 220611.00
638796.00 220609.00
638799.00 220520.00
638653.00 220519.00
638653.00 220609.00
= 1 facility
```

The last line in the file defines the category number and a label for the area.
To define the imported raster extent and resolution it is important to set the
current region: we use a pre-defined rural sub-area at 1m resolution stored in
a region file **rural_1m**. Then we import the ASCII file with **r.in.poly** and
display it on top of a previously imported aerial imagery:

```
# list available regions and set the current region
g.list region
g.region rural_1m -p

# import and display the data
r.in.poly in=newfacility.txt out=newfacility_1m
r.info newfacility_1m
d.erase
d.rast ortho_2001_t792_1m
d.rast -o newfacility_1m
```

The resulting raster map **newfacility_1m** contains the desired area labeled as
"1 facility".

Import MrSID files   High resolution imagery data are often distributed in
the proprietary MrSID format. To import this format using **r.in.gdal**, the
GDAL library has to be compiled with MrSID support after downloading the
necessary Software Development Kit (SDK). Alternatively, you can convert
the file to GeoTIFF using the proprietary converter **mrsidgeodecode** as we
show for a TIFF file available in **ncexternal/**:

```
mrsidgeodecode -i lkwhee4.sid -o geotifflkwhee4.tif -of tifg
# verification
gdalinfo geotifflkwhee4.tif

# flag -o since projection info is missing but the map matches
r.in.gdal geotifflkwhee4.tif out=orthoIR1998 -o
g.region rast=orthoIR1998.blue -p
d.erase
d.rgb b=orthoIR1998.blue g=orthoIR1998.green r=orthoIR1998.red
```

The resulting image is a 1998 infrared Digital Orthoimagery Quarter Quad-
rangles (DOQQ) at 1m resolution.

Import binary arrays and raster formats not supported by GDAL
The module `r.in.bin` reads numerous binary array grids such as GTOPO30
DEM (worldwide elevation data at 30 arc-seconds resolution provided by
USGS), Etopo-2 DEM (worldwide elevation data at 2 arc-minutes resolution), Globe DEM (worldwide elevation data at 30 arc-seconds resolution, provided by NOAA), BIL, AVHRR (Advanced Very High Resolution Radiome-
ter), and GMT (Generic Mapping Tool). Please refer to the related manual
page (`g.manual r.in.bin`) for encoding details. Many of these formats are
much easier to import using `r.in.gdal`, but we include a few examples[2] of
import into `mync_11` for comparison (see `ncexternal/` subdirectory for sample data). To avoid reading the GTOPO30 data incorrectly, you can add a
new line `PIXELTYPE SIGNEDINT` in the `.HDR` to force interpretation of the file as
signed rather than unsigned integers. For this, open the `W100N40.DEM` in a text
file. Open GRASS in `mync_11` LOCATION and import the data as follows:

```
# extract and import GTOPO30 DEM
tar -xvzf GTOPOw100n40.tar.gz

# edit W100N40.DEM as described above, then import
r.in.gdal W100N40.DEM out=gtopo30_usa
g.region rast=gtopo30_usa -p

# check range for strange maximum value
r.info -r gtopo30_usa
 min=1
 max=6710
r.colors gtopo30_usa col=terrain
d.mon x0
d.erase
d.rast gtopo30_usa

# alternative import method
# note: add 'anull=-9999' to set the sea level to NULL
r.in.bin -sb input=W100N40.DEM out=gtopo30_usa2 bytes=2 \
         north=40 south=-10 east=-60 west=-100 \
         r=6000 c=4800 anull=-9999
r.info -r gtopo30_usa2
r.colors gtopo30_usa2 col=terrain
d.rast gtopo30_usa2
```

As another example, we import ETOPO2v2 DEM data:[3]

```
# import ETOPO2v2 DEM for entire NC - includes bathymetry
r.in.bin NCETOPO2_4604.g98 out=ncetopo_2min bytes=2 \
         north=38 south=32 east=-72 west=-85 r=180 c=390
```

---

[2] GTOPO30 tile covering North Carolina,
  http://edc.usgs.gov/products/elevation/gtopo30/w100n40.html
[3] ETOPO2v2 DEM data Web site,
  http://www.ngdc.noaa.gov/mgg/fliers/06mgg01.html

```
r.colors ncetopo_2min col=terrain

# import a GMT type 1 (float) binary array (use -b to
# swap bytes if needed)
r.in.bin -hf input=your_map.grd out=gmtmap
```

GRASS includes several modules for importing raster maps in specialized formats that are not currently supported by GDAL. These include gridatb files formatted for TOPMODEL supported by the module r.in.gridatb, MATLAB files by r.in.mat, and SRTM HGT files from the NASA server by r.in.srtm. The module r.in.aster performs rectification, georeferencing and import of ASTER imagery, it requires compilation of GDAL with HDF support.

Import raster data directly from Web Sources GRASS 6 supports direct download of raster data from Open Geospatial Consortium's Web Map Service (OGC WMS) compliant Web servers by the module r.in.wms. The module downloads the requested data layers from the given map server for the current or given region, reprojects them and patches the tiles together. You can use it, for example, to download a USGS Digital Raster Graphics file (digitized topographic map) from MS Terraserver into the nc_spm LOCATION:

```
# list available layers
r.in.wms -l output=terraserver-drg  \
mapserver=http://terraserver.microsoft.com/ogccapabilities.ashx

g.region rural_1m -p
r.in.wms output=terraserver-drg \
    mapserver=http://terraserver.microsoft.com/ogcmap6.ashx \
    layers=DRG region=rural_1m format=tiff
```

Note that to run this module you need to have several standard Unix and Web tools installed, such as wget, sed, grep, see the manual page for more details.

To download LANDSAT pseudo-color pan-sharpened map (15m) into the nc_spm LOCATION, run:

```
# enforce 15m resolution
g.region swwake_30m res=15 -p
r.in.wms layers=global_mosaic \
        mapserver=http://wms.jpl.nasa.gov/wms.cgi \
        output=wms_global_mosaic_nc
d.erase
d.rast wms_global_mosaic_nc
d.vect roadsmajor col=yellow
d.barscale at=0.0,0.0 tcol=white bcol=none
```

```
# visualize in NVIZ (set its viewing resolution to 1)
nviz elevation col=wms_global_mosaic_nc vect=roadsmajor
```

Figure 4.4 shows the resulting map in the GRASS monitor.

Import of raster data without ancillary georeferencing files If you obtain a raster map in a common format such as TIFF, but without the related TFW file, you can update the geocoding manually. Of course, you have to get the related georeference information from the data provider. You can import the map by overriding the projection check with **r.in.gdal** -o. The lower left corner coordinates of the imported map will be at the origin of the LOCATION coordinate system, which is usually outside the study area and when you try to display the imported map you often don't see anything because the map is outside the current region. To include the georeferencing information the imported raster map header needs to be modified using the module **r.region**, in our case, we use the header information (spatial extent) of another raster map which perfectly matches our imported map:

```
r.in.gdal -o NonGeoIMG792.tif out=NonGeoIMG792
r.info NonGeoIMG792
[...]
      Projection: Lambert Conformal Conic (zone 0)
            N:       3048   S:          0   Res:      1
            E:       3048   W:          0   Res:      1
  Range of data:    min = 53  max = 188
```

Fig. 4.4. LANDSAT pseudo-color pan-sharpened map of South-West Wake County (NC) imported from JPL/NASA WMS (northern part of map shown)

If you use **d.rast NonGeoIMG792** nothing will be displayed. In this case the map header needs to be modified:

```
r.region NonGeoIMG792 rast=ortho_2001_t792_1m
r.info NonGeoIMG792
[...]
  Projection: Lambert Conformal Conic (zone 0)
            N:      222504    S:      219456    Res:      1
            E:      640081    W:      637033    Res:      1
  Range of data:     min = 53   max = 188
d.rast NonGeoIMG792
```

You can also use a vector map, a named region, a 3dview file or coordinate values to define the missing spatial extent of the imported map. Alternatively, the header can be modified using **r.support**. After starting it, specify the map name and go through the following dialog:

1. "Edit the header?" <y>. Rows and columns can be checked now. The values should be correct.
2. Pressing <ESC><ENTER> changes into the coordinates menu which looks similar to the LOCATION definition screen.
3. Now you have to update the boundary coordinates. Enter the correct coordinates and GRID RESOLUTION for this map by moving around with the cursor keys. Afterward hit <ESC><ENTER> to proceed.
4. The additional questions can be skipped with <ENTER>.

Then you can use the **r.support** module to assign labels for the raster map categories, add the map title, or change the color table, see Section 5.1.9 for details.

Import of 3D raster data (voxel) You can import 3D ASCII raster data in GRASS format using r3.in.ascii, see the manual page for the format description. You can also create a volume raster model based on 2D raster data by converting 2D raster slices into 3D raster or a 3D volume map based on 2D elevation and value raster maps, see examples in the volume data processing section in Section 5.6.

4.1.4 Import and geocoding of a scanned historical map

In this section, we explain rectification and georeferencing of a scanned map (e.g. a historical map). For this procedure, it is important to understand the relation between on ground distance, scale and spatial extent. This general cartographical relation is also needed when transforming a hardcopy map into a digital map. When scanning maps, keep in mind the proper handling of copyright restrictions.

Determining the scanning parameters The relationship between the
distance on the ground and the corresponding length of a raster cell is determined by the scanning resolution. When working with toposheets, scanning
resolution between 150 and 300dpi is recommended. Of course, the text labels
on the map should stay readable. Depending on the number of colors in the
map, the image can be scanned as color image with 256 colors. Assuming a
scanning resolution of 300dpi, we first calculate its equivalent in centimeters:

$$300dpi = 300\frac{lines}{2.54cm} = 118.11\frac{lines}{cm} \tag{4.1}$$

Suppose that the scale of the scanned map is 1:25,000. Thus, one centimeter
on the map is equivalent to 25,000cm on the ground. Now we can calculate
the on-ground distance that corresponds to the length of a raster cell:

$$\frac{distance\ on\ the\ ground}{scanned\ lines\ per\ cm} = \frac{25,000cm}{118.11lines} = 211.6\frac{cm}{line} = 2.12\frac{m}{line} \tag{4.2}$$

The resulting value of 2.12m is the spatial resolution of the scanned map at
the 300dpi scan resolution. If you want the spatial resolution to be an integer,
do the inverse calculation and adjust the scanning resolution accordingly.

Geocoding of scanned maps After scanning the map, store it in a file. If
needed, you can convert it to a GRASS supported format using `gimp`, `display`
(ImageMagick package) or `xv` which are available for many operating systems,
as well as the netpbm tools[4] which can be run on command line. General
procedure for geocoding a scanned map has the following five steps:

- create or use an existing xy LOCATION and import the map into it using
  `r.in.gdal -e`;
- restart GRASS with the projected LOCATION. If GIS manager did not
  open, start it with `gis.m`. Display reference map(s);
- from the menu, launch File ⤳ Georectify which opens a convenient graphical user interface to the geocoding tools:
  - select the map type you want to geocode (raster or vector map);
  - select a source LOCATION/MAPSET (usually the xy LOCATION);
  - add the map(s) to a group if not done by `r.in.gdal` import (internally
    launches `i.group` and `i.target`);
  - select the group;
  - select the unreferenced map to display for interactive georectification
    (normally a raster in the group you want to georectify);
  - start searching for ground control point in both the unreferenced and
    the reference map (internally launches `i.points`);
  - rectify the unreferenced map into the target LOCATION (internally
    launches `i.rectify`).

---

[4] Netpbm tools, `http://sourceforge.net/projects/netpbm/`

This task is using image processing tools for geocoding.

As an example, we geocode a "historical" city map of Raleigh (original scale 1:62,500) from the U.S. Geological Survey 1951[5]. First, start GRASS with the nc_spm LOCATION and import the scanned map by r.in.gdal into the new xy LOCATION:

```
g.region n=232700 s=220400 w=632100 e=648700 res=10 -p
r.in.gdal -e raleigh_nc_1951.jpg location=nc_xy \
          out=raleigh_nc_1951
```

The command will create the new LOCATION nc_xy. The flag -e automatically ensures that the new xy LOCATION has a region large enough for the imported map.

Make sure that the GIS manager gis.m is running. The first step is to display reference maps, for example streets_wake and lakes vector maps (use the Add vector layer icon, the maps are actually displayed by clicking on the Display active layers button in the Map Display 1 window). Zoom to the city center of Raleigh (approximately centered at 642700, 225400 NC State Plane coordinates) or to the current region using the Zoom to button. Choose blue color filling for the lakes and redraw the map display.

Next we set up the map(s) to geocode. From the menu of the GIS manager gis.m, launch File ⤳ Georectify for the geocoding tools. For 1. Select mapset, we select through the file manager LOCATION nc_xy and therein MAPSET PERMANENT (the latter will be shown in the selection field). Then we use 2. Create/Edit group just to verify, that r.in.gdal created the group raleigh_nc_1951. We select this existing group in 3. Select group. With 4. Select map, we choose one of the color channels, e.g. raleigh_nc_1951.blue. Now we are ready to search ground control points (GCPs) in both the reference maps and the unreferenced maps by clicking on 5. Start georectifying. You can zoom in both maps using the zoom tool in each map display and set GCPs by activating the GCP icon (in the georeferenced map display window you have to activate the Pointer icon to set a point there). The GCP coordinates are automatically added to the Manage ground control points (GCPs) window table. To set additional points, always click into a new table row. Finding common points is not very easy, it may be best to search a lake and then pan within the maps. To make it easier we have added a map with suggested GCPs to the book's Web site. Once you have sufficient and well distributed GCPs, you can calculate a RMS error.[6] It should not be larger than half of the true resolution of the scanned map as we have calculated above. The overall RMS error is computed from the errors for individual matching points. If it is too large, you can delete a point from the GCPs table (toggle the checkbox to

---

[5] City map of Raleigh 1951 download,
http://www.lib.utexas.edu/maps/north_carolina.html

[6] The RMS error is computed from the distance of the set matching point towards the accurately placed matching point. It is calculated as:
$$rms = \sqrt{(x - x_{orig})^2 + (y - y_{orig})^2}$$

switch a point on and off) and select a new point. Once sufficient GCPs are selected and assigned properly, leave i.points, by clicking on the disk icon in the GCPs table and the points will be saved.

Finally, we perform the transformation of the unreferenced map. We select a 1st order polynomial (as "order of transformation") since we don't expect too many spatial distortions. This will perform linear transformation (stretching and rotating). To start the transformation, click on the Rectify maps in group icon. Internally, the module i.rectify is launched to transform the map. This may require some time, depending on map size, resolution, and hardware, but you can run it in background and continue working with GRASS. After the transformation has finished, you can look at the new map in the target LOCATION and add it to the GIS manager map list for display. All three bands (red,green,blue) were geocoded and to create a map with the original colors, you can run r.composite. The above described procedure can also be used for georeferencing vector maps.

4.1.5 Raster data export

GRASS raster data can be exported to more than 20 formats using r.out.gdal. In addition, there is a set of export modules designed for specific formats, most of them were already mentioned in the import section: GRASS ASCII (r.out.ascii), ARC/INFO ASCII GRID (r.out.arc), BIL (r.out.bil), BINARY ARRAY (r.out.bin), PPM (r.out.ppm), MPEG (r.out.mpeg), TIFF (r.out.tiff), Matlab's MAT (r.out.mat), Topmodel's GRIDATB (r.out.gridatb).

As mentioned above, only the portion of the map that falls within the current region will be exported. The export modules can be used on command line as well as interactively with menus. Several export modules, such as r.out.ascii or r.out.ppm can be used with UNIX piping, i.e. redirecting the data stream to another module, as illustrated by the following example that creates an 8 bit GIF image:

```
r.out.ppm elevation out=- | ppmquant 256 | ppmtogif > elev.gif
```

The result of r.out.ppm is directly sent to ppmquant to quantize the elevation categories to 8 bit (256 colors), and then to ppmtogif. The data transfer is done through "standard output" (stdout) indicated by - (dash). The GIF data stream resulting from ppmtogif is written to the elev.gif file. The produced GIF file is stored into the current directory.

Several commands support export of GRASS raster data for external visualization tools, for example, r.out.pov exports elevation file into a height field in Targa (TGA) format that can be used with the Persistence of Vision (POV) raytracer[7] and r.out.vtk, r3.out.vtk export the 2D and 3D raster

---

[7] POV-Ray Web site, http://www.povray.org

data as VTK-ASCII file that can be used for visualization with Paraview or MayaVi[8]. You will learn more about these options in the visualization chapter.

Export to XYZ ASCII format A common format for raster data exchange is the plain XYZ ASCII format (i.e. x, y coordinates with the z value or category number). Unlike the GRASS ASCII raster export with `r.out.ascii` and `r3.out.ascii` (which exports the data as an ASCII matrix), the following command produces a file with one line for each cell, each line containing three columns (easting, northing, z):

```
r.out.xyz elev_ned_30m out=elev_ned_30m_xyz
```

To export category labels (attributes of the raster cell, if they are numerical) instead of category values or floating values, you can use `r.stats` with the following flags:

```
r.stats -1lg landuse96_28m nv="-9999" > luse96_28m.asc
```

```
more luse96_28m.asc
 629994.2531 228513 18 Mixed Hardwoods/Conifers
 630022.7531 228513 18 Mixed Hardwoods/Conifers
 630051.2531 228513 15 Southern Yellow Pine
 [...]
```

The optional `nv` parameter will replace the NULL value with a different character or string, in our case -9999.

4.2 Vector data

Point, line, area and 3D features can be represented in GRASS by a vector data model. It stores the feature's geometry, topology, and attributes. Different vector object types are used to store points (*vector points*), lines (*vector lines*), and polygons (called *vector areas*). GRASS 6 has a topological vector data support that includes attribute management handled by a database management system. Old vector and site data in the pre-GRASS 6 formats must be converted into the new format using `v.convert` for vector data and `v.in.sites` for old site data.

4.2.1 GRASS vector data model

Vector geometry GRASS stores vector data geometry using *vector object types* (graphic elements or primitives) such as a point, line, area boundary, and centroid (label point for an area). Points defining a line are called *vertices*, end points of a vector line are special vertices called *nodes* (see Figure 4.5).

---

[8] Paraview Web site, `http://www.paraview.org`

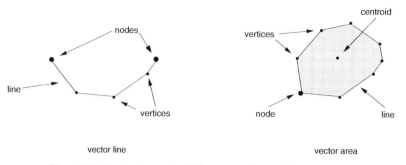

Fig. 4.5. Vector types in GIS: vector line and vector area

Vector lines are represented as a directed sequence of points and may consist of a single line (arc) or multiple connected lines (polylines). A closed ring of line segments defines an area boundary; an area boundary with a centroid inside defines a vector area. An area inside an area is called *island*. A 3D area is called *face*. Point data are represented as nodes that are not connected.

In GRASS 6, the following vector object types are defined:

- point: a point;
- line: a directed sequence of connected vertices with two endpoints called nodes;
- boundary: the border line to describe an area;
- centroid: a point within a closed boundary;
- area: the topological composition of centroid and boundary;
- face: a 3D area (see Fig. 4.6);
- kernel: a 3D centroid in a volume (not yet used in GRASS 6);
- volume: a 3D corpus, the topological composition of faces and kernel (not yet used in GRASS 6).

Note that all lines and boundaries can be polylines (with nodes in between). Conversion between several vector types is possible. Vector data are stored in a portable way in the GRASS DATABASE so that they can be exchanged directly across different platforms and architectures.

GRASS vector data map is stored in several separate files The geometry part comprises:

- vector map ASCII *header* with information about the map creation (date and name), its scale and threshold;
- binary *geometry* file which includes the coordinates of graphic elements (primitives) that define the vector feature;
- binary *topology* file describes the spatial relationships between the map's graphic elements;
- *history* ASCII file with complete commands that were used to create the vector map, as well as the name and date/time of the map creation;

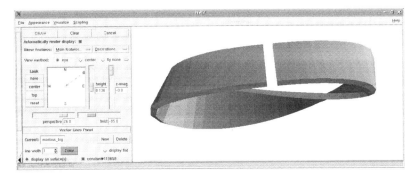

Fig. 4.6. Möbius strip visualization to illustrate the 3D vector faces capabilities of GRASS 6 (imported from 3D DXF)

- binary *category index* file which is used to link the vector object IDs to the attribute table rows;
- ASCII file which contains *link* definition(s) to attribute storage in database (DBMS): table name, database, key column and more are defined here.

Topology   The GRASS vector data model includes topology describing spatial relations between the graphic elements that define the feature location and geometry. GIS topology describes the spatial relationships between connecting or adjacent geographic components. With topological information, it is possible to efficiently perform the following tasks (Bartelme, 1995; Curtin, 2007):

- find neighborhood relationships between objects;
- analyze if one object contains another object (island areas);
- find intersections of objects;
- analyze vicinity of two objects;
- perform vector network analysis based on graph theory.

In a topological GIS like GRASS, a valid common border between two adjacent areas is stored only once and is shared between these two areas. Also shared nodes do not have to be duplicated. In a non-topological GIS this border would be digitized and stored twice. Topological representation of vector data helps to produce and maintain vector maps with clean geometry and supports certain analyses that can not be conducted with non-topological or spaghetti data. In GRASS, topological data are refered to as level 2 data and data without the built topology is referred to as level 1.

Sometimes topology is not necessary and the additional memory and space requirements are burdensome to a particular task. Therefore a few modules allow us to work directly on level 1 (non-topological), especially for import and processing of large point data sets (e.g., as with lidar data). Point data without attributes can be imported without topology but the support is then rather limited, we will discuss this issue in more detail in the following vector data import section. Most vector modules require level 2 that is needed to build spatial index for efficient vector data processing.

Detailed rules for digitizing vector data in a topological GIS are given in the digitizing section 6.3.1. For discussions on general computational geometry, see the book of O'Rourke (1998).

Attributes in DBMS

Vector points, lines and areas usually have attribute data that are stored in DBMS. The attributes are linked to each vector object using a *category number* (attribute ID, usually the "cat" integer column). The category numbers are stored both in the vector geometry and the attribute table. The category numbers can be printed or maintained using v.category.

A vector object can have zero to several categories. The vector objects can be linked to one or more attribute tables, each link to a distinct attribute table represents a vector data "layer". For example, multiple layers can be used to link the vector objects with multitemporal attributes, with each layer representing a single time snapshot. Multiple layers are also useful for integration of thematically distinct but spatially related vector objects. In this case, each table represents attributes only for objects with a given theme, creating thus separate layers for each theme. Although the layers make combination of points, lines and areas in a single vector map possible, it is generally recommended to store different vector types in separate vector maps. The first layer is active by default, i.e. the first table corresponds to the first layer. Additional tables are linked to subsequent layers. Map layers can be listed or maintained using v.db.connect.

The default database driver used by GRASS 6 is DBF. It provides only limited Structured Query Language (SQL) support; therefore other DBMS are commonly used, for example, PostgreSQL, MySQL, or SQLite. By introducing the full DBMS support, GRASS 6 handles multiattribute vector data by default. The db.* set of commands provides basic SQL support for attribute management, while the v.db.* set of commands operates on the vector map.

4.2.2 Import of vector data

GIS vector data are available in many different formats. There is no single standard available but some formats, such as ESRI SHAPE, are widely used for vector data exchange. Because of the complex data structure, exchange of vector data is often more complicated than for raster data. In GRASS 6, the import has been greatly simplified by introduction of the v.in.ogr module, a vector analog to r.in.gdal.

If you just want to display external vector data provided in any of the OGR[9] (OGR is part of the GDAL library) supported formats along with

---

[9] OGR library, http://www.gdal.org/ogr/

your GRASS data you can link these data into your MAPSET by running
v.external command, for example:

```
# make sure that we have the right region set
g.region swwake_10m -p

# unzip the SHAPE data and link them to GRASS from curr. dir.
unzip geod_pts_spm.zip
v.external dsn=. layer=gdc output=ext_geodetic_pts
d.vect ext_geodetic_pts
```

It is important to know that the linked data are "read-only" and that they use
the OGR/OGC simple feature model that does not include topology (some
GRASS modules may not work properly with such data). It is therefore recommended to import the data into native GRASS vector format if operations
beyond data viewing are performed.

Many common vector data formats, such as SHAPE, Arc/Info Coverage, MapInfo, DGN, SDTS, PostGIS, or TIGER can be imported using the
v.in.ogr module (see its manual page for a complete, up-to-date list of supported formats). It provides options for importing subsets of the original vector data by defining the desired spatial extent, minimum area or data with
certain attribute values defined by SQL statement. For example, we can import a subset of statewide soil map data provided as a SHAPE file set in the
current directory that covers only our study area as follows:

```
# get the coordinates of the study region
g.region swwake_10m -p

# import the NC state soil data associations from a SHAPE file
# clipped to the current region (W,S,E,N)
v.in.ogr gslnc.shp out=soils_nc \
        spatial=630000,215000,645000,228500

# check the imported data, there will be only two associations
v.info soils_nc
v.info -c soils_nc
d.vect soils_nc
```

Topology is built automatically for each imported data set. Note that in the
GRASS 6 implementation topology building for large data sets requires substantial memory and disk space so it may be necessary to split large vector
data sets into subregions when importing. Attribute data for SHAPE files are
stored by default in a DBF database (or you change the DBMS settings, see
Section 6.2.1 for details).

If you observe that polygons are not built correctly (overlapping areas,
area without category due to lost centroids, etc.), use the snap parameter
of v.in.ogr to polish the topology while importing. This snapping threshold for boundaries must be selected with care depending on the scale of the

vector map. Additionally, `min_area` can be used to suppress small areas by defining the minimum size of an area to be imported (given in square map units). Smaller areas and islands will be ignored then. The value should be greater than the square of the snapping threshold. Snapping problems often occur when importing "Simple Features" data like SHAPE files that were not generated in a topological GIS. Such data sometimes contain gaps or slivers which need to be corrected. If the topology is still not built without problems, use the advanced tools of `v.clean` to polish it or manually edit the map with `v.digit`. More examples of vector data import are described in the manual page for `v.in.ogr`. See Section 6.5 for details on cleaning topology.

Vector data in the ESRI E00 format are imported with `v.in.e00`. It requires the `avcimport` and `e00conv` programs to be installed.[10] The E00 format is preferred to the SHAPE files because it includes projection information and keeps better data structure; however, it is much less common because of its complexity. The import module re-builds the topology and stores the attributes in a database table. The module can read E00 files from any directory. If no path is specified, the current directory is used.

You can generate a vector area map that represents the rectangle defining the current region as follows:

```
v.in.region out=myregion
d.vect myregion
```

The resulting vector map can be used for clipping subsets of your vector data for a given region using the vector overlay tools. MATLAB vector data can be imported using the command `v.in.mapgen`.

Import of GRASS ASCII vector files and ASCII point data The standard GRASS ASCII vector maps are usually generated by `v.out.ascii`, manually or by custom applications, and can include all supported vector objects (points, lines, boundaries, etc.). We can modify the text file `newfacility.txt` (used in the raster ASCII import example in Section 4.1.3 and provided in `ncexternal/`) to GRASS ASCII vector format by modifying a line indicating the vector object type (primitive). We use B for boundary rather than A for area, followed by 5 for number of points and 1 for a single category. Then we include the coordinates of the boundary and the layer number and the category ID on the last line:

```
B 5 1
 638656.00  220611.00
 638796.00  220609.00
 638799.00  220520.00
 638653.00  220519.00
 638653.00  220609.00
 1 1
```

---

[10] AVCTools Web site: http://avce00.maptools.org

We save the file as **newfacility.asc** and then convert it to a GRASS binary vector map using **v.in.ascii** (the -n flag indicates that the ASCII file does not include the standard vector map header):

```
g.region rural_1m -p
v.in.ascii -n input=newfacility.asc out=my_facility \
           format=standard

# display the imported map over orthophoto
d.erase
d.rast ortho_2001_t792_1m
d.vect my_facility col=red
```

You can learn more about GRASS ASCII format by exporting different types of GRASS vector maps (points, lines, areas) using **v.out.ascii** and then exploring the result (see more in the next section).

Point data in ASCII format can be imported using **v.in.ascii**. For example, you can import the bare earth lidar point data provided as x, y, z text file (find the data within **ncexternal/**) as follows:

```
v.in.ascii -ztb BE3720079200WC20020829m.txt out=mylidar_pts \
           x=1 y=2 z=3

# set the region to imported data to see all of them
g.region vect=mylidar_pts -p
d.erase
d.rast ortho_2001_t792_1m
d.vect mylidar_pts siz=1 col=yellow
```

The flags mean that the data will be imported as 3D vector (-z), without creating an attribute table as there is no attribute (-t) and no topology will be built (-b). Without topology the data can only be displayed by **d.vect** or interpolated by **v.surf.rst**, so if you want to do more, you either need to skip the relevant flag or build topology after the file is imported using **v.build** (as we have mentioned, this may be problematic for large data sets with millions of points).

Another common example of ASCII point data are the data exported from a spreadsheet as a comma separated variable (.csv) file. Such data can be easily imported by **v.in.ascii** whith the separator parameter set to **fs=","**. The manual page for **v.in.ascii** includes many excellent examples of various vector ASCII data imports, including tips for generating such data.

Import of GPS, Gazetteer and DBMS point data   Waypoints, routes and tracks can be directly imported from a GPS device or a text file using commands **v.in.garmin** or **v.in.gpsbabel**. Both commands work in a similar way, the first is using **gpstrans** to convert the data and the second one is based on **gpsbabel**. Both modules automatically project the data in geographic coordinates to the current coordinate system unless you use flag -k. For example, to read GPS tracks from a USB connected Garmin device, run

```
v.in.gpsbabel -t in=/dev/ttyUSB0 format=garmin output=mytracks
```

Depending on the number of waypoints, lengths of tracks or routes, the data transfer can take some time. You can learn more in Section 10.3.

US-NGA GEOnet Names Server[11] (GNS) country files can be imported as GRASS vector point data using v.in.gns. The data include geographic and UTM coordinates, so you need to import them into a lat-long LOCATION such as our nc_ll and then project into State Plane from nc_spm using v.proj:

```
# open GRASS in nc_ll LOCATION and run
v.in.gns nc_names.txt vect=nc_names_ll

# open GRASS in nc_spm LOCATION and project
v.proj nc_names_ll out=nc_names locat=nc_ll mapset=PERMANENT

g.region vect=nc_names -p
d.erase
d.vect nc_names displ=shapes,attr attrcol=Feature_Name size=3
```

You can also import vector point data stored in a database system using the module v.in.db, and select a subset to be imported using a category column name and SQL query.

Import of DXF files GRASS supports import and export of 2D and 3D multiple layer vector maps in *DXF format*. As an example we show import of a DXF vector map with multiple layers representing map planimetry (roads, building footprints, fences, etc.) The DXF data can be imported as a vector map with a single layer that will include all DXF layers (useful for a preview of what is in the file, but difficult to work with) or you can choose only selected layers, as in the following example run from the nc_spf LOCATION:

```
# set region to the small rural area and display airphoto
g.region rast=IMG_airBW_79200WC_3ft -p
d.rast IMG_airBW_79200WC_3ft

# list available layers
v.in.dxf -l P079216.DXF

# import all layers as one map
v.in.dxf -1 P079216.DXF
d.vect P079216
```

---

[11] US-NGA GEOnet Names Server, http://earth-info.nga.mil/gns/html/

```
# import only the layers with buildings
v.in.dxf P079216.DXF out=buildings79216 \
        layer=BLDG_COMMER_BL,BLDG_RESID_BL
d.vect buildings79216 col=red
```

To reproject these local government data provided in State Plane coordinate system, units feet into our sample data set in State Plane system, units meters, open GRASS with nc_spm LOCATION and use v.proj to reproject the imported DXF layers.

DXF polyface meshes can be imported as 3D wireframe using flag -f (without flag it will be imported as vector faces type):

```
v.in.dxf -f building.dxf out=building_wire
```

Buildings and other 3D vector features can be visualized using the module nviz that we will describe in detail in the visualization chapter 7.

Import vector data directly from Web Sources GRASS 6 supports direct download of vector (feature) data from Open Geospatial Consortium's Web Feature Service (OGC WFS) compliant Web servers using v.in.wfs. As an example, we download the registered GRASS users from the community mapserver portal (run command in latitude-longitude LOCATION nc_ll):

```
# enter URL in single row
v.in.wfs \
  wfs="http://mapserver.gdf-hannover.de/cgi-bin/\
grassuserwfs?REQUEST=GetFeature&SERVICE=WFS&VERSION=1.0.0" \
  out=grass_users
v.db.select grass_users \
            where="name ~ 'Helena' OR name ~ 'Markus'"
 cat|ogc_fid|name|company
 467|42|Helena Mitasova|NCSU
 636|2|Markus Neteler|ITC-irst (Povo)
 [...]
```

The WFS request downloads a XML file from the WFS server which is then converted into a GRASS vector map with attribute table.

4.2.3 Coordinate transformation for xy CAD drawings

CAD drawings, usually provided in the DXF format, are sometimes delivered in non-georeferenced xy coordinates. To use them along with other GIS data, these coordinates have to be transformed to the coordinate system of the current LOCATION. You can do it using the geocoding tool in the GRASS GIS manager gis.m as explained in the section 4.1.4 or using the module v.transform. The later requires a table of ground control points (GCPs, also called tie points) to perform this transformation. It is a table of points with xy coordinates and their corresponding georeferenced coordinates. To generate this table, coordinates of points such as road intersections etc. are identified in the DXF map and another corresponding reference map. Also, GPS measurements can be used. With v.transform, the map can be shifted and stretched in x, y, and z direction, as well as rotated around the z axis. It maintains 3D structures in case they are present in the CAD drawing.

In GRASS, the DXF map geocoding process requires three steps:

- DXF data are imported into the projected LOCATION, although the DXF map keeps its xy coordinates reference;
- the ground control points are identified within the imported DXF map as well as the reference map or taken from GPS measurements and stored in a text GCPs table;
- finally, the imported map is transformed to the current LOCATION coordinate system by shifting and rotating the input DXF map using the GCPs table.

To illustrate the procedure we will import the DXF map Masterplanxy.dxf (xy coordinates) and transform it to the map Masterplan stored in a georeferenced LOCATION, in our case the State Plane NAD83 with units meters. To start, we import the DXF map into the nc_spm LOCATION as a GRASS vector map in xy coordinates. To view it, you will have to change the region as shown below:

```
# import the dxf file in xy coordinates
v.in.dxf Masterplanxy.dxf out=Masterplan_xy

# change the region to view the imported map
g.region vect=Masterplan_xy -p
d.mon x0
d.erase
d.vect Masterplan_xy
```

If you do not expect internal map distortions, only four ground control points (GCP) will be needed for the transformation. In our example, the easy to identify points are at highway interchanges. To get the (x, y) coordinates for the GCPs from the imported DXF map, you can use d.where and then store the coordinates in an ASCII file (e.g. called GCPforDXF.txt) in your home directory. You can use any text editor to create the file and write or rather paste each (x, y) coordinate pair on a separate line. The next step is to find the corresponding four GCPs in the georeferenced map or to assign corresponding coordinates from a GPS measurement. When using GPS data, just enter the georeferenced coordinate values. When using a reference map in the current LOCATION, for example the vector map streets_wake, you can open a new monitor, reset the current region and display the map (for convenience we show the zoomed-in region for the streets_wake that includes the Master plan area):

```
d.mon x1
g.region -p n=228396.0 s=222148.0 w=634206.0 e=641883.0
```

```
d.erase
d.vect streets_wake
```

Again, use **d.where** to get the corresponding four GCP coordinate pairs from
the reference map. Store these values in the ASCII text file **GCPforDXF.txt**
next to the related xy coordinates as EASTING NORTHING with all values
delimited by a space. The GCP text file will look as follows (the first two
values are the X and Y coordinates, the second two values are EASTING and
NORTHING):

```
10547.28 19180.92   636087.08 227883.50
12681.62 11905.81   636728.77 225661.66
22963.16  2289.81   639873.03 222725.95
27782.63   683.32   641340.88 222252.71
```

Now you can transform the imported map in xy coordinates to the State Plane
NAD83 using the module **v.transform**:

```
v.transform Masterplan_xy out=Masterplan points=GCPforDXF.txt
 [...]
 Residual mean average: 4.581719
```

The module transforms the coordinates of the nodes and vertices of the DXF
map and prints out the error for each GCP and the overall RMS (root mean
square) error. If the RMS error is too high for the given application you can
try to improve the result by more accurate selection of GCPs. To verify the
transformation result, reset the current region to the transformed map and
display it along with the reference map:

```
g.region vect=Masterplan -p
d.erase
d.vect Masterplan col=red
d.vect streets_wake
```

Both maps should match well.

4.2.4 Export of vector data

GRASS vector data can be exported in various formats using **v.out.ogr** and
several format-specific commands.

Export to OGR-supported formats   You can export GRASS vector maps
into more than 15 formats including SHAPE, KML, GML2, PostGIS and DNG
using the **v.out.ogr** command that is based on the OGR Simple Features
Library (part of GDAL). For example, to export the **roadsmajor** vector file
from our sample data set to ESRI SHAPE format, including an appropriate
projection information stored in ESRI-style .**prj** format (indicated by the flag
-e) run:

```
# get the info about the vector map
v.info roadsmajor
v.out.ogr -e roadsmajor dsn=roadsmajor type=line
```

Several files are written into the **roadsmajor** subdirectory of the current directory: **roadsmajor.shx**, **roadsmajor.shp**, **roadsmajor.prj** and **roadsmajor.dbf**. Note that you did not have to specify the output format or feature type because the SHAPE format and line features are default. The written DBF file may be inspected with OpenOffice (**http://openoffice.org**), a free office software. Additional examples of export of areas, 3D faces and points into different formats are below:

```
# export soils map to PostgreSQL/PostGIS
v.out.ogr soils soils \
     dsn="PG:host=localhost,dbname=postgis,user=postgres" \
     type=area format=PostgreSQL

# export 3D faces into KML; facility.kml needs to be
# reprojected to LatLong/WGS84 for WorldWind or Google Earth
v.out.ogr facility_3d dsn=facility.kml olayer=facility \
         format=KML type=face
# export points into CSV
v.out.ogr schools dsn=schools type=point format=CSV
```

To export to DXF (and some other formats) GDAL/OGR needs to be compiled with additional libraries, but you can also use the module **v.out.dxf** to export data into the DXF format:

```
v.out.dxf roads out=roads.dxf
```

The DXF file is written to the current directory or to a specified path.

Export to GRASS ASCII vector format   You can use the module **v.out.ascii** to export a GRASS vector map to GRASS ASCII format. Only geometry and category numbers are exported in this format. You can export the line and point vector maps as follows:

```
# export to GRASS ASCII format
v.out.ascii roadsmajor out=roadsmajor.asc format=standard

# alternate way to export to CSV table format
v.out.ascii schools out=schools.txt fs=" "
```

Note that the format "point" is default so we did not have to specify it for the vector map **schools**.

Export to visualization and graphics formats: POV, VTK and SVG
You can export GRASS vector maps for sophisticated visualization in POVRay using **v.out.pov** or VTK-based tools such as **Paraview** or **MayaVi** using **v.out.vtk**. See Chapter 7 on visualization for more details. An important

addition in GRASS 6 is the possibility to export vector maps in the Scalable
Vector Graphics (SVG) format that can be used for graphically rich web appli-
cations, hardcopy output, and as exchange format for numerous applications.
For example, you can export the soils map as follows:

```
v.info -c census_wake2000
v.out.svg census_wake2000 out=census_wake2000.svg \
          type=poly attrib=TRACTID
```

The resulting map can be displayed in Web browsers with SVG support.

5

Working with raster data

In this chapter, we explain processing of raster data, including examples of spatial analysis and modeling. We provide basic description of the GRASS raster tools accompanied by numerous practical examples; you can learn more details from the manual pages, tutorials and publications provided on the GRASS Web site. For description of the GRASS raster data model, as well as raster data import and export, please refer to the Sections 4.1.1 and 4.1.3.

5.1 Viewing and managing raster maps

In this section, we continue to use the North Carolina sample data set to illustrate our examples. Please refer to Section 3.1.4 on how to start GRASS with the nc_spm LOCATION.

5.1.1 Displaying raster data and assigning a color table

As mentioned before, display of GRASS data is easy to handle using your favorite GRASS GUI, especially when you need to combine various raster and vector maps. The available GUIs evolve rapidly, therefore, we use the more stable underlying commands in this book. The command line interface (CLI) is also faster than GUI if you need to work over the Internet.

We have already displayed raster maps using the **d.rast** module (see Subsection 3.1.5, so here we just repeat the procedure. First set the current region to the map of interest, then open the GRASS monitor, and run the raster display module (see Section 4.1.2 for region management and Section 3.1.5 for GRASS monitor usage):

```
g.region rast=elevation -p
d.mon x0
d.rast elevation
```

The module **d.rast** offers two useful optional parameters, `catlist` and `vallist`. When using `catlist` you can selectively display categories for integer maps, while `vallist` applies to floating point maps. For example, we display selected land use categories with the display background set to black and selected elevation values over the geology map:

```
# display only categories 1, 2 (developed) for land use map
d.rast landuse96_28m cat=1,2 bg=black

# check the range of elevation values of the entire map
r.into -r elevation

# display only elevation higher than 100m over the geology map
d.rast geology_30m
d.rast -o elevation val=100-160
```

The flag -o allows you to display the raster map (or its selected subset of categories or values) over the map already displayed in the monitor. Remember, from Section 3.1.2, that you need to type only the first letters of each keyword (e.g., `cat` instead of `catlist`).

To get a list of the maps currently displayed in the GRASS monitor, run:

```
d.frame -l
```

The GRASS monitor can be erased with **d.erase**. You can display all raster maps in the selected MAPSETS in a single monitor as follows:

```
d.erase
d.slide.show mapsets=PERMANENT
```

Optionally, you can define a name **prefix** to see only selected maps. To learn how to view raster data in 3D using the **nviz** module, see Chapter 7.

Color tables and legends   Each raster map has its own color table. When no color table file is present, the rainbow color coding will be used (not always the best choice). You can assign a color table for a raster map using the **r.colors** module. It provides a range of pre-defined color tables. To give it a try, change the color table of the elevation map to various pre-defined color schemes and then display it again using **d.redraw**:

```
d.rast elevation
r.colors help
r.colors elevation col=gyr
d.redraw
r.colors elevation col=elevation
d.redraw
```

The parameter **color** uses GRASS internal color tables (in `$GISBASE/etc/colors/`). The color tables typically used with elevation raster maps are available under the names **elevation** and **terrain**. If the parameter **color** is set to **rules**, you can build customized color tables by color list, by category values, and by "percent" values or their combination. The colors can be defined by name or by the numbers 0-255 for red-green-blue triplets. The following example illustrates how to combine different methods for color table specification:

```
r.colors elevation col=rules
 Enter rules, "end" when done, "help" if you need it.
 fp: Data range is 55.578... to 156.329...
 > 0% aqua
 > 20% green
 > 100 255 255 0    #yellow
 > 120 orange
 > brown
 > end


d.redraw elevation

# put back the original color table
r.colors elevation col=elevation
```

The parameter **rules** allows you to define the path to an existing color table file and the parameter **raster** can be used to copy a color table from an existing raster map. You will find additional examples of color table definitions throughout the book and in the manual page for this command.

You can add a legend and a scale to your displayed map as follows:

```
d.barscale at=5,90
d.legend elevation at=70,15,5,10
```

You will get the barscale drawn at the lower left corner of the graphical window (at 5% from the left, at 90% of the window from the top) and the legend at lower left corner in a box starting at 15% from the bottom, 5% from the left extending to 70% from the top and 10% from the left. If you want to interactively define the location and extent of the legend using the mouse, include the -m flag. Then click the left mouse button to "Establish a corner", drag the mouse over the area where you want your legend and finish by clicking the right button to "Accept box for legend". The same flag is available for barscale. The menus for these interactive display commands are always explained in the terminal window. If you have selectively displayed only certain categories or values using the **catlist** or **vallist** options for the **d.rast** command, you can selectively display the relevant colors in your legend using the **use** option for a list of categories and the **range** option for a range the map values. Try the graphical user interface for more sophisticated options to display maps with legends, scales, and titles. A simple way to show a raster map with legend is to use **d.rast.leg** (see Section 7.1.1 for changing the legend font):

```
g.region rast=landuse96_28m -p
d.erase
d.rast.leg landuse96_28m
d.erase -f
```

To remove the map including the legend frame, use **d.erase -f**. You can use the command **d.text** to add text to your image.

5.1.2 Managing metadata of raster maps

Information about the data source, accuracy, producer, date of mapping or image acquisition, date of map production, and eventual modifications, is called metadata (data describing data). Data documentation is crucial for GIS work, especially for evaluation of data quality and suitability for a given task. This is particularly important for long-term projects or where GIS data are shared with other users. Most geospatial data provided by the US government agencies include a comprehensive, standard metadata file. You can find several examples in the external data sets provided in within **ncexternal/** subdirectory. For example, see the text file **gdc.txt** stored within **geod_pts_spm.zip** for the NC geodetic points data.

We have already used the command **r.info** to get basic information about a raster map. The module **r.describe** provides a simple output of category numbers or range of values, while **r.cats** prints a list of category numbers and associated labels:

```
r.info landuse96_28m
r.describe landuse96_28m
r.cats landuse96_28m
```

Modifying metadata   GRASS offers an option for maintaining a "history file" that stores basic raster map documentation. Many analytical modules save their calculation steps into the history file automatically. You can add additional information, including raster map units and vertical datum, by modifying the "map history" using the command **r.support**:

```
# make you own copy of the raster map so that you can modify it
g.copy rast=elevation,myelevation
r.info myelevation
r.support myelevation history="Downloaded from USGS,\
  projected to spm" units=meter vdatum=NAVD88
r.info myelevation
```

or by running **r.support** in interactive mode:

```
r.support myelevation
```

This will guide you through series of questions that give you the opportunity to edit the header, update histogram, edit categories or color table; you may proceed with <ENTER> until you reach the question: Edit the history file for [myelevation]?. Confirming with <y> opens an input screen showing the metadata for this map.

You can see that map date, title, creator, a description containing the map creating method "generated by r.proj" and a few more entries are already stored there. You may fill the field "Data source" with "USGS NED". With <ESC><ENTER> you reach the next screen with comments that include the full command used to create the map and a comment that we have added above. You may add more comments here for example, "Based on 2001 NC Flood mapping program survey". Another <ESC><ENTER> takes you back to the questionnaire mode of r.support; you can skip the rest of the questions with <ENTER>, and leave the module. You can check your modification by running:

```
r.info myelevation
```

This will display the updated data description, boundary coordinates, and data range.

Raster map timestamps Mapping and monitoring often produce time series of spatial data that require to store the relevant temporal information. We can store it separately from the history file using the module r.timestamp. This command has two modes of operation. If no date argument is supplied, then the current timestamp for the raster map is printed. If a date argument is specified, then the timestamp for the raster map is set to the specified date or date range. Absolute timestamp that includes date, time and time zone can be added as follows (you need to open GRASS with the MAPSET PERMANENT or copy the raster maps into your own mapset to run the examples below):

```
r.timestamp lsat7_2002_10 date="24 May 2002 09:30:00 -0000"
```

Check the defined timestamp by:

```
r.timestamp lsat7_2002_10
```

Time range is represented by two comma-separated timestamps (e.g. date="07 october 1999, 24 march 2000"). Also relative timestamps can be specified:

```
r.timestamp elevation_shade \
            date="17 hours 25 minutes 35.34 seconds"
r.timestamp cfactorgrow_1m date="130 days"
```

Timestamps can be removed by:

```
r.timestamp elevation_shade date=none
r.timestamp elevation_shade
```

5.1.3 Raster map queries and profiles

One of the most common tasks when working with digital maps is querying the values at a given location. This can be done either interactively using a mouse, or on the command line for a location given by its coordinates. To query single or multiple maps interactively use **d.what.rast**, the mouse button menus are explained in the terminal window:

```
g.region rast=landuse96_28m -p
d.erase
d.rast landuse96_28m
d.what.rast
 Buttons
  Left:  what's here
  Right: quit
 637637.66225962(E)  221085.80408654(N)
 landuse96_28m in PERMANENT   (15)Southern Yellow Pine

d.what.rast elevation,landuse96_28m
 ...
 641350.22271715(E)  220066.25835189(N)
       elevation in PERMANENT, quant   (102)
       elevation in PERMANENT, actual  (102.465187)
 landuse96_28m in user1    (1)High Intensity Developed
```

The command **d.what.rast** will work without specifying a map name when a raster map is already displayed in the GRASS monitor. By clicking with the left mouse button at a certain location, you will get the coordinates at that point as well as the category values and labels in the given map(s). For the floating point maps you will get the actual value and an integer quantized value (category number). When you are finished with the query, don't forget to end the request mode using the right mouse button within the GRASS monitor. To get only the coordinates at a point by mouse click, use **d.where**.

Non-interactive queries of one or more raster maps can be performed either at individual points or along a transect (profile along a line). You can query a single or multiple maps at a point given by its coordinates or at a set of points stored in a file with the module **r.what**. In the following example, we use the coordinates of schools stored in the file **schools.txt** (provided in the sample data set subdirectory **ncexternal/**, generated in Section 4.2.4):

```
# query at a single point: provides category number and label
r.what -f landuse96_28m east_north=638650,220610
 638650|220610||4|Managed Herbaceous Cover

# query multiple maps at multiple points
# 'more' lists the file contents, leave with <q>
more schools.txt
r.what -f elevation,landuse96_28m \
       null=-9999 < schools.txt > schools_el_lu.txt
```

```
# note: coordinates shortened here
more schools_el_lu.txt
 633649.29|221412.94|1|145.07||2|Low Intensity Developed
 628787.13|223961.62|2|-9999|-9999
 [...]
d.rast.leg landuse96_28m
d.vect schools_wake icon=basic/circle fcol=yellow
```

Category labels (in our case the type of land use) are included in the output
when running the command with the -f flag. The result in the second example
is written into an ASCII file schools_el_lu.txt. The schools located outside
our area have NULL for elevation and land use, defined as -9999 by the null
parameter. Note that the decimal digits printed beyond *millimeters* are due
to printing format and do not reflect the data precision. Additional interesting
examples of r.what usage are given in its manual page.

Profiles You can interactively draw profiles within the GRASS monitor
with the command d.profile. You may try it with the elevation map:

```
d.profile elevation
```

Then follow the menu provided within the monitor.
    Numeric terrain profiles can be generated using r.profile. It outputs
raster map values located on user-defined line(s) and writes the result to a file
or to standard output (stdout). For example, you can extract a profile along
a line specified by coordinates of two GPS tracking points or a profile along
a series of points stored in a file road_profile_xy.txt (see the subdirectory
ncexternal/ of the NC sample data set to get this text file):

```
g.region rast=elevation -p
r.profile elevation out=- profile=641373,221962,641272,219354 \
         res=200
 Using resolution 200
 Output Format:
 [Along Track Dist.(m)] [Elevation]
 Approx. transect length 2609.955078 m.
 0.000000 74.698303
 200.000000 78.065613
 400.000000 82.880783
 [...]

cat road_profile_xy.txt
cat road_profile_xy.txt | r.profile -g elevation \
    out=road_profile_xyz.txt
more road_profile_xyz.txt
 641373.134328 221962.686567 0.000000 74.698303
 641373.134328 221952.686567 10.000000 74.768227
 641373.134328 221942.686567 20.000000 74.704346
 [...]
```

The output list for the first example has two columns. The first column is the cumulative transect length, the second column is the raster value found at the corresponding grid cell resampled to 200m resolution. The second example takes the coordinate input from a file and stores output in a file road_profile_xyz.txt that includes easting, northing (due to the flag -g) as well as the cumulative transect length and the raster value at that position. You can use the flag -i to select the profile points interactively in the GRASS monitor. Profiles defined by a starting point, direction and distance can be extracted using r.transect.

5.1.4 Raster map statistics

Univariate statistics for a raster map can be computed with r.univar. It computes the number of cells, minimum, maximum, range, arithmetic mean, variance, standard deviation, variation coefficient and sum of all values. As an example, we apply it to the map elevation:

```
g.region rast=elevation -p
r.univar elevation
 [...]
 Of the non-null cells:
 --------------------
 n: 2025000
 minimum: 55.5788
 maximum: 156.33
 range: 100.751
 mean: 110.375
 mean of absolute values: 110.375
 standard deviation: 20.3153
 variance: 412.712
 variation coefficient: 18.4057 %
 sum: 223510266.5581016541
```

Extended statistics that includes median and 90th percentile can be computed by running the command with the -e flag. Univariate statistics is applicable to raster maps representing continuous fields and even those require caution when interpreting the results, we discuss these issues in more detail in the Section 10.2. Please refer to Appendix A.1 for equations used in this module to compute the univariate raster statistics.

You can use the modules r.report and r.stats to report summaries for raster maps such as area sizes for each category with units given by parameter units (cells, hectares, percent, etc.):

```
r.report landuse96_28m unit=c,h,p
r.stats -pl zipcodes,landuse96_28m
```

The r.report example outputs a table that shows category number and label, number of cells, area in hectares and percent for each land use category.

The r.stats example computes percent of land use type for each zipcode. In addition, you can use the command r.stats with the flag -1 to output individual cell values (left to right, top to bottom) in a format that can be further customized using additional flags. Often, the result is piped to other tools such as awk for further processing (see an example in Section 5.1.9).

5.1.5 Zooming and generating subsets from raster maps

An easy way to create spatial subsets is to zoom and save a raster map. You can interactively zoom into a selected location within a map displayed in the GRASS monitor using the command d.zoom (see Section 3.1.5 for details). You can now save the selected region using the g.region command:

```
g.region save=myzoomregion
```

Besides zooming, you can use g.region also to adjust the current region settings to well defined region boundary values as we have shown in Section 4.1.2. You must run d.redraw (or the sequence d.erase; d.rast map) after using g.region to give the GRASS monitor the information about the changed current region. The module d.erase which is internally used by the d.redraw script sends the updated coordinates to the monitor while erasing its contents. Therefore, a redraw is required to get the map(s) back.

Generating map subsets   If you have a large raster map, but you want to work only on its smaller subset, you can select it and store it into a separate raster map. This saves processing time, especially when you want to try a more complex calculation before applying it to the full map.

To create a raster map subset, we will use the GRASS feature that makes the GRASS raster computations limited to the current region at current resolution. After defining the area of interest with g.region or d.zoom, you can use the module r.mapcalc to extract the map portion into a new map:

```
# change region to the airphoto tile 792 and resolution 10m
g.region rast=ortho_2001_t792_1m res=10 -ap
d.erase
d.rast ortho_2001_t792_1m

# create a subset of the map elevation for this subregion
r.mapcalc "elevation_792_10m=elevation"
d.rast elevation_792_10m

# zoom out to see that it is a subset, click middle button
d.zoom
```

Through this simple map algebra expression the map portion defined by the current region is saved as elevation_792_10m, by copying the cell values from the larger original map elevation to the new map (nearest-neighbor method).

5.1.6 Generating simple raster maps

GRASS has several modules that allow users to compute various types of simple raster maps. To create a new raster map using on-screen digitizing you can use **r.digit**. For example, you can create an additional area, circle and line for a planned new building with parking, circular fountain and a service road as follows:

```
# change region to the high resolution, small rural area,
# display the orthophoto and proposed facility area
g.region rural_1m -p
d.erase
d.rast ortho_2001_t792_1m
d.rast -o facility

# digitize an additional raster map area and raster line
# using the orthophoto as a background
r.digit myfacility_1m

# this menu appears
 Please choose one of the following
    A define an area
    C define a circle
    L define a line
    X exit (and create map)
    Q quit (without creating map)
```

Now select the appropriate letter and digitize the desired polygon(s), circle(s) or line(s) using the middle button, then follow the directions to create a raster map from the drawn objects. Try to digitize the facilities in the north-east as areas (<A> for area, category number 1, category label "house"). To save the map, leave with <X>. In Section 6.3 we show a more sophisticated digitizing tool.

You can then display the result on top of the orthophoto and check the information about the new map:

```
d.rast -o myfacility_1m
r.info myfacility_1m
```

When running **r.info** you will see that the raster map was in fact created by the command **r.in.poly** that is used to import raster maps defined by ASCII vector data (see Section 4.1.3).

A raster map representing set of concentric circles around given points can be created using the **r.circle** command. For example, you can create a set of circles around a selected school given by a vector point starting at the distance of 200m growing up to 1000m with a step equal to the current resolution:

```
g.region swwake_10m -p
d.erase
```

```
# display streets and schools, find the coordinates
# of the selected school: Centennial
d.vect streets_wake
d.vect schools_wake col=red disp=shape,attr attrcol=NAMESHORT
d.what.vect -x schools_wake

# create raster map circle and display it over the vector data
r.circle out=school_circle coor=637768,222962 min=200 max=1000
d.rast -o school_circle
```

You can then use such a raster map for various types of analysis such as identifying spatial relationship between the given point and landscape properties
as function of the distance from this point.

Random sampling   Analysis that involves random sampling may require
a raster map with randomly distributed cells that are within a given minimum distance. You can create such map using the module r.random.cells
(we use low resolution of 100m in our example to make the results easier to
understand):

```
g.region rast=elevation res=100 -ap
r.random.cells out=randomcells_200m distance=200
d.erase
d.rast randomcells_200m
r.info randomcells_200m
d.what.rast
```

Each random cell has a unique non-zero value ranging from 1 to the number
of cells generated, in our case it was 2866 (you may get a slightly different
number, depending on how the random cells are distributed). The unfilled
cells are assigned the value zero. You can use map algebra, explained later in
this chapter, to replace the zeroes by nulls and the random values by a single
value, if necessary.

Random surface, useful for modeling uncertainty and analysis of sensitivity
of various algorithms to errors in the modeled surface, can be generated by
r.random.surface:

```
r.random.surface out=rand_surf_100m
r.random.surface out=rand_surf_100m_d1000 distance=1000
d.rast rand_surf_100m
d.rast rand_surf_100m_d1000
```

The first surface generated with the default spatial dependence set to
distance=0 generates a surface where the cells are independent of each other.
The second surface has the range of influence (spatial dependence) set to
1000m, essentially creating a random surface with features that are around
1000m in size (Ehlschlaeger and Goodchild, 1994). The values in the resulting

map range between 1 and the value given by the parameter `high` (in our case it was 255 given as the default value). Alternatively, random surface without spatial dependence can be created by module `r.surf.random` where you can choose both the minimum and maximum value. You can find more detailed explanation and relevant publications about the above described modules in the manual pages.

In the next set of examples, we generate a raster map representing a plane that passes through the point e=638789, n=219570, z=113 with inclination (dip) of 25 degrees and orientation (azimuth) of 120 degrees. We also show how to create a fractal surface (see Wood, 1996) and a Gaussian surface:

```
r.plane name=plane dip=25 azimuth=120 easting=638789 \
        northing=219570 elevation=113 type=float
d.rast plane

# fractal surface
r.surf.fractal surf_fractal
d.rast surf_fractal

# Gaussian surface
r.surf.gauss surf_gauss mean=100 sigma=10
d.rast surf_gauss
```

The fractal surface resembles natural topography and its structure depends on the fractal dimension (Mandelbrot, 1983). The fractal dimension (D) lies between the Euclidian dimensions 2 (plane) and 3 (volume), the closer D is to 3 the more rugged is the generated relief. Our surface uses the default dimension 2.05. The Gaussian surface represents random values around the mean value of 100 with standard deviation 10.

5.1.7 Reclassification and rescaling of raster maps

Reclassification of a raster map creates a new map based on the transformation of existing categories in the original map to a new set of categories. Usually, ranges of categories are grouped into a new category using the module `r.reclass`. Those category numbers which are not explicitly reclassified to a new value will be reclassified to NULL. Before using `r.reclass` you need to know:

- transformation rules (reclass table) describing which old categories will be assigned to which new categories;
- optionally, names for the new categories (category labels).

We recommend using the module on the command line and storing the reclass table in a file. This is convenient in the case that additional modification of the reclass table is required. The file containing the reclass rules is read from standard input (i.e., from the keyboard, redirected from a file, or piped through another program). The following examples illustrate the concept.

First, we reclassify the raster map `zipcodes`, which includes 13 categories (you may check that with `r.report zipcodes`). The new map will include 6 categories representing the area towns and cities. We store the reclass rules into a text file `zipreclass.txt`:

```
27511 27513 27518 = 1 CARY
27529 = 2 GARNER
27539 = 3 APEX
27601 27604 27605 27608 = 4 RALEIGH-CITY
27603 27606 27610 = 5 RALEIGH-SOUTH
27607 = 6 RALEIGH-WEST
```

To apply the rules to the `zipcodes` map and create a new reclassified map **towns**, we run:

```
# display the original map
g.region rast=zipcodes -p
d.erase
d.rast zipcodes

# display with simple legend, omit entries with missing label
d.rast.leg -n zipcodes

# check reclass table and then pipe it to the reclass command
cat zipreclass.txt
cat zipreclass.txt | r.reclass zipcodes out=towns \
                     title="Cities and towns"

# check the result as percent area table and as map
r.report towns unit=p
d.rast.leg -n towns
d.vect roadsmajor
```

The first `cat` command just shows the table contents. We then send the table to the GRASS module using the pipe into the second command. You can check the resulting map **towns** by reporting the percent area for each new category (town or city section) and displaying the map with major roads for better orientation. The reclassification rules can be displayed with `r.info towns`. To remove the legend frame use `d.erase -f`.

The next example explains reclassification of a continuous field map. We want to reclass the aspect map into 4 aspect ranges reflecting the terrain orientation. Aspect angle can range from 0 to 360 degrees (origin is east and rotation counterclockwise) so we create a reclass rules table `aspectreclass.txt`:

```
0 thru 45 = 4 East
45 thru 135 = 1 North
135 thru 225 = 2 West
```

```
225 thru 315 = 3 South
315 thru 360 = 4  East
```

This table is then applied to the map:

```
cat aspectreclass.txt
cat aspectreclass.txt | r.reclass aspect out=orientation \
                           title="Terrain orientation"
d.rast.leg orientation
```

As another example, we show a mixture of the reclass rules described above.
The mixed rules table reclasses the landuse96_28m map to a reduced number
of categories:

```
0 = NULL
1 2 = 1 developed
3 = 2 agriculture
4 6  =  3 herbaceous
7 8 9 = 4 shrubland
10 thru 18 = 5 forest
20 = 6 water
21 = 7 sediment
```

As in the previous examples, we store this table into a file landuserecl.txt
and run r.reclass on the landuse96_28m map:

```
cat landuserecl.txt
cat landuserecl.txt | r.reclass landuse96_28m out=landclass96\
                           title="Simplified landuse classes"
r.report landclass96 unit=p
r.colors landclass96 col=ryg
d.rast.leg landclass96
```

A hint: To minimize typing efforts, you can start from the category table of
the landuse96_28m map, store it to the file landuserecl.txt and modify it
accordingly. The module r.cats outputs the category table; we can redirect
it to an initial reclass table:

```
r.cats landuse96_28m > landuserecl.txt
```

You can then use a text editor to prepare the reclass table from the file
landuserecl.txt. Be cautious with reclassed maps. Since r.reclass internally
generates a table referencing the original raster map rather than creating a
real raster map, a reclass map will not be accessible if the original raster map
upon which it was based is removed. If such a case occurs, g.remove prints a
warning. You can use r.mapcalc to convert a reclass map to a regular raster
map:

```
g.region rast=landclass96 -p
r.mapcalc "landclass_simpl=landclass96"
```

Filter by area size In case that you need to filter areas by size, use
r.reclass.area. In our example, we eliminate all areas that are smaller than
1 hectare:

```
r.report landuse96_28m unit=h
r.reclass.area in=landuse96_28m out=landuse96_28m_1 great=1
r.report landuse96_28m_1 unit=h

# assign the same colors as in the original map and compare
r.colors landuse96_28m_1 rast=landuse96_28m
d.rast.leg -n landuse96_28m
d.rast.leg -n landuse96_28m_1
d.vect roadsmajor
```

The script **r.reclass.area** will generate a new map landuse96_28m_1 where
the minimum area size is 1 hectare, setting the omitted small areas to NULL.
These no-data areas could be filled again with surrounding values using
r.neighbors with the mode parameter, see Section 5.4.1. Use **d.erase -f** to
remove the legend frame.

Rescaling You can also reclassify a raster map by rescaling it to a different
range of values using **r.rescale** or **r.rescale.eq**; both modules are based on
r.reclass. To show the difference between the two rescaling modules, we will
apply them to the slope map where most cells in our region have relatively
low values (up to 6 degrees) and only few represent steep slopes:

```
g.region rast=slope -p
# remove above d.rast.leg frame, too
d.erase -f
r.rescale slope from=0,90 to=1,6 out=slope_c6
r.rescale.eq slope from=0,90 to=1,6 out=slope_ceq6
d.rast slope_c6
d.rast slope_ceq6
r.info slope_c6
r.info slope_ceq6
```

The histogram equalized reclassification has more evenly distributed categories throughout the region as opposed to standard rescaling where most
area falls into the categories 1 and 2.

5.1.8 Recoding of raster map types and value replacements

More complex reclassification, conversion between different raster map types
and value replacement can be performed using the module **r.recode**. It has
routines for conversion between every possible combination of raster type (e.g.,
CELL to DCELL, DCELL to FCELL, see Section 4.1.1 for more detailed
explanation of different raster data types). The recoding is based on rules
that are read from standard input (i.e., from the keyboard, redirected from

a file, or piped through another program). The default output floating point raster data precision is FCELL, to write DCELL precision raster use the flag -d. The general form of a recoding rule is:

```
oldmin:oldmax:newmin:newmax
```

In the following example, we use a special UNIX method to direct data, such as the recode rules, into the command r.recode. This method is very convenient for script programming because it skips printing out messages about the data range and help. We specify EOF (end of file) in the first line and the module reads input until the second EOF appears. If you do not specify EOF, the module prints the data range (both floating point and integer, if applicable) and prompts you for the rules. Alternatively, you can also store your recode rules in a text file and redirect it or pipe it as an input to r.recode, as we did in our r.reclass examples. To simply convert a raster map between types, for example from DCELL to CELL, run the command with the following rule:

```
r.info zipcodes_dbl
r.recode zipcodes_dbl out=zipcodes_int << EOF
 27511.0:27610.0:27511:27610
EOF
r.info zipcodes_int
```

This will convert a DCELL raster map to a new CELL raster map with the same range of values. The same task can be accomplished using r.mapcalc as we will show in the Section 5.2. To convert a raster map from FCELL to CELL while simultaneously changing the range of values in the aspect raster map from 0.0-360.0 (infinite number of directions) to 1-8 (D8 – eight directions often used for flow routing), run the command with the following rule (in this example, we skip EOF to illustrate how it works):

```
r.info aspect
r.recode aspect out=aspect_d8
 Data range of aspect is 0 to 360 (entire map)
 Integer data range of aspect is 0 to 360
 Enter the rule or 'help' for the format description:
 > 0.:360.:1:8
 > end

r.info aspect_d8
```

To illustrate the conversion of a raster map from CELL to DCELL while simultaneously replacing the values, we replace the landuse categroy in a recently created raster map landclass96 (see the section above for land use associated with each number) with associated land cover erosion factor (C-factor):

```
g.region rast=landclass96 -p
r.recode -d landclass96 out=cfact96_30m
 Integer data range of landclass96 is 1 to 7
```

```
[...]
> 1:1:0.0:0.0
> 2:2:0.1:0.1
> 3:4:0.01:0.01
> 5:5:0.0001:0.0001
> 6:6:0.0:0.0
> 7:7:0.9:0.9
> end
```

The values appear twice because the module expects both the minimum and maximum for the old and new values. When we display the resulting map, only small spots with high erosion factor are visible, so we need a new color table to cover the values that change over several magnitudes:

```
d.rast cfact96_30m
r.colors cfact96_30m color=rules
 > 0 grey
 > 0.0001 green
 > 0.01 yellow
 > 0.1 orange
 > 0.9 red
 > end
d.rast cfact96_30m
d.vect roadsmajor
```

You will find additional alternatives for defining the rules in the r.recode manual page. The value replacement method using recoding is often faster than formulating complex "if" conditions with raster map algebra module r.mapcalc.

5.1.9 Assigning category labels

Raster maps may include category labels that are internally stored in a special category table. Sometimes, this table does not exist, for example, when the map has been created by r.mapcalc or when it was imported without labels. In other cases you may want to modify or update the existing labels.

Modifying existing category labels for a raster map   The module r.support can be used to update existing category labels. As an example, we modify existing labels of the geology_30m map by replacing some of the geology type acronyms with the full description. First, let us look at the original map:

```
r.report geology_30m
```

Because the map is stored in the read-only PERMANENT MAPSET we cannot modify it (see Section 3.1.2 for explanations), and we first have to create a copy within our MAPSET:

```
g.copy rast=geology_30m,mygeology_30m
```

The category label editor is included in the module **r.support**. Start it with the name of your map:

```
r.support mygeology_30m
```

Answer the questions with <n> until it asks you if you want to "Edit the category file for [mygeology_30m]?" Enter <y> here. You will get to the first screen where the highest category number should be defined. Because you are not going to change the number of categories, you can accept the current value and continue with <ESC><ENTER>. This takes you to the category table where you can move around with cursor keys. The table displays the first 10 categories, the others are on the next page(s). If you want to change just few categories, you can get directly to it by entering it's number into the field "Next category ". As an example, proceed to the category number 24: Enter <24> in that bottom line and hit <ESC><ENTER>. Now you have the requested category number on top of the table which should read "Ccl". Go to the line and enter the full name "Lower Chilhowee".[1] You can then go to the category number 10 and then to 20 to learn how it is working. To leave this mode, either scroll through the full table with <ESC><ENTER> or just type <end> into the line "Next category ". Now you get back to the questionnaire mode of **r.support**, you can skip the rest of the questions with <ENTER> to exit the module. Finally, check the updated table with **r.cats** or **r.report**. This procedure looks a bit old-fashioned, but you can use it even remotely through low-bandwidth (wireless) network access because no graphical user interface is required.

If you have a derivative map which is based on category labels of another map, you can easily transfer category labels (the category numbers have to correspond of course):

```
r.support newmap rast=landuse96_28m
```

Note that you can create categories for floating point maps representing continuous fields such as elevation or precipitation. To do this, you can run **r.support** and create categories for ranges of FP data.

Assigning new attributes to a raster map   The next application is a bit more sophisticated because we want to automatically assign new attributes to a raster map based on calculations. For illustration, we create a new raster map **towns_area** by substituting the category labels that represent the town names in the map **towns** to labels representing town areas (**towns** was created in Section 5.1.7). The procedure will read the map **towns**, calculate the areas for the individual areas, output the results as reclass rules and reclass the **towns** map according to the rules.

---

[1] NC geological map legend,
  http://www.geology.enr.state.nc.us/usgs/blueridg.htm

To get the area information, we can use **r.report** or **r.stats**. Compare the results of both (we use -**h** to suppress page headers):

```
g.region rast=towns -p
r.report -h towns units=h
+------------------------------------------------------------+
|                   Category Information        |            |
| #|description                                 | hectares|
|------------------------------------------------------------|
| 1|CARY . . . . . . . . . . . . . . . . . . . . |   2608.490|
| 2|GARNER . . . . . . . . . . . . . . . . . . . |   1415.720|
| 3|APEX . . . . . . . . . . . . . . . . . . . . |    254.440|
| 4|RALEIGH-CITY . . . . . . . . . . . . . . . . |   1605.140|
| 5|RALEIGH-SOUTH. . . . . . . . . . . . . . . . |12,276.320|
| 6|RALEIGH-WEST . . . . . . . . . . . . . . . . |   2089.890|
|------------------------------------------------------------|
|TOTAL                                          |20,250.000|
+------------------------------------------------------------+
```

```
# output in map units (here: square meters)
r.stats --q -an towns
 1 26084900.000000
 2 14157200.000000
 3 2544400.000000
 4 16051400.000000
 5 122763200.000000
 6 20898900.000000
```

The -**a** flag allows us to print the area values in square meters related to a category, while -**n** suppresses NULL values. The flag --**q** suppresses printing of percent complete messages to standard output. Note that the area calculation depends on the raster resolution. The results from **r.report** and **r.stats** should be comparable.

However, we cannot use the output of **r.stats** as rules for reclassification of the **towns** map directly as it is not properly formatted for such task. Since we want to store the area values as category labels for each town we need to modify the output of **r.stats**, using a UNIX tool called **awk**. It is a "pattern scanning and processing language" which is very useful for modification of character strings and simple calculations with data stored in text files (see more details on how to use **awk** in the GRASS Wiki[2]). It allows us to modify a data stream on the fly:

```
r.stats --q -an towns | awk \
  '{printf "%d=%d %.2fha\n", $1, $1, $2/10000.}'
1=1 2608.49ha
2=2 1415.72ha
3=3 254.44ha
[...]
```

---

[2] GRASS Wiki, GNU text tools,
http://grass.gdf-hannover.de/wiki/GNU_text_tools

```
r.stats --q -an towns | awk \
   '{printf "%d=%d %.2fha\n", $1, $1, $2/10000.}' \
   | r.reclass towns out=towns_area
r.report -h towns_area
 [...]
 |1|2608.49ha                                                    |
 |2|1415.72ha                                                    |
 |3|254.44ha                                                     |
 [...]
```

The redirection is done with UNIX piping which sends the output of r.stats directly to awk to do some formatting, then further on to r.reclass for generating the new map. If desired, you can copy the original color table from towns to towns_area with r.colors (see Section 5.1.1).

Clumping raster area features   For some applications, we may need to create an individual category number for each raster area (polygon). For example, when the raster map includes several areas with the same surface water type (assigned the same category number), we may need to distinguish each area, in case we are interested in computing the size of each area using r.report.

The module r.clump finds all areas of contiguous cell category values in the input raster map and assigns a unique category value to each such area ("clump") in the resulting output raster map. This can be used to calculate statistics based on the clumps instead of individual cells (also called "zonal statistics" in other GIS).

Assume that we have a lake map with 3 lake types (3 category numbers) in 22 polygons. After "clumping", we will have all 22 polygons numbered individually and we can report areas for each of them individually:

```
g.region rast=lakes -p
r.report lakes units=a
 [...]
 |    #|description                                | acres|
 |---------------------------------------------------------|
 |34300|Dam/Weir  . . . . . . . . . . . . . . . |    10.922|
 |39000|Lake/Pond . . . . . . . . . . . . . . . |   867.296|
 |43600|Reservoir . . . . . . . . . . . . . . . |    11.614|
 |    *|no data. . . . . . . . . . . . . . . |49,147.918|
 |---------------------------------------------------------|
 |TOTAL                                     |50,037.750|

r.clump lakes out=lakes_individual
r.report lakes_individual units=a
 [...]
 | #|description                                 | acres|
 |---------------------------------------------------------|
```

```
| 1|  . . . . . . .. . . . . . . . . . . . .|      5.164|
| 2|  . . . . . . . . . . . . . . . . . . . .|      5.535|
[...]
|20|  . . . . . . . . . . . . . . . . . . . .|    533.439|
|21|  . . . . . . . . . . . . . . . . . . . .|     10.922|
|22|  . . . . . . . . . . . . . . . . . . . .|      7.611|
| *|no data. . . . . . . . . . . . . . . . .|49,147.918|
|----------------------------------------------------|
|TOTAL                                    |50,037.750|
```

While most raster polygons have assigned the same category (39000 lake/pond) in the original **lakes** map, all raster polygons in the **lakes_individual** map are numbered individually. This is useful for calculating area size for each individual lake.

5.1.10 Masking and handling of no-data values

Raster MASK allows the user to block out certain areas of a map from analysis by hiding them from other GRASS raster modules. MASK is a raster map which contains the values 1 and NULL. Those cells where the MASK map shows value 1 are available for display and processing while those assigned NULL are hidden. The map name **MASK** is a reserved filename[3] for raster maps. If you have a **MASK** map in your MAPSET, it will be used as for all raster operations when reading raster data. Any raster data falling outside of the MASK are treated as if its value were NULL.

To create a MASK, you need a base map that is used to select which values will represent the hidden and the active areas. As an example, we may decide to work only with urban areas defined in the raster map **urban**. We can set the MASK that will exclude all areas outside the urban category from operations (display, analysis) as follows:

```
# find the category number for urban areas
d.rast urban
r.cats urban
 55

# set the mask and check its effect
d.rast elevation
r.mask urban maskcats=55
d.rast elevation
```

Now, for any map that you display, in our example it was **elevation**, you will see only the areas designated as urban while the rural areas are left out. If you are using "bash" shell, the command line prompt will change to [Raster MASK present] as long as a MASK is present. Note that this raster MASK does not apply to vector maps.

---

[3] Note that MS-Windows does not distinguish between "mask" and "MASK". We prefer to use "MASK" throughout the book.

You can also generate MASK directly by r.mapcalc as we explain later in Section 5.2, or you can rename an existing binary raster map to MASK with g.rename. Finally, you can remove MASK either with

```
r.mask -r MASK
```

or g.remove rast=MASK, or by renaming it to another name for later re-use; for example, g.rename rast=MASK,mymask.

NULL value management   GRASS distinguishes between 0 (zero) and no-data (NULL). While zero may represent a true value, such as temperature, NULL is used where no value is available. You can specify the values to be considered as NULL when importing raster data. This is important because other systems may have a different NULL encoding. Sometimes you may want to modify the current values, such as setting a specific value to NULL or setting NULL to a true value. Here, we explain how to exchange NULLs with a single other value using the module r.null. You can change a certain value (e.g. -9999 or 0) to NULL, using the setnull parameter:

```
# generic example for setting cells with -9999 value to NULL
r.null mymap setnull=-9999

# create a copy of SRTM DSM map
# replace cells with elevation=0 by NULL
g.copy rast=elev_srtm_30m,myelev_srtm_30m
g.region rast=myelev_srtm_30m -p
d.rast myelev_srtm_30m
r.null myelev_srtm_30m setnull=0.
d.rast myelev_srtm_30m
```

The example changes zero to NULL in the myelev_srtm_30m raster map. Note that when displaying the resulting map and checking the data range with r.info or r.univar we still find unrealistic negative elevation along the lake borders. Later on, when talking about r.mapcalc, we introduce more complex replacement methods that will allow us to change these negative elevation values to NULL too. For some applications, we may need to fill in the areas with missing values, such as the surface water elevation in our SRTM DSM example above. We will explain this after you learn more about raster data processing, in the section about raster data interpolation.

To replace the NULLs by a true value, you can use the null parameter:

```
# replace NULLs with 0 and check the values in the white area
g.copy rast=urban,urban_0
r.null urban_0 null=0
d.rast urban_0
d.what.rast urban,urban_0
```

The example will change NULL to zero, we will show in the section about r.mapcalc why this may be needed for some map algebra operations.

5.2 Raster map algebra

Raster map algebra is a powerful tool for spatial analysis and modeling using raster data. In GRASS, map algebra is performed with r.mapcalc using the following general command:

```
r.mapcalc "newmap=expression(map1, map2, ...)"
```

where expression is any legal arithmetic expression involving existing raster maps, integer or floating point constants, and functions known to the calculator. The expression should be enclosed within quotes in command line mode. You can also type r.mapcalc, then enter one or more expressions at a prompt (quotes are not necessary), and the expressions are executed after you type end:

```
r.mapcalc
  mapcalc> newmap1=expression1(map1, map2, ...)
  mapcalc> newmap2=expression2(map2, map3, ...)
  mapcalc> end
```

The expression is stored in the history file of the newly computed raster map and can be displayed using r.info. The following **operators** (listed in the order of precedence) are available in r.mapcalc:

-	negation
!	not
~	one's complement (bit operator)
^	exponentiation
%	modulus (remainder upon division)
/	division
*	multiplication
+	addition
-	subtraction
<<	left shift
>>	right shift
>>>	right shift (unsigned)
>	greater than
>=	greater than or equal
<	less than
<=	less than or equal
==	equal
!=	not equal
&	bitwise and
		bitwise or
&&	logical and
&&&	logical and*

		logical or
|||	logical or*
#	color separator into R, G, and B color portions

   * The &&& and ||| operators handle NULL values differently compared to other operators (see Section 5.2.4).

The following **functions** are available in r.mapcalc:

abs(x)	return absolute value of x
acos(x)	inverse cosine of x (result is in degrees)
asin(x)	inverse sine of x (result is in degrees)
atan(x)	inverse tangent of x (result in degree)
atan(x,y)	inverse tangent of y/x (result is in degrees)
cos(x)	cosine of x (x in degree)
double(x)	convert x to double-precision floating point
eval([x,y,...,]z)	evaluates the values of the given expression, pass results to z
exp(x)	exponential function of x
exp(x,y)	x to the power of y
x^y	alternative for x to the power y
float(x)	converts x to single-precision floating point
graph(x,x1,y1[x2,y2..])	convert the x to y based on points in a graph
if	decision operator
if(x)	1, if x does not equal 0, otherwise 0
if(x,a)	a, if x does not equal 0, otherwise 0
if(x,a,b)	a, if x does not equal 0, otherwise b
if(x,a,b,c)	a, if x > 0, b if x equals 0, c if x < 0
int(x)	converts x to integer [truncates]
isnull(x)	1, if x equals "no data" (NULL)
log(x)	natural log of x
log(x,b)	log of x base b
max(x,y[,z...])	largest of the listed values
median(x,y[,z...])	median of the listed values
min(x,y[,z...])	smallest of the listed values
mode(x,y[,z...])	mode of the listed values
not(x)	1 if x is zero, 0 otherwise
pow(x,y)	x to the power y
rand(low,high)	generates random number between the values *low* and *high*
round(x)	rounds x to the nearest integer
sin(x)	sine of x (x in degree)
sqrt(x)	square root of x
tan(x)	tangent of x (x in degree)
xor(x,y)	exclusive-or (XOR) of x and y

`r.mapcalc` provides some additional, internal **variables**, which are related to the "moving window" used for calculations:

x()	current x-coordinate of moving window
y()	current y-coordinate of moving window
col()	current col of moving window
row()	current row of moving window
nsres()	current north-south resolution
ewres()	current east-west resolution
null()	NULL value

The value NULL (no-data) is specified with `null()`. As denoted before, NULL differs from 0 (zero).

5.2.1 Integer and floating point data

The resulting raster type in map algebra operations is defined by the type of the input raster maps and constants. The result of an expression including integer maps and constants will be an integer map; it will be a floating point map if at least one of the constants or input maps is floating point. For example, when dividing two integer maps, it is important to use multiplication by 1.0 to store the result as a floating point map and preserve the decimal values. To illustrate this rule, we will add a constant to an integer map:

```
g.region rast=ortho_2001_t792_1m -p
r.mapcalc "img_int = ortho_2001_t792_1m + 123"
r.info img_int
 [...]
 Range of data:    min = 176  max = 311

r.mapcalc "img_fp = ortho_2001_t792_1m + 123."
# use -r to only get the range
r.info -r img_fp
 min=176.000000
 max=311.000000
```

The resulting map `img_int` is stored as an integer map, while `img_fp` is stored as a floating point map. To transform an integer map into a floating point map, simply multiply it by 1.0 or use the float() or double() functions. The calculation of the Normalized Difference Vegetation Index (NDVI from LANDSAT-TM5) is a good example of an application where the function of integer maps needs to be stored as a floating point map. In the next example, we have used short generic names, but you can try it with the maps `lsat7_2002_40` (substitute for `tm4`) and `lsat7_2002_30` (substitute for `tm3`):

```
g.region rast=tm4 -p
r.mapcalc "ndvi1 = 1.0 *  (tm4 - tm3) / (tm4 + tm3)"
```

```
r.info -r ndvi1

r.mapcalc "ndvi2 = float(tm4 - tm3) / float(tm4 + tm3)"
r.info -r ndvi2
```

The maps **ndvi** and **ndvi2** are the new floating point raster output maps, **tm3** and **tm4** are LANDSAT channels used as integer input maps (see also Section 8.3.1). Without the multiplication by 1.0, the result would be saved as integer and important information would be lost. When using the `float()` function, it is important to apply it before doing the division.

5.2.2 Basic calculations

Cell-wise addition or subtraction of two or more raster maps is one of the common map algebra tasks. For example, we can compute the difference between the SRTM DSM **elev_srtm_30m** and the USGS DEM **elev_ned_30m**:

```
g.region rast=elev_ned_30m -p
r.mapcalc "srtm_ned30_dif=elev_srtm_30m - elev_ned_30m"

# create a custom color table to distinguish the negative
# and positive values
r.colors srtm_ned30_dif col=rules
 fp: Data range is -142.24... to 86.19...
 > -145 blue
 > -10 aqua
 > 0 white
 > 10 orange
 > 90 red
 > end

d.erase
d.rast srtm_ned30_dif
d.what.rast elev_srtm_30m,elev_ned_30m,srtm_ned30_dif
d.vect streets_wake
```

You can see that we have many areas where the USGS DEM **elev_ned_30m** is around 5 to 30m lower than the SRTM DSM **elev_srtm_30m** (positive values, orange areas) mostly due to the vegetation and buildings captured by SRTM.

To illustrate a more complex arithmetic expression, we can calculate a weighted average of two maps (here decimal points are used to ensure that the resulting **elev_avg** is stored in floating point format):

```
r.mapcalc "elev_avg=(3.*elev_srtm_30m + 5.*elev_ned_30m)/8."
d.rast elev_avg
```

5.2.3 Working with "if" conditions

Various logical operations can be performed with raster data by combining operators with if() functions. We can create a new raster map by applying an if() function to a set of raster maps or values as illustrated by the following examples:

- *if map = a, then b, else c*, for example, set all cells in landclass96 with category 1 (developed) to the same category 1 in the new map calcmap1 and all other categories to category 2:
  ```
  g.region rast=landclass96 -p
  r.mapcalc "calcmap1=if(landclass96 == 1, 1, 2)"
  d.erase
  d.rast.leg landclass96
  d.rast.leg calcmap1
  ```
- *if map is not equal a, then b, else c*, will result in the opposite classification (2 for developed, 1 for non-developed):
  ```
  r.mapcalc "calcmap2=if(landclass96 != 1, 1, 2)"
  d.rast.leg calcmap2
  ```
- we can use *if map >= a, then b, else c* to create the same map as the previous example, because all non-developed categories are equal or greater than 2:
  ```
  r.mapcalc "calcmap3=if(landclass96 >= 2, 1, 2)"
  d.rast.leg calcmap3
  ```
- *if map >= a and map <= b, then c, else d*, leads to category 1 vegetated and category 2 other:
  ```
  r.mapcalc "calcmap4= \
             if(landclass96>=3 && landclass96<=5,1,2)"
  d.rast.leg calcmap4
  ```

The if() functions can be combined to define more complex logical operations:

- Select the values 1 and 2 from landclass96 and save them in calcmap5 while setting all other values to 0:
  ```
  r.mapcalc "calcmap5a= \
             if(landclass96==1,1,0) + if(landclass96==2,2,0)"
  d.rast.leg calcmap5a

  # alternatively with logical 'or' operator
  r.mapcalc "calcmap5b= \
           if(landclass96==1 || landclass96==2,landclass96,0)"
  d.rast.leg calcmap5b
  ```
- Select the values 1 and 2 from landclass96 and save them in calcmap6 as a category 1 while setting all other values to 0, leading to a binary map:
  ```
  r.mapcalc "calcmap6= \
             if(landclass96==1,1,0) || if(landclass96==2,1,0)"
  d.rast.leg calcmap6
  ```

5.2.4 Handling of NULL values in r.mapcalc

The basic rule to remember when working with NULL data in map algebra is that operations on NULL cells lead to NULL cells. For example, if one of the input maps included in the r.mapcalc expression has NULLs in the given area, the resulting map will have NULLs in this area, too (both for addition and multiplication functions). In this way, NULL behaves differently from zero, which will have in this area zeroes for multiplication but not necessarily for addition. Therefore, if we want to do operations with NULL data, we need to use a special function isnull(). For example, if we want to fill the NULLs in the raster map lakes with values from landclass96 (in other words, when cell value in lakes is NULL, then write corresponding value of landclass96, otherwise use value in lakes divided by 1000), so we run:

```
r.mapcalc "luselakes=if(isnull(lakes),landclass96,lakes/1000)"
d.rast.leg luselakes
```

The result will be a land cover map with more detailed representation of lakes and ponds than the original landclass96 map. We have divided the map lakes by 1000 because its categories are on the order of ten thousands and we want the categories in the new map to be at approximately he same order of magnitude as the landclass96 map. You can check the categories in the input maps and in the result using r.cats. If you don't use the isnull() function, the NULL values will remain in the output map (NULL propagation). To illustrate how to avoid NULLs when adding maps, we will compute a new SRTM DSM elev_avg_nonull where NULLs in the SRTM map myelev_srtm_30m (in locations with larger lakes; see Section 5.1.10 for creation of the SRTM map) are replaced by elevations from the USGS NED elev_ned_30m. We use the function isnull() to replace NULLs by 0 (zero):

```
r.mapcalc "elev_avg_nonull=elev_ned_30m + \
           if(isnull(myelev_srtm_30m), 0, \
           elev_srtm_30m-elev_ned_30m)"
```

We had to subtract USGS DEM in the if statement to avoid adding it to the valid values of the SRTM DSM. The above examples show that it is important to carefully evaluate the use of the function isnull() when applying map algebra to raster maps containing NULLs.

If you want to change all cell values that fulfill a certain condition (for example, all values greater than 1) into NULL value, and all other values to a given value or a map (we use value 1 in our example), you need to use a function null():

```
r.mapcalc "developed = if(landclass96 > 1, null(), 1)"
```

The result will be a map with developed areas represented by the category 1 and all other land use types set to NULL.

There are also special AND (&&&) and OR (|||) operators that behave differently with NULL data compared to the general rule stating that NULL in the input map always leads to NULL in the resulting map. For example, we want to create a map where developed areas with SRTM-based elevation less than 100m are set to the category 1 while all other areas are set to category 2:

```
r.mapcalc "dev_low = \
           if(landclass96<3 &&  myelev_srtm_30m<100.,1,2)"
r.mapcalc "dev_low2= \
           if(landclass96<3 &&& myelev_srtm_30m<100.,1,2)"
```

The first example results in a map with NULLs in the areas with lakes that were NULLs in the SRTM map `myelev_srtm_30m`. In the second example, the NULLs are replaced by category 2 (the condition is treated as false).

5.2.5 Creating a MASK with r.mapcalc

You can use `r.mapcalc` as an alternative way to create a MASK, especially if it is based on more than one map or you want to use more complex rules than those offered by `r.mask`. As an example, we will create a MASK that will allow us to perform operations only in developed areas with a given range of elevations. We compute a raster map MASK using an expression which assigns the value 1 to the cells that have elevations between 60m and 100m in the given `elevation` map and land use categories 1 and 2 (high and low intensity developed) in the given map `landuse96_28m`, while NULLs will be assigned elsewhere:

```
# set the region, display the input maps and create a MASK
g.region rast=elevation -p
d.erase
d.rast elevation
d.rast landuse96_28m
r.mapcalc "MASK=if((elevation > 60 && elevation < 100) \
           && (landuse96_28m==1 || landuse96_28m==2),1,null())"

# display elevation again to see the MASK effect
d.rast elevation

# rename MASK to disable it, and display elevation again
g.rename rast=MASK,maskfile
d.rast elevation
```

You will see that the display command run after the MASK is computed shows only the selected elevations in the developed areas. All raster map operations performed after setting a MASK are performed only in the non-masked areas. After the MASK is renamed, the entire elevation map is displayed again.

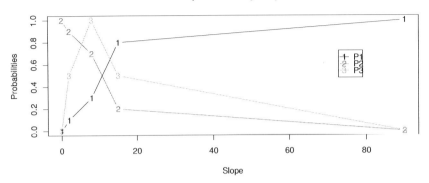

Fig. 5.1. Linear functions between two consecutive points to transform slope into probability values (e.g. for a soil property) as input for **r.mapcalc** (graph created with R)

5.2.6 Special graph operators

You can use a special graph operator to transform values in an input map representing a continuous field to new values in an output map defined by a series of linear relationships applied to a sequence of interval values. In other words, if the relationship between two variables is defined by a discrete set of (v_i, w_i) points, where v_i is a value in input map and w_i is a corresponding value in the output map, you can use the **graph** operator to perform the transformation using linear function between two consecutive points. We illustrate the concept by transforming a **slope** map to a map representing probability values (e.g. for a soil property) that increase linearly with the slope value, but the rate of increase changes at given threshold values (in our example at 2., 8. and 15. degrees). The second example shows computation of probability values that decrease with slope value, and in the third example the probability values increase up to 8 degrees slope and then they decrease (see Fig. 5.1):

```
r.mapcalc "probability1 = graph(slope, 0.,0.,2.,0.1,8.,0.3, \
          15.,0.8,90.,1.)"
r.colors probability1 col=byr
d.rast.leg probability1
r.mapcalc "probability2 = graph(slope, 0.,1.,2.,0.9,8.,0.7, \
          15.,0.2,90.,0.)"
r.colors probability2 col=byr
d.rast.leg probability2
r.mapcalc "probability3 = graph(slope, 0.,0.,2.,0.5,8.,1., \
          15.,0.5,90.,0.)"
r.colors probability3 col=byr
```

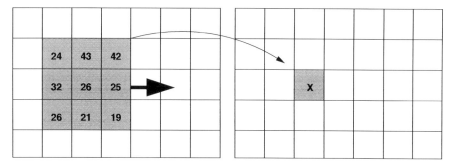

Fig. 5.2. Square "moving window" method for neighborhood operations in raster map algebra. The raster cell value X of the new map is calculated from a 3×3 matrix of the old map

```
d.rast.leg probability3

# check and compare the transformed values
d.what.rast slope,probability1,probability2,probability3
```

When defining the corresponding values, make sure that the values for the input map v_i are increasing, the values for the output map w_i can change arbitrarily.

5.2.7 Neighborhood operations with relative coordinates

The map algebra module r.mapcalc can be used for analysis and modeling that involves neighborhood operations by providing an option for working with relative coordinates of a **moving window** (Figure 5.2). The offset format is map[r,c], where r is the row (y) offset and c is the column (x) offset. For example, map[1,2] refers to the cell one row north (above) and two columns to the west (left) of the current cell, map[-2,-1] refers to the cell two rows south (below) and one column to the east (right) of the current cell, and map[0,1] refers to the cell one column to the west (left) of the current cell. Neighboring cells can be used in calculations, and larger, possibly asymmetrical moving windows (beyond the common 3×3 matrix) can be defined. We will use a simple smoothing average applied to the noisy SRTM DSM elev_srtm_30m to illustrate the concept:

```
g.region rast=elev_srtm_30m -p
d.erase
r.mapcalc "elev_srtm30m_smooth=(elev_srtm_30m[-1,-1]    \
          + elev_srtm_30m[-1,0] + elev_srtm_30m[-1,1] \
          + elev_srtm_30m[0,-1] + elev_srtm_30m[0,0]  \
          + elev_srtm_30m[0,1]  + elev_srtm_30m[1,-1] \
          + elev_srtm_30m[1,0]  + elev_srtm_30m[1,1])/9."
```

```
# assign the resulting map the same color table as the original
r.colors elev_srtm30m_smooth rast=elev_srtm_30m
d.rast elev_srtm_30m
d.rast elev_srtm30m_smooth
```

You can check the range of values in the original and smoothed maps using **r.info -r** or **r.univar** and verify that the range of values in the smoothed map is lower than in the original.

The current row and column values of the moving window can be integrated into expressions using the functions **row()** and **col()**, for current coordinates use **x()** and **y()**. To illustrate the functionality, we can generate a map mosaic using the following rules. If the x and y coordinates are smaller than the given values, then use the values from the map **elevation** in the resulting **elev_combined** map, otherwise use values from **elev_srtm_30m**:

```
r.mapcalc "elev_combined=if(x()<637455. && y()<224274.,\
                          elevation, elev_srtm_30m)"
d.rast elev_combined
```

In the above example, the variable containing the current coordinates of the moving windows was used. Note that color values of the original maps are not transferred. As another example of an application with internal variables we can generate a tilted plane, dipping to the northwest with starting altitude 100m (we have to specify 98m, as row and column each start with 1):

```
r.mapcalc "tiltplane = 98 + row() + col()"
r.info -r tiltplane
d.rast tiltplane
```

Evaluation of internal temporary variables   To perform multiple steps within one expression, we can use the function **eval()**. It evaluates temporary variables without the need to store them in a raster map. The intermediate steps are written within the **eval()** function, delimited by comma and the last result is saved as a raster map. As an example, we select a range subset from rounded values of a floating point SRTM DSM **elev_srtm_30m**, while the NULL values will be kept. We use the variables "t1" and "t2" to store the temporary results representing the rounded map and the map with rounded out of range values replaced by the values from the **elevation** map and then put back the NULLs:

```
g.region rast=elev_srtm_30m -p
r.mapcalc "elev_srtm_sub = eval (t1=round(elev_srtm_30m), \
    t2=if(t1 >= -33 && t1 <= 55,elevation,elev_srtm_30m), \
    if(isnull(elev_srtm_30m), elev_srtm_30m, t2))"
d.rast elev_srtm_sub
```

A useful alternative for value replacement is the module **r.recode** that we have explained earlier (see Section 5.1.8). For complex value replacements this

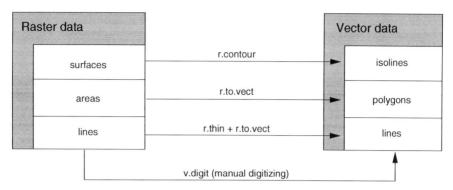

Fig. 5.3. Modules for transformation of different types of raster data to vector representation

may be more convenient than writing lengthy "if" statements in r.mapcalc. Many interesting examples of r.mapcalc applications can be found in Shapiro and Westervelt (1992).

5.3 Raster data transformation and interpolation

GRASS provides a range of capabilities for transformation of raster maps to vector maps. The approach depends on the type of raster data and the application. In this section, we also explain how to perform interpolation and approximation with raster data.

5.3.1 Automated vectorization of discrete raster data

If the raster data represent linear features, homogeneous areas, or points they can be transformed to vector data using the module r.to.vect (see Figure 5.3).

Vectorizing lines   Raster lines often need to be thinned (skeletonized) to a single pixel width using the module r.thin before they can be transformed to vector data. The lines are then vectorized using the r.to.vect module with the default feature type option set to line. For example, to vectorize the raster map **streams_derived** that represents stream network derived from the 10m resolution DEM by r.watershed, we run:

```
g.region rast=streams_derived -p
d.erase
d.rast streams_derived
r.thin streams_derived out=streams_derived_t
r.to.vect streams_derived_t out=streams_derived_t
```

```
d.vect streams_derived_t col=blue
d.vect streams col=red
```

You can compare the vectorized stream network with the vector map **streams**
provided by the Wake county government. This streams map was digitized
from 1:1200 orthophotos using additional information from photogrammetri-
cally derived contour maps and is considered to be the most accurate stream
data source for this area. The vector map **streams_derived_t** should fit well
with the Wake county **streams** vector map although there are some significant
differences within lakes and around highways.

Vectorizing raster polygons   A different method is needed for vectorization of raster polygons. Such polygons may be generated by reclassification of a
raster map (see Section 5.1.7) or by several modules, for example **r.watershed**.
Using the area borders, we can convert the raster polygons to vector areas.
Note that vectorizing areas do not require thinning and can be done directly
with **r.to.vect**. As an example, we vectorize the watershed map **basin_50K**
derived from the 10m resolution DEM by **r.watershed**:

```
g.region rast=basin_50K -p
d.erase
d.rast basin_50K

# convert to vector map while assigning each area
# a unique category number,
# store raster value as attribute "value" in attribute table
r.to.vect -s basin_50K out=basin_50K feature=area
d.vect -c basin_50K

# alternatively convert to vector map while
# using raster value as category number (flag -v)
r.to.vect -sv basin_50K out=basin_50Kval feature=area
d.vect -c basin_50Kval

# compare how raster values are stored in the new vector maps
d.what.vect -x basin_50K,basin_50Kval
 [...]
 table: basin_50K
 key column: cat
 cat : 5
 value : 20
 [...]
 table: basin_50Kval
 key column: cat
 cat : 20
```

The flag -s smoothes corners when generating vector lines or boundaries.
The vector topology is built automatically. The flag -v allows us to use

the raster map values as category numbers, otherwise the category numbers
are generated by the module and the values are stored as vector map attribute called `value`. The randomly color-filled polygons are displayed with
`d.vect -c mapname`.

The module `r.to.vect` also supports transformation of a raster map into
a gridded vector points map. The density of points is controlled by the current
region GRID RESOLUTION which you can adjust with `g.region`. To see how
it works, you can generate a vector point map (with the points 50m apart)
from the raster map `elev_ned_30m` as follows:

```
g.region swwake_10m res=50 -p
d.erase
d.rast elev_ned_30m

# transform to points with values stored as attributes
# and build topology
r.to.vect elev_ned_30m out=elev_ned50m_pts feature=point
v.info -c elev_ned50m_pts
d.vect elev_ned50m_pts size=1

# transform to points with values stored as z-coordinate,
# do not build topology
r.to.vect -zb elev_ned_30m out=elev_ned50m_ptsz feature=point
v.info -c elev_ned50m_ptsz
d.vect elev_ned50m_ptsz size=1
```

By default, the raster map values (in our case elevation) are stored in the
attribute table in a column called `value`, along with a category number (column `cat`) generated by the module for each point. Alternatively, the raster
values can be stored as z-coordinate rather than an attribute using `-z` flag, as
shown by our second example above. If the current resolution is lower than
the resolution of the raster map, the nearest neighbor cell value is used. The
computation of topology requires a lot of resources (memory and disk space)
and it can be skipped for vector point data output by running the module
with the `-b` flag. A vector map without topology is not supported by most
modules and should be considered as a temporary map that is to be converted
to raster or exported as ASCII file using `v.out.ascii`.

A raster map can be transformed to a set of random points using `r.random`.
For example, we can transform the elevation map used in the previous example; however, this time the resulting set of points (we select 6000) will be
randomly distributed:

```
g.region rast=elev_ned_30m -p
d.erase
r.random elev_ned_30m rast=elev_rand vect=elev_randpts n=6000
d.rast elev_ned_30m
d.vect elev_randpts size=2
```

The output vector points are the centers of randomly selected grid cells.

5.3.2 Generating isolines representing continuous fields

Continuous fields (e.g. elevation, temperature) are often represented by iso-lines, or in the case of an elevation surface, contours. Isolines can be derived from raster data using the module r.contour. The module determines the minimum and the maximum isoline values for a given raster map, but at least the isoline/contour interval should be provided by the step parameter. For example, we can generate contours with a 10m interval from the raster map elevation by running:

```
g.region rast=elevation -p
d.erase
d.rast elevation
r.contour elevation out=elev_contour_10m step=10
d.vect elev_contour_10m
```

The vector map topology is built automatically and the resulting contour lines are stored in the vector map elev_contour_10m.

The contour interval should be carefully selected so that the contours do not hide the underlying map, but at the same time, they should be dense enough to provide a good representation of the surface. In case that there are no additional requirements on the contour interval, its optimal value (step parameter) can be computed by the following formula, developed by Imhof (in Hake and Grünreich, 1994:382):

$$A = n * \log n * \tan \alpha \quad \text{with} \tag{5.1}$$

$$n = \sqrt{\frac{M}{100} + 1} \tag{5.2}$$

where $A[m]$ is the contour interval (difference in elevation), $\alpha[deg]$ is the slope angle class for the given relief type, and M is the map scale denominator. The value of α is selected based on the relief type:

- mountains: $\alpha = 45°$
- rolling hills: $\alpha = 25°$
- plains: $\alpha = 10°$

This leads to a contour interval of 15m for a map scale of 1:50,000 for a region with rolling hills ($\alpha = 25°$). In our example, the map displayed on the screen is approximately 1:100,000 scale, leading to $n = 31$. We can estimate the average slope needed to select the appropriate α value, by computing the univariate statistics of the slope map:

```
r.univar slope
```

The mean slope is 3.86°, so we can use the slope angle for plains ($\alpha = 10°$) leading to the rounded contour interval value of 8m, close to what we have

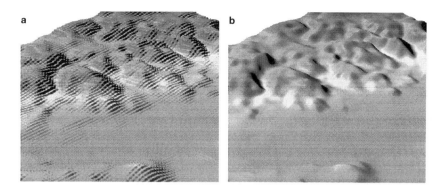

Fig. 5.4. Difference between a) nearest neighbor resampling and b) interpolation to higher resolution by RST

used in our example. More detailed, but still readable elevation surface representation is obtained by using the rounded mean slope of $4°$ leading to the contour interval of 3m that we will use in `r.contour` as the `step` parameter:

```
r.contour elevation out=elev_contour_3m step=3 min=55 max=154
d.rast elevation
d.vect elev_contour_3m
```

We have also defined the minimum and maximum contours to ensure that they are generated at easy to handle rounded values.

5.3.3 Resampling and interpolation of raster data

Transformation of a raster map to different resolution is performed automatically whenever the region resolution settings are changed. It is done by simple resampling designed for rasters representing discrete categories. Changing resolution of raster maps that represent continuous fields requires interpolation (Figure 5.4). Interpolation is also needed when filling gaps in merged raster data or when a raster map is patchy and contains NULL values that need to be replaced to achieve continuous coverage.

Nearest neighbor resampling, bilinear and bicubic interpolation GRASS uses automatic resampling when the actual region resolution is different from the resolution of the given raster map. When resampling from lower to higher resolution, the high resolution cells are assigned the same values as the cell within which they are located. When resampling from higher to lower resolution, the low resolution cell is assigned the value of the high resolution cell which is located the closest to its center (nearest neighbor). Resampling can be also applied to a raster map by the module `r.resamp.interp` that includes the nearest neighbor, bilinear and bicubic interpolation. Resampling is designed for raster data which represent geometrical features, such as lines and areas (raster maps with categories, e.g. the map `landuse96_28m`). If applied to raster maps representing continuous fields, such as elevation, the resulting surface may have a checkerboard pattern as you can see in the following example and in the Figure 5.4:

```
g.region rast=elevation -p
r.resamp.interp landuse96_28m out=landuse96_10m_nn met=nearest
r.resamp.interp elev_ned_30m out=elev_ned10m_nn method=nearest

# check the resampled landuse map
d.erase
d.rast landuse96_10m_nn
# check the resampled elevation surface using the aspect map
r.slope.aspect elev_ned10m_nn aspect=aspect_ned10m_nn
d.rast aspect_ned10m_nn
```

Nearest neighbor resampling works for the landuse map (areas with category values) but creates artifacts for the elevation map (continuous surface) which requires interpolation.

Continuous surfaces can be re-interpolated to a different resolution using either the bilinear or the bicubic method that are both implemented in the module `r.resamp.interp`. The bilinear interpolation uses a product of linear functions in x and y direction derived from 4 neigboring points (see the mathematical formulation in the Appendix A.1). Due to the small number of points used for the interpolation this method is very fast but for larger changes in resolution, a checkerboard structure may appear. The module has a special option for "wrap-around" interpolation of latitude-longitude rasters. The bicubic method uses a product of 3rd order polynomials approximated from 16 points defining the cell centers in a given rectangular area (see the equations in the Appendix A.1). The method produces smoother surface and can be applied to larger differences in resolution. You can compare these methods with the result of the nearest neighbor resampling after running:

```
g.region rast=elevation -p
r.resamp.interp elev_ned_30m out=elev_ned10m_bil meth=bilinear
r.resamp.interp elev_ned_30m out=elev_ned10m_bic meth=bicubic

# check the interpolated elevation surface using aspect maps
r.slope.aspect elev_ned10m_bil aspect=aspect_ned10m_bil
r.slope.aspect elev_ned10m_bic aspect=aspect_ned10m_bic
d.erase
d.rast aspect_ned10m_bil
d.rast aspect_ned10m_bic
```

Bilinear or bicubic methods are better alternatives for resampling of continuous data than the nearest neighbor. However, for both methods, the cell values at the edges of the region and edges of nodata areas that do not have the required 3×3 neighbors are set to NULL.

Inverse distance weighted average (IDW) interpolation The module
r.surf.idw computes the value at a grid point as a weighted average of a
given number of neighboring grid points (default number of points is 12). The
weight depends on a distance between the computed grid point and the given
point (see the equation in Appendix A.1). The module is intended to generate
a surface through a set of irregularly spaced data values. It does not perform
well for regular grids, therefore we discuss its use in the vector data processing
chapter, Section 6.8.1.

Regularized spline with tension (RST) interpolation If there is a
large difference in resolution (1:3 or more) or there are larger gaps in the raster
data that r.resamp.interp cannot handle due to its use of relatively limited
3×3 neighborhood, it is advisable to use splines for the interpolation. If appropriate parameters are selected, the result will be better than with the IDW
method (output raster map is floating point and should have fewer artifacts).
The spline interpolation is available for raster and vector data in the modules
r.resamp.rst and v.surf.rst respectively. As an example, we resample the
30m USGS DEM elev_ned30m to 10m resolution using r.resamp.rst. The
handling of resolution by this module is different than in the previous cases,
with the input resolution set to the input raster file elev_ned30m and resulting
resolution given as parameters ew_res, ns_res:

```
g.region rast=elev_ned_30m -p
r.resamp.rst elev_ned_30m elev=elev_ned10m_rrst \
             ew_res=10 ns_res=10


# first change the region to the resampled map
# check the interpolated elevation surface using aspect map
g.region rast=elev_ned10m_rrst -p
r.slope.aspect elev_ned10m_rrst aspect=aspect_ned10m_rrst
d.erase
d.rast aspect_ned10m_rrst
```

The resulting surface is similar to the bicubic interpolation, but the entire region is interpolated, including the border that was set to NULL by
r.resamp.interp. Besides the interpolation, the r.surf.rst module optionally calculates topographic parameters such as slope, aspect and curvatures
(profile, tangential and mean). The raster implementation r.resamp.rst of
the RST interpolation is less general than the vector version and if the results
are not satisfactory you can transform the raster map into random vector
points, using r.random (see Section 5.3.1) and use the v.surf.rst module
to interpolate to desired resolution, as we will describe in the Section 6.8.
For details about the spline interpolation including the equations see the Section 6.8.6, Appendix A.1 and papers by Mitas and Mitasova (1999); Mitasova
and Mitas (1993); Mitasova and Hofierka (1993).

As we have shown, when using interpolation, it is important to check
the quality of the resulting map (see for example, Web document by Dylan

Beaudette)[4] We have used aspect maps for visual comparison but you can also compute the difference between the input and output raster maps and quantitatively evaluate the differences. We will explain the spatial interpolation of raster maps from rasterized contours (or isolines) using `r.surf.contour` and `r.surf.nnbathy` in the vector data processing chapter (Section 6.8.1).

Filling data holes in a raster map   Sometimes raster maps may include areas with NULL values (e.g., due to clouds or water in case of lidar data). These areas can be filled using two approaches:

- replacing the NULL values with a given value;
- filling the NULL value areas using their boundary values by interpolation (for continuous fields) or by using neighborhood operations (for the discrete category maps).

The first approach can be done with `r.null` as explained in the previous sections. For the second option, you can use the script `r.fillnulls`. It will fill the holes with interpolated values based on the values at the no-data area boundaries using internally the `v.surf.rst` module. The script stores the hole boundaries in a separate temporary map which forms a set of "NULL lakes". The values for the "lakes" are interpolated and merged back into the original map. Only the holes are filled with the new values, the original non-NULL values remain unchanged. We can use the script to fill the holes in the SRTM-V1 DSM (see Section 5.1.10 for creation of the SRTM map `myelev_srtm_30m`):

```
g.region swwake_30m -p
# first replace the garbage values along the lake boundaries
# to NULL and extract the null areas to new map
r.mapcalc "lakes_srtm=if(isnull(myelev_srtm_30m),1,null())"
d.erase
d.rast lakes_srtm

# create a buffer around them that covers the wrong boundary
# elevations, buffering 3-4 pixels
r.buffer lakes_srtm out=lakes_srtm_buff dist=120

# change the buffered lakes to NULL and the NULLs to 0
r.null lakes_srtm_buff null=0 setnull=1,2

# find the correct lowest elevation in the region
r.info -r elev_ned_30m
min=55.173603
max=156.386520
```

---

[4] Beaudette, D. E., Image matrix,
   `http://169.237.35.250/~dylan/gdalwarp/image_matrix.html`

```
# set all cells with elevation less than 56m and within
# the buffered lakes to NULL (NULL propagation)
r.mapcalc "elev_srtm30m_null=if(elev_srtm_30m<56, null(), \
          elev_srtm_30m+lakes_srtm_buff)"

# fill the lakes using interpolation
r.fillnulls elev_srtm30m_null out=elev_srtm30m_filled
d.rast elev_srtm30m_filled

# compare with USGS 30m NED that has the lakes filled
r.mapcalc "srtmfillned30_dif=elev_ned_30m-elev_srtm30m_filled"
r.colors srtmfillned30_dif col=rules
 [...]
> 0% 0 0 126
> 30% 0 0 255
> 0 white
> 70% 255 0 0
> 100% 126 0 0
> end
d.histogram srtmfillned30_dif
d.rast.leg srtmfillned30_dif
```

It is important to realize that depending on the shape of the NULL data area(s), problems may occur due to an insufficient number of input cell values for the interpolation process or, as in our case, if the boundary values do not represent very well the values in the filled areas. In our example, the SRTM elevations around the lakes often represent top of the trees, therefore the interpolated elevation of the lakes is higher than in the USGS 30m NED. Additional problems may occur if an area containing NULLs reaches the map boundary. You have to carefully check the result using **r.mapcalc** to generate a difference map to the input map and/or **d.what.rast** to query individual cell values.

Resampling using aggregation You can resample a raster map from higher to lower resolution using several statistical methods included in the module **r.resamp.stats**. We will again use the elevation and landuse maps as examples, this time resampling from 30m to 100m resolution:

```
g.region rast=elev_ned_30m res=100 -p
# continuous field
r.resamp.stats elev_ned_30m out=elev_new100m_med method=median
# discrete categories
r.resamp.stats landuse96_28m out=landuse96_100m method=mode
```

For more accurate computation, you can use flag -**w** that weights the values from the input map according to the proportion of the input cell area that is located within the output map cell.

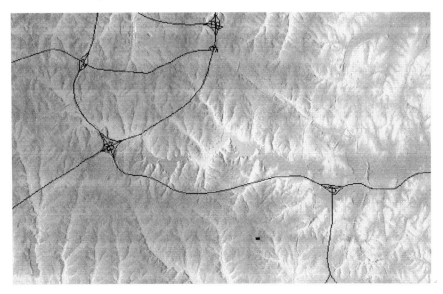

Fig. 5.5. Map composite of roads, lakes map, planned facility, and elevation model created with **r.patch** (only northern half of the map shown)

5.3.4 Overlaying and merging raster maps

As we have already mentioned, it is possible to overlay raster maps visually in the GRASS monitor using **d.rast -o**. To store such a raster map overlay in a new map, the two or more maps need to be merged into a single new map using **r.patch**. The module requires the names of input maps and a name for the new output map. The specified input map order determines the result: The NULL areas in the first map (which is on top of the virtual map stack) are filled with values from the second map and so forth for additional maps. It is important to properly set the region, the best approach is to use the option that sets the region extent and resolution based on the set of the maps to be patched. For example, multiple adjacent digital elevation models can be merged as follows (we use the lidar based 6m resolution DEM from the NC Flood mapping program that is distributed in small tiles):

```
# set region to include all imported maps
g.region rast=el_D793_6m,el_D783_6m,el_D782_6m,el_D792_6m -p
r.patch in=el_D793_6m,el_D783_6m,el_D782_6m,el_D792_6m \
        out=elevlidD_6m
r.colors elevlidD_6m rast=elevation
d.erase
d.rast elevlidD_6m
```

In another example, we will compose several raster maps within the same area. All the input maps contain some NULL values which are filled by the underlying map(s) (in case that they contain non-NULL values in these particular cells):

```
g.region rast=elev_ned_30m -p
r.patch roadsmajor,facility,lakes,elevation out=composite
r.colors composite col=rules
 [...]
 fp: Data range is 1 to 43600
 > 1 black
 > 55 green
 > 90 yellow
 > 130 orange
 > 160 brown
 > 30000 aqua
 > 50000 aqua
 > end

d.erase
d.rast composite
```

The roads network is on top in the patched map, followed by the small facility and lakes. The DEM is filling all areas not being filled by other maps (see Figure 5.5). Map overlays can also be done with r.mapcalc as we have done in the Section 5.2. The "r.mapcalc tutorial" describes several examples (Shapiro and Westervelt, 1992).

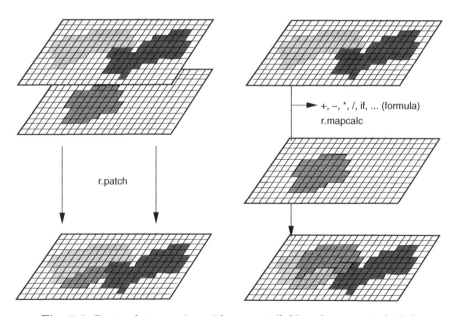

Fig. 5.6. Raster data merging with r.patch (left) and r.mapcalc (right)

Figure 5.6 shows the differences between a map merge using `r.patch` and `r.mapcalc`. The module `r.patch` patches on basis of overlays, while `r.mapcalc` combines the raster maps based on a user defined expression, as described in the Section 5.2.

For validation, the module `r.cross` checks the plausibility of merged maps against the input maps by enumerating all existing combinations which occur. In this example, we only check the categorical maps:

```
g.region rast=roadsmajor,facility,lakes -p
r.cross in=roadsmajor,facility,lakes out=composite_cross
r.cats composite_cross
d.erase
d.rast composite_cross
d.what.rast
 635893.57262104(E)  215484.55759599(N)
 composite_cross in user1 (116)category 0;category 0;Lake/Pond
```

The resulting map can be displayed and checked using `d.what.rast` to see whether the results are equivalent to the source data or not. In case that you want to display only the table with category labels (attributes), use `r.cats` in conjunction with the `r.cross` output map.

5.4 Spatial analysis with raster data

GRASS raster modules can be used to perform a wide range of spatial analysis tasks. Map overlay, generation of buffers, finding shortest paths, and deriving topographic parameters can be combined to analyze complex spatial relationships. We describe many of the available modules in this section.

5.4.1 Neighborhood analysis and cross-category statistics

The neighborhood operators determine a new value for each cell as a function of the values in its neighboring cells. All cells in a raster map, except for the cells at the map boundaries, become the center cell of a neighborhood as the neighborhood window moves from cell to cell throughout the map. The following neighborhood operators, with the user defined size of the moving window, are available in `r.neighbors` (see Appendix A.1 for equations):

- average: the average value within the neighborhood;
- diversity: the number of different cell values within the neighborhood;
- interspersion: the percentage of cells containing categories which differ from the category assigned to the center cell in the neighborhood, plus 1;
- maximum: the maximum cell value within the neighborhood;
- median: the value found half-way through a list of the neighborhood's cell values, when these are arranged in numerical order;

- minimum: the minimum cell value within the neighborhood;
- mode: the most frequently occurring cell value in the neighborhood;
- stddev: the statistical standard deviation of cell values within the neighborhood (rounded to the nearest integer);
- sum: sum of cell values within the neighborhood;
- variance: the statistical variance of cell values within the neighborhood (rounded to the nearest integer).

When using the neighborhood operator, you need to carefully consider whether you are working with a category map or a map that represents a continuous field based on measured values. The map type is reported with r.info -t mapname (compare Section 4.1.1). As an example of use with a category map, we can compute a simple landuse diversity map as follows:

```
g.region rast=landuse96_28m -p
r.info -t landuse96_28m
r.neighbors landuse96_28m out=lu_divers method=diversity size=7
d.erase
d.rast.leg lu_divers
d.vect streets_wake
r.report lu_divers unit=p
```

You can see that one of the largest areas with the smallest diversity (a single category 1) is downtown Raleigh while most of the region has a combination of 2-3 landuse categories. You can experiment with different neighborhood window size (we have used 7×7 cells, close to an approximate city block size of 200m $\times$ 200m) to see its impact on the resulting map.

To show how the operator can be applied to a continuous field map, we will use it to smooth the SRTM DSM. We already smoothed it using a 3×3 neighborhood in the section illustrating the neighborhood operator in r.mapcalc, here we will compute the average elevation using a 5×5 neighborhood and then display the original and smoothed DEMs:

```
g.region rast=elev_srtm_30m -p
r.neighbors elev_srtm_30m out=elev_srtm30m_smooth5 \
          method=average size=5
r.colors elev_srtm30m_smooth5 rast=elev_srtm30m_smooth
d.erase
d.rast elev_srtm_30m
d.rast elev_srtm30m_smooth
d.rast elev_srtm30m_smooth5
```

For better comparison, we transfer the color table from elev_srtm30m_smooth to the new map. You can see that with 5×5 neighborhood the smoothing effect is quite substantial. Additional examples can be found in the manual page of r.neighbors.

Alternatively, r.mapcalc can be used to perform neighborhood operations with relative coordinates, see Section 5.2.7 for details. In the broader sense, also the `r.li` suite (see Section 5.4.5) and `r.texture` (see Section 8.8) belong to neighborhood analysis.

Category-based statistics of base and cover map   To find an average, median, mode, standard deviation, minimum, maximum, sum, and several additional statistical measures for values in a given cover raster map within areas assigned the same category numbers in a base raster map, you can use `r.statistics`. This command integrates and extends the functionality of `r.average`, `r.median`, and `r.mode`. For example, you can compute average and standard deviation of elevation (continuous field) based on each landuse category. The resulting maps are computed as reclassified base raster maps (landuse) with the computed values (elevations) stored as category labels. You can then use `r.mapcalc` to convert them to raster maps with the computed values stored as raster values:

```
# convert elevation to integer: update to FCELL not yet done
g.region rast=elevation -p
r.mapcalc "elevint=round(elevation)"
r.statistics base=landuse96_28m cov=elevint out=elevstats_avg \
             method=average
r.cats elevstats_avg
 [...]
 1        104.7486264124
 2        106.2974427995
 3        116.0606444918
 [...]

# convert cat. labels to values, assign colors and display map
r.mapcalc "elev_avg_bylanduse=@elevstats_avg"
r.colors elev_avg_bylanduse rast=elevation
d.rast.leg elev_avg_bylanduse
d.what.rast elevstats_avg,elev_avg_bylanduse,landuse96_28m

# compute and display the standard deviation map
r.statistics base=landuse96_28m cover=elevint \
             out=elevstats_stddev method=stddev
r.mapcalc "elev_stddev_bylanduse=@elevstats_stddev"
d.rast elev_stddev_bylanduse
d.what.rast elev_stddev_bylanduse,landuse96_28m
```

The resulting maps show that the water bodies have the lowest average elevation (as one would expect), followed by high density development while the evergreen shrubs category has the highest average elevation. The standard deviation in elevation is again the lowest for water bodies, but it is the highest for high density development.

You can also use the module to find the most frequently occuring category in a cover raster map for the areas defined by the base raster map. The result will be a table that shows the distribution of categories in the cover map for each category in the base map. As an example, we will compute the most frequent landuse category and distribution of landuse categories for each zipcode:

```
g.region rast=zipcodes -p
r.statistics base=zipcodes cover=landuse96_28m \
            out=landusestats_mode method=mode
r.mapcalc "landuse_mode_byzip=round(@landusestats_mode)"

# copy the category labels
r.support landuse_mode_byzip rast=landuse96_28m
r.cats landuse_mode_byzip

d.erase
d.rast landuse_mode_byzip
d.what.rast landusestats_mode,landuse_mode_byzip

# compute distribution of land use categories
r.cats landuse96_28m
r.statistics base=zipcodes cover=landuse96_28m out=dummy \
            method=distribution
  [...]
  27601            1 59.518782
  27601            2 36.515791
  27601            4 2.540248
  27601           11 1.411982
  27601           18 0.013196
  [...]
```

The downtown zipcode 27601 has as the most common landuse category 1 "High Intensity Development", most of the suburban zipcodes (27511, 27518, ...) are in southern pine forests. The second example with the method distribution outputs a table that shows for each zip code percent covered by the landuse category (for example, the zipcode 27601 has 59.5% of its area in category 1). This output is useful for scripting, we will show below how to get more readable result using r.report.

To compute the statistical measures of values which are stored as category labels, use r.statistics -c. In this case, the category label for each category in the cover map must be a valid number (integer or decimal).

Cross-category reports We can easily generate reports for two or more maps which include occurrence of categories in the second map for each category in the first map. As an example, we can create a report which includes the size of the areas in each land use category listed for each zipcode:

```
g.region rast=landuse96_28m -p
r.report zipcodes,landuse96_28m units=h,p
  [...]
```

```
|MAPS: Zipcode areas derived from vector map (zipcodes in PER|
|        NC Land Use 1996 clipped (landuse96_28m in PERMANENT|
|------------------------------------------------------------|
|                  Category Information   |         |   %   |
|       #|description                     |  hectares| cover|
|------------------------------------------------------------|
[...]
|----------------------------------------|----------|------|
|27608|RALEIGH                           |   439.915|  2.17|
|     |----------------------------------|----------|------|
|     |  1|High Intensity Developed . . . .|    52.471| 11.93|
|     |  2|Low Intensity Developed. . . . .|   254.153| 57.77|
|     |  4|Managed Herbaceous Cover . . . .|     1.706|  0.39|
|     |15|Southern Yellow Pine . . . . . .|   131.585| 29.91|
|----------------------------------------|----------|------|
|27610|RALEIGH                           |  1334.446|  6.59|
|     |----------------------------------|----------|------|
|     |  1|High Intensity Developed . . . .|   242.457| 18.17|
|     |  2|Low Intensity Developed. . . . .|   515.535| 38.63|
[...]
```

As you can see, this produces a nicely formatted version of the **r.statistics**
output with method distribution. If you want to build a table using a set of
maps, you can select them with **g.mlist** and send the list to the module in
one line:

```
# validate selection
g.mlist type=rast pattern="land*9*" sep=, mapset=PERMANENT
# store selection into shell variable for easier re-use
MAPS=`g.mlist type=rast patt="land*9*" sep=, mapset=PERMANENT`
echo $MAPS
r.report -n map=$MAPS
```

The backtick <` `> characters have a special meaning in a shell (on the command line or in scripts). A command enclosed by these characters is executed
and the message sent by the command can be stored in an environmental variable (in our case **$MAPS**) as above. More details on shell script programming
are given in Section 9.2.

   You can also compute size or area percentage of mutual occurence of categories for two raster maps and output it as a table. Tabulating such coinci-
dence is usually useful for maps with small number of categories, so we will use
the reclassified maps **landclass96** and **towns** that we have created in previous
sections to illustrate the output:

```
g.region rast=landclass96,towns -p
r.cats landclass96
r.cats towns
r.coin -w map1=landclass96 map2=towns unit=p
```

The resulting table shows percent of each landclass for each town as well as the summaries. For example, 60% of our region is in South Raleigh and about half of the area is covered by forest and 30% is developed.

Covariance and correlation matrix   It is also possible to compute a category raster map that is a cross product of the category values from multiple raster maps using the module `r.cross`. We illustrated its functionality in Section 5.3.4 to verify patching of raster maps. The covariance and correlation matrix of several raster maps can be computed using `r.covar`. As an example, we will print the correlation matrix for our different elevation data sets:

```
g.region rast=elev_ned_30m -p
r.covar -r elevation,elev_ned_30m,elev_srtm_30m
 [...]
1.000000 0.996847 0.897588
0.996847 1.000000 0.894038
0.897588 0.894038 1.000000
```

The result shows that the 10m and 30m DEMs are practically identical (both were derived from lidar data), while the SRTM DSM has slightly lower correlation (we have already explored its problems around lake boundaries and noise).

Linear regression analysis   We can also perform a more detailed linear regression analysis for the elevation data:

```
g.region rast=elev_srtm_30m -p
r.regression.line map1=elevation map2=elev_srtm_30m
 y = a + b*x
 [...]
     a       b        R        N         F
 -2.7344 1.0479 0.897692 225000 -0.805848

   medX    sdX     medY     sdY
 109.964 20.2356 112.491 23.6204

r.regression.line map1=elevation map2=elev_ned_30m
 [...]
     a       b        R        N         F
 0.29376 0.996681 0.996815 225000 -0.993636

   medX    sdX     medY     sdY
 109.964 20.2356 109.892 20.2329
```

The result includes the following coefficients: offset (a) and gain (b), residuals (R), number of elements (N), medians (medX, medY), standard deviations (sdX, sdY), and the F test for testing the significance of the regression model as a whole (F).

Surface and volume calculation The area of a surface represented by
a raster map can be computed by r.surf.area which calculates both the
area of the horizontal plane for the given region and the area of the 3D surface estimated as a sum of triangle areas created by internally splitting each
rectangular cell by a diagonal (in map units):

```
g.region rast=elev_ned_30m -p
r.surf.area elev_ned_30m
 [...]
 Current Region plan area: 2.025000e+08
 Estimated Region Surface Area: 2.031663e+08

# current region plan area can also be calculated from
g.region -e
```

Volume of an object defined by a surface (or its subsections defined by
clumps) and a horizontal plane can be computed using r.volume. In our example, we assume that a facility will be built in a rural subarea of our study
region. We have already defined the footprint of the facility using r.digit and
stored it as a raster map myfacility_1m (see Section 4.1.3); it is also available
in the mapset PERMANENT as a raster map facility. The construction will
require grading that includes excavation to the depth of 4m. To calculate the
costs of the excavation work, we have to find the corresponding earth volume.

First we set MASK defined by the raster map facility so that only the
facility area is included in the computation. Then we can find the minimum
elevation for the facility area by running e.g. r.univar:

```
# set the region, display facility on top of orthophoto
g.region rural_1m -p
d.erase
d.rast ortho_2001_t792_1m
d.rast -o facility

# set MASK to facility map and find min elevation
r.mask facility
r.univar elevation
 [...]
 minimum: 123.521
 maximum: 129.026
 [...]
```

The minimum elevation is 123.521m, and the elevation of the bottom of the
excavation area will then be 123.52m-4m=119.52m. We subtract this elevation
from the masked portion of the DEM and store the result into a new map
excavation that defines the volume that we want to compute:

```
r.mapcalc "excavation=elevation-119.52"
d.rast excavation
r.univar excavation
```

```
[...]
minimum: 4.00057
maximum: 9.50554
[...]
r.volume excavation
 Cat     Average  Data    # Cells        Centroid        Total
 Number  in clump Total   in clump  Easting  Northing  Volume
 1        7.88      103489 13130    638725.50 220565.50 103489.00

# remove the mask after we are done
r.mask -r MASK
```

The earth volume to be excavated for the facility is roughly 103,489m³. You can re-run the computation with a higher resolution, lidar-based DEM elev_lid792_1m, you will see that the difference in volume estimate is around 3%, so our 10m resolution DEM is sufficiently accurate. For a more interesting (and realistic) example you can set the bottom elevation of the facility to 125m and use r.mapcalc and r.volume to compute the cut and fill volumes and estimate the amount of soil that will be left.

5.4.2 Buffering of raster features

Investigation of an impact of a linear or area feature on its neighborhood belongs to common tasks in spatial analysis. The areas within a given distance from the studied feature can be generated using a buffering function implemented in the r.buffer module.

As an example, we want to find the noise impact along highways in our study region. We use a simplified model based on the assumption that the potential noise impact is a function of the distance from the road. Our goal is to find which developed areas are influenced by different levels of noise. The result may determine whether noise protection walls have to be installed. The buffer zones (isophones) may be 250m (high impact), 500m (moderate impact) and more than 500m (low impact). We have selected the last buffer to be 2,500 meters in order to define a distance beyond which the impact can be considered negligible.

We apply the buffer zones to the rasterized version of the major roads map roadsmajor and then use map algebra to find the developed areas that are located within these buffers:

```
# set region and create buffers along major roads
g.region rast=landuse96_28m -p
r.buffer roadsmajor out=roads_buffers dist=250,500,2500
d.erase
d.rast roads_buffers

# intersect developed areas from landuse map with road buffers
r.cats landuse96_28m
r.mapcalc "noise=if(landuse96_28m==1 || landuse96_28m==2,\
          roads_buffers, null())"
```

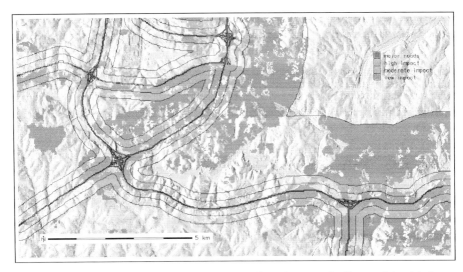

Fig. 5.7. Noise impact map from interstate (simple noise buffer model with iso-phones)

```
# transfer the category labels
r.support noise rast=roads_buffers

# convert the buffers to vector and display the result
r.to.vect -s roads_buffers out=roads_buffers feature=area
r.colors noise col=ryg
d.rast elevation_shade
d.rast -o noise
d.vect roads_buffers type=boundary
d.barscale at=5,90

# find total area for each level of impact
r.cats roads_buffers
d.what.rast noise,landuse96_28m
r.report -n noise units=h
 [...]
 |#|description                                          |  hectares|
 |----------------------------------------------------------------|
 |1|distances calculated from these locations. .| 136.53923|
 |2|0-250 meters . . . . . . . . . . . . . . . .|1045.93432|
 |3|250-500 meters . . . . . . . . . . . . . . .| 716.89185|
 |4|500-2500 meters. . . . . . . . . . . . . . .|2505.62880|
 [...]
```

The resulting map **noise** shows only those residential areas which are influenced by the major roads (see Figure 5.7). We have used **r.support** to assign labels to the map categories to indicate the level of impact. We have then queried the map directly and printed a report of affected developed area sizes (while filtering out all NULL values by running it with flag -n). Buffers can be applied in a similar way to raster areas.

Some applications may require to keep the original areas and the buffer in the same category and use a special metrics for creating the buffer. The module **r.grow** includes maximum and Manhattan metric along with Euclidean distance for computation of buffered areas. For example, we need to compute the total area covered by lakes and their protective 100m buffer (given as radius 10 cells wide) in the region with the zipcode 27606:

```
# set region, restrict analysis to selected zipcode
g.region rast=zipcodes -p
r.mask zipcodes maskcats=27606

# create the buffer, change categories to lake=1, buffer=2
# assign color table to resulting map
r.grow lakes out=lakes_grow radius=10 metric=manhattan \
             old=1 new=2
r.colors lakes_grow col=wave

# display the results and compute the protected area
d.erase
d.rast zipcodes cat=27606
d.rast lakes_grow -o
r.report -n lakes_grow unit=h
r.mask -r MASK
```

The result shows that the protected lake area in this zipcode is 502ha. Refer to Wikipedia[5] for explanation of Manhattan metric, measured along the edges of a grid or lattice, here we have used it just for illustration (it results in a more blocky area).

5.4.3 Cost surfaces

Cost surfaces are raster maps showing the cumulative costs of moving between different geographic locations in an input raster map. The value assigned to each cell in the input raster map represents the cost of traversing that cell. The module **r.cost** will produce an output raster map in which each cell represents the lowest total cost of traversing the space between each cell and the user-specified points.

We will explain the functionality using a simple application. Assume that there was an accident with fire on the major highway at coordinates 634886E, 224328N (NC State Plane meters coordinate system used in our nc_spm LO-CATION). Our task is to identify the fire station(s) that can provide the

---

[5] Wikipedia "taxicab" article on metric geometry,
   http://en.wikipedia.org/wiki/Taxicab_geometry

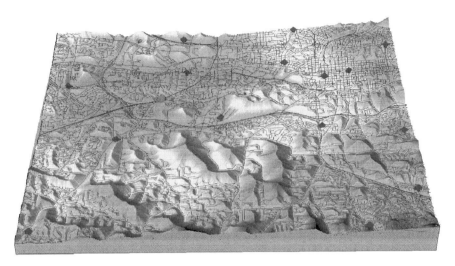

Fig. 5.8. Cost surface displayed in nviz with point symbols representing the location of the accident (sphere) and firestations (pyramids)

fastest help. The decision is influenced by two parameters: the potential speed and the distance to the fire. To solve the problem, we have to calculate a "cost surface".

A detailed roads map is available as a vector map **streets_wake**. It includes speed limit for each road segment as an attribute. We transform the vector map to a raster map **streets_speed** at 30m resolution while using the speed attribute for the raster map values. We then fill the NULLs in the resulting map with 5 mi/hr speed so that neither the target point (fire location) nor the starting point (firestation) falls within a NULL area that cannot be included in the cost surface computed by **r.cost**:

```
g.region swwake_30m -p
d.erase
v.info -c streets_wake
v.to.rast streets_wake out=streets_speed use=attr col=SPEED

# replace NULL in the street_speed map with 5 mi/hr speed
r.null streets_speed null=5
r.info streets_speed
r.colors streets_speed col=gyr
d.rast streets_speed
```

The resulting map shows speed limits ranging from 5 miles per hour in off-street areas, through 25 miles per hour on small residential streets, to 65mph on the Raleigh beltline and interstate highways. To store the location of the fire in a vector map, we import it as a vector point:

```
echo "634886 224328 1"|v.in.ascii out=fire_pt fs=space
d.vect fire_pt col=red icon=basic/marker siz=20
```

The echo command is a UNIX command to display a line of text. Here, we pipe the string of coordinates to the module v.in.ascii which is faster but effectively the same as storing the coordinates string into a text file and subsequently importing it.

Next we have to transform the street speed map into a map representing potential travel time per mile. Although the fire trucks will travel at speeds above the speed limit, they still have to adjust to the type of road and the traffic, so we can use the inverse value of speed limit as a relative measure of time it takes to pass through each cell. The reason to use the relative travel time rate instead of the speed is that r.cost considers high values as costly. We can use map algebra to generate a new map streets_travtime in [hr/mi]:

```
r.mapcalc "streets_travtime=1./streets_speed"
r.info streets_travtime
d.rast streets_travtime
d.vect fire_pt col=red icon=basic/marker siz=20
```

The new map of relative travel time provides the input to the cost surface module, which also requires the coordinates of the fire location. The costs to travel along the roads based on the inverse speed are then calculated from this location. Finally, we display the results along with the location of fire and existing firestations (Figure 5.8):

```
r.cost -k streets_travtime out=streets_cost coor=634886,224328
d.rast streets_cost
d.vect firestations col=red siz=10
d.vect firestations displ=attr attrcol=LOCATION
d.vect fire_pt col=red icon=basic/marker siz=20
d.what.rast
 [...]
 3.274   12 SE Maynard
 635933.17819149(E)  225959.77393617(N)
 2.008   20 Western Blvd
 633186.50265957(E)  221417.8856383(N)
 3.069   52 Holly Springs
nviz streets_cost vect=streets_wake point=firestations,fire_pt
```

You can find the fire stations that can reach the fire the fastest by visual inspection of the resulting cost map and by querying the resulting cost map at the locations of the fire stations using d.what.rast. The result shows that the fire brigade coming from the Western Blvd station (category 20) can reach the fire slightly faster (costs are lower) than those from the other stations. You can also use r.what to find cost for each firestation by providing their coordinates obtained from v.out.ascii:

```
# get firestation coordinates; several are outside the region
d.vect firestations displ=cat col=black
v.out.ascii firestations
[...]
630879.21198056|224876.55413017|12
[...]

# find cost for stations with categories 20, 52
r.what streets_cost east_north=635940,225912,633178,221353
635940|225912||2.0080178411
633178|221353||3.2176276305

# find the cost for all
v.out.ascii firestations fs=' ' | r.what streets_cost
```

The results confirm that the station category 20 provides the fastest access.

Of course we are also interested in getting the optimal route through the road network. The optimal routing module r.drain, which analyzes the cost surface, needs the coordinates of the fire stations located on the street_cost network, we have identified them using v.out.ascii. Additionally we specify the flag -n to count the number of cells along the path (a simple indicator for the distance):

```
# compute the optimal path
r.drain -n streets_cost out=route_20Western coor=635940,225912
r.drain -n streets_cost out=route_52Holly coord=633178,221353

# display the results, overlaying various maps
d.erase
d.vect streets_wake col=grey
d.vect fire_pt col=red icon=basic/marker size=20
d.vect firestations col=red icon=basic/box size=4
d.vect firestations displ=cat col=black xref=right lsize=15 \
       lcol=red
d.vect firestations displ=attr attrcol=LOCATION
d.rast -o route_20Western
d.rast -o route_52Holly

# print the length of the path in cells
r.describe route_20Western
r.describe route_52Holly
```

The results are two path maps with accumulated numbers of cells indicating the distance. Precise distance measure could be done by converting the raster lines to vector lines and generating a related line length report as we will show in the vector analysis chapter (see Section 6.4.1). The results should be used with caution, because we did not take into account the fact that some roads that appear connected on the raster map do not cross in reality. For example, the route route_52Holly makes turn onto the highway in a location where there is an overpass over the highway and such turn is not possible. Using 3D vector data with true overpasses solves this problem (see vector network analysis in Section 6.6).

Computing a distance map   To compute the shortest distance of each grid cell in the study area from rasterized lines, we can use cost surfaces with cost value 1. For example, we can compute a raster map that will represent the shortest distance to a major road for each grid cell in our study area as follows:

```
# set region, convert vector road map to raster
g.region swwake_30m -p
v.to.rast roadsmajor out=roadsmajor use=val

# compute the distance map
r.mapcalc "area_one=1"
r.cost -k in=area_one out=dist_toroad start_rast=roadsmajor
r.mapcalc "dist_meters=dist_toroad * (ewres() + nsres())/2."
d.rast.leg dist_meters
```

We have used the map **area_one** as a cost map and the distances were calculated from the major roads network. The resulting **dist_toroad** map represents distance information in number of cells to the closest road (it can be queried with **d.what.rast**). To calculate an approximate distance map in meters **dist_meters**, we have multiplied the cell values by the cell resolution (we have used average of resolution in north-south and west-east directions). Keep in mind, that this won't compute the exact Euclidian distance, but a polygonal approximation with accuracy dependent on the resolution.

Walking over terrain   A modification of cost map specially designed for a walking person is **r.walk**. We will show its functionality using the following example: A child got lost in the forested park in the southern part of our region. To make the search effort more efficient, we compute a map that will show how far the lost child could get (the distribution of cost) from the point where she was last seen while taking into account the topography and landcover (see Appendix A.1 for the equations used in the model, Aitken, 1977; Fontanari, 2001):

```
g.region swwake_30m -p

# create friction map based on land cover
r.cats landclass96
r.recode landclass96 out=friction << EOF
  1:3:0.1:0.1
  4:5:10.:10.
  6:6:1000.0:1000.0
  7:7:0.3:0.3
EOF
```

```
# import coordinates of the starting point
# compute the cost map and generate isochrones
echo "635576,216485"|v.in.ascii out=lostperson fs=","
r.walk -k elev=elev_ned_30m fric=friction out=walkcost \
      coord=635576,216485 lambda=0.5 max=10000
r.contour walkcost out=walkcost step=1000

# display the results
d.erase
d.rast lakes
d.rast -o walkcost
d.vect lostperson
d.vect walkcost col=red
d.vect streets_wake
```

The result shows that the lost person can get farther towards the north where
there are lower slopes and some developed areas, while the lake to the south,
along with steep bluffs and forest minimize the movement in that direction.
The result is highly dependent on the choice of friction parameters. The flag
-k indicates the use of Knights move for computation of distances, you can
learn more about the algorithm and additional parameters in the manual page
for the r.walk module.

5.4.4 Terrain and watershed analysis

Topographic analysis (geomorphometry, terrain analysis, or the more general surface analysis) provides methodology for estimation of parameters that
describe geometrical properties of a studied surface (Hofierka et al., 2007),
including:

a) summary parameters and profiles, including volumes, surface areas, roughness indexes, and fractal dimension (we have worked with some of them
   already in Section 5.1.4);
b) point parameters describing the geometry of surface in a given point, such
   as slope, aspect, and different types of curvatures;
c) flow parameters based on integration along flowlines, such as slope length,
   flow accumulation, upslope contributing area, watershed (basin) boundaries, stream networks;
d) ray tracing parameters based on lines (rays) emitted from or towards the
   surface, such as line of sight or insolation;
e) landforms, process-related indexes (wetness, topographic erosion potential) and other combined measures or features.

GRASS provides a comprehensive set of tools for surface analysis, we provide
their overview with examples in the following section, see also Hofierka et al.
(2007).

Point parameters: slope, aspect and curvatures   These parameters represent measures of change in elevation (gradient) and measures of rate of this change (curvatures) and are usually expressed as the following parameters (find the exact mathematical definitions and equations in the Appendix A.1):

- slope [deg]: steepest slope angle, function of gradient magnitude;
- aspect [deg]: slope orientation, direction of gradient, the steepest slope direction, direction of flow;
- profile curvature $[m^{-1}]$: curvature in the direction of the steepest slope (perpendicular to contour lines), measures rate of change in gradient magnitude, convex curvature leads to accelerated and concave to decelerated flow;
- tangential curvature $[m^{-1}]$: curvature in the direction of contour tangent (perpendicular to the steepest slope direction), measures rate of change in gradient direction, convex curvature leads to dispersal and concave to convergent flow.

You can compute these parameters using the module r.slope.aspect that generates raster maps of slope, aspect, curvatures and partial derivatives from a raster map of true elevation values. The slope values are calculated in degrees by default, or you can change the units to percentage using the parameter format. In case you are working in a coordinate system that uses feet (or units other than meter), you need to be aware of the fact that the module automatically converts the horizontal distances to meters. You must use the parameter zfactor to convert elevation to meters to obtain correct values of slope, curvatures and derivatives. The aspect values represent the direction of flow (pointing downslope), measured in degrees from east, increasing counter-clockwise: $90°$ is north, $180°$ is west, $270°$ is south and $360°$ is east. The aspect value of 0 is assigned to the cells with slope equal zero by default where aspect is undefined. You can define the minimum slope for which aspect is undefined by parameter min_slp_allowed.

You can also calculate profile and tangential curvature maps. Curvatures are expressed in m^{-1}, that means curvature of 0.01 corresponds to a 100m curvature radius; negative values represent concave shapes (valleys) while positive values indicate convex shape (ridges, Mitasova and Hofierka, 1993). You can also output the raster maps of first and second order partial derivatives and use them to compute additional parameters, such as slope in a given direction or special types of curvatures, such as mean or Gaussian. We will use this option in erosion model later in this chapter. For general explanation of curvatures see, for example Rigon et al. (2006); Alexandrov et al. (1989), or Mitasova and Hofierka (1993).

The module computes topographic parameters based on approximation of the terrain surface by a second order polynomial. The partial derivatives needed for estimation of slope, aspect, and curvatures are then computed as weighted averages of elevation differences in the 3×3 neighborhood of the given grid point (see Appendix A.1).

To illustrate the analysis, we derive slope, aspect and curvature maps from the elevation map as follows:

```
g.region rast=elevation -p
r.slope.aspect el=elevation slope=slope asp=aspect\
            pcurv=profcurv tcurv=tancurv
d.erase
d.rast.leg slope
d.rast.leg aspect
d.rast.leg profcurv
d.rast.leg tancurv
```

The `elevation` DEM is based on lidar surveys, and the topographic parameters highlight the level of detail captured by these data. The curvature maps are relatively noisy, we will learn how to compute smoother curvatures at a desired level of detail using `v.surf.rst` in the next chapter, Subsection 6.8.6. Although most DEMs are now provided as floating point with *cm* or better precision, special care needs to be taken when working with DEMs represented by integer values in meters: artificial flat areas may be present and the aspect is biased in the cardinal directions. Reinterpolation to floating point DEM is then recommended (see e.g. Section 5.3.1 or .ection 5.3.3). You can see the effect of integer precision by computing a polar diagram for the 30m DEM represented with *cm* and *m* precision:

```
g.region rast=elev_ned_30m -p
# simulate integer DEM
r.mapcalc "elev_ned_30m_int=int(elev_ned_30m)"
r.slope.aspect elev_ned_30m asp=aspect_ned30m
r.slope.aspect elev_ned_30m_int asp=aspect_ned30m_int
d.erase -f
d.polar aspect_ned30m
d.polar aspect_ned30m_int
```

The aspect derived from a *m* integer precision DEM is highly dominated by spikes at multiples of 45° (0°, 45°, 90°, etc.). See Hofierka et al. (2007) for a discussion. A map related to aspect is shaded terrain that can be computed and displayed with elevation color map as follows:

```
g.region rast=elevation -p
r.shaded.relief elevation shadedmap=elevation_shade
d.erase
d.his h_map=elevation i_map=elevation_shade
```

The module `r.shaded.relief` provides parameters defining the altitude and azimuth of the light source (sun) position as well as z-scale that allows you to highlight different types of topographic features. Even more options for creating colored, shaded relief maps provides an add-on module `r.csr` available through the GRASS AddOns Wiki site.[6]

---

[6] GRASS AddOns Web site, `http://grass.gdf-hannover.de/wiki/GRASS_AddOns`

Flow parameters and watersheds   Flow parameters are derived by flow tracing (also refered to as flow routing) and are "integral" rather than "point" parameters: The value at each cell is computed as a summary function of values from a set of other cells – in our case along the flowpaths. Flow tracing is based on flow paths (also refered to as flow lines or steepest slope lines), which are curves perpendicular to contours with direction given by aspect (minus gradient). The basic flow parameters are:

- flow path length (including hillslope length);
- flow accumulation (also flow line density, used to compute upslope contributing area);
- stream network;
- watershed (basin) areas (upslope area for a given outlet).

Several algorithms have been implemented in GRASS to support for flow routing, using different methods for estimation of the flow direction and distribution of "water" to the downslope cells. Direction of flow (gradient direction or aspect) can be approximated using the following methods:

- D8 algorithm uses 8 directions representing aspect discretized to 0, 45, 90, ... degrees and estimated from elevation differences between the given grid cell and its 8 neighboring cells; this approach is implemented in r.watershed, r.terraflow (Figure 5.9a,c);
- D-infinity or vector-grid algorithm uses a floating point value of aspect, that means practically infinite number of directions; this approach is applied in r.flow (Figure 5.9c).

Flow can be routed at various levels of complexity:

- single flow direction (SFD) moves flow into a single downslope cell, it is used with D8 by r.watershed and with Dinf by r.flow (Figure 5.9a,b);
- multiple flow direction (MFD) partitions flow into two or more downslope directions, it is used by r.terraflow, r.topmodel, and the r.mapcalc flow implementation described in the "r.mapcalc tutorial" (Shapiro and Westervelt, 1992, see Figure 5.9c);
- bivariate (2D) flow used by an experimental module r.sim.water see Figure 5.9d.

Watersheds and stream networks can be extracted from raster DEMs using the module r.watershed. The module outputs a raster map that represents partitioning of the given area into watersheds (basins) with the minimum size (in number of cells) given by the parameter threshold. It also outputs flow accumulation map, flow directions (D8) and several additional terrain or stream network related parameters (see the manual for more details). For display purposes, it is useful to transform the watershed areas and stream networks into vector format. In the following example, we derive streams and watershed boundaries from the 30m resolution DEM:

```
g.region rast=elev_ned_30m -p
r.watershed elev_ned_30m thresh=10000 accum=accum_10K \
          drain=draindir_10K basin=basin_10K stream=rivers

# extract more detailed streams from flow accumulation
r.mapcalc "streams_der_30m=if(abs(accum_10K)>100,1,null())"

# convert to vector format and display results
r.to.vect -s basin_10K out=basin_10K feature=area
r.to.vect -s streams_der_30m out=streams_der_30m
d.erase
d.his h_map=basin_10K i_map=elevation_shade
d.vect basin_10K type=boundary
d.rast -o lakes
d.vect streams_der_30m col=blue
d.vect streams col=red
```

Incomplete basins are omitted (assigned NULL value) in the resulting watersheds raster map basin_10K. Cells with flow accumulation that originated outside the given region are assigned negative values in the output map accum_10K and we had to use absolute values of flow accumulation to extract detailed stream network streams_der_30m using map algebra. The river segment values in the output map rivers correspond to the watershed basin numbers. We have set the threshold parameter to the number of cells representing the smallest watershed to be extracted – we have used g.region -p to get the number of cells in the given region and made sure that the threshold is a reasonably large number. Small thresholds lead to small, strip shaped basins and can stall the program. If smaller hydrologic units are needed, you could further divide the study area into half-watersheds using the r.watershed option half.basin for this purpose.

The r.watershed module does not require filling of depressions (pits, sinks) in DEM prior to its application because it uses the least-cost algorithm to traverse the elevation surface to the outlet. When applied to the new type of DEMs that are based on lidar or radar surveys, this often leads to more accurate results compared to standard methods that rely on filled depressions (Kinner et al., 2005). When compared with the provided stream data that were digitized from aerial photos, the automated stream extraction works relatively well in natural areas; however, it cannot trace the flow properly in locations with man-made structures, such as highways and built areas, with bridges, culverts or subsurface drainage.

Sink filling Flow accumulation and drainage direction analysis in the r.watershed module provides an option for a binary depression (pits, sinks, lakes) input map which contains depressions that are large enough to store surface runoff. Such a depression map can be generated with the r.fill.dir module by filling the depressions in the DEM and then subtracting the filled DEM from the original elevation map using r.mapcalc. The result can be converted to a binary raster map on the fly using if-condition and provided as input depression map for **r.watershed**. In some cases, it is necessary to run **r.fill.dir** repeatedly (using output from one run as input to the next run) before all depressions are filled:

```
g.region rast=elevation -p
r.fill.dir elevation  el=elev_fill1 dir=dir1 areas=unres1
r.fill.dir elev_fill1 el=elev_fill2 dir=dir2 areas=unres2
r.fill.dir elev_fill2 el=elev_fill3 dir=dir3 areas=unres3
r.mapcalc "depr_bin=if((elevation-elev_fill3)< 0., 1, null())"
d.erase
d.rast elevation
d.vect roadsmajor
d.rast -o depr_bin
d.vect lakes type=area fcol=aqua
```

The results show large number of depressions, mostly due to the roads acting as dams because the culverts under roads cannot be represented in standard raster DEMs. Smaller depressions are in the forested areas along the streams, some are real, others are due to errors in data or artifacts of the procedure used to extract bare ground.

Often we need to find a watershed associated with a given outlet, for example to compute a contributing area upstream from a monitoring station. To find such watershed, we first run **r.watershed** to produce drainage (flow) directions map and then use these maps as input for the module **r.water.outlet** along with the coordinates of the outlet. We use **d.what.rast** to identify the outlet coordinates, or if the coordinates are given, to check whether it is located on the stream derived by **r.watershed** (if not, we use the closest point on the derived stream as outlet). Then we can delineate the desired watershed as follows:

```
# set region to the high resolution study area
# derive flowaccumulation and flow direction
g.region rural_1m -p
r.watershed elev_lid792_1m thresh=5000 accum=accum_5K \
          drain=draindir_5K basin=basin_5K

# display extracted streams over aerial image
d.erase
d.rast ortho_2001_t792_1m
d.rast -o accum_5K cat=1000-1000000

# identify outlet on the extracted stream
d.what.rast
 [...]
 638872.62954796(E) 220042.58544653(N)
 accum_5K in user1  (238474)
echo "638872.6 220042.6 1" | v.in.ascii out=outletA30 fs=space
d.vect outletA30 col=yellow
```

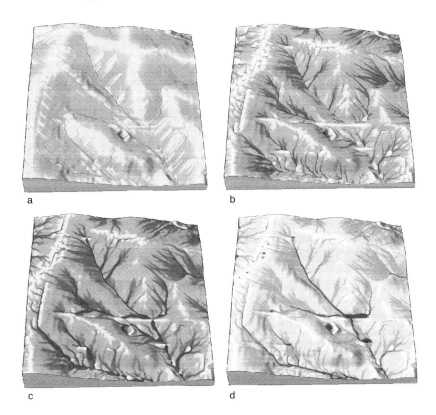

Fig. 5.9. Flow accumulation maps based on a) D8 method (**r.watershed**), b) vector-grid (D-infinite, **r.flow**), c) multiple directions flow (**r.terraflow**), d) steady state water depth (**r.sim.water**. Note that flowaccumulation is expressed as number of contributing cells (a,b), upslope area in map units (c) or water depth in |m| (d)

```
# delineate the watershed and convert it to vector format
r.water.outlet drainage=draindir_5K basin=basin_A30 \
               east=638872.6 north=220042.6
r.to.vect -s basin_A30 out=basin_A30 feature=area

# display watershed boundary along with contours
d.vect basin_A30 type=boundary col=green width=2
r.contour elev_lid792_1m out=elev_lid792_cont_1m step=1 min=104
d.vect elev_lid792_cont_1m col=white

# compute the watershed area
r.report basin_A30 unit=h,a
```

You can visually check the watershed boundaries against contours or a DEM and compute the size of the contributing area by running r.report.

Spatial distribution of water flow To analyze the spatial distribution of water flow, you can modify the color table for the flow accumulation map accum_5K and display it with d.rast. Due to the D8 algorithm, the flow pattern on hillslopes is artificial (biased towards the 8 directions, see Figure 5.9a); therefore, we will use r.flow to derive a refined representation of overland water flow patterns. This module allows us to trace flow upslope or downslope and compute raster maps representing flow accumulation, flow path length, and a vector map representing flow lines. The module uses the vector-grid (D-infinite) algorithm (Mitasova and Hofierka, 1993; Mitasova et al., 1996); therefore, it can better handle flow routing at high resolutions. First, we set the region to the input elevation map, then we use r.flow to compute raster maps of flow accumulation and flow path length as well as flow lines vector map:

```
g.region rast=elev_lid792_1m -p
d.erase
r.flow elev_lid792_1m flout=flowlines lgout=flowlg_1m \
      dsout=flowacc_1m
r.flow -u elev_lid792_1m lgout=flowlgup_1m dsout=flowaccup_1m

# display maps along with contours to see relation to terrain
d.rast flowacc_1m
d.vect elev_lid792_cont_1m col=red
d.rast flowaccup_1m
d.vect elev_lid792_cont_1m col=red
d.vect flowlines
```

When displaying the flow accumulation along with contours, you can see that downslope flow accumulation concentrates in valleys, while upslope flow accumulation merges on ridges. You can also display the flowlines to verify that they are perpendicular to contours. To compute upslope contributing area, you need to multiply the flow accumulation map produced by r.flow by cell area using r.mapcalc. The flow path length map flowlgup_1m provides information about the length of the flowline drawn from each cell. It can be used to compute the longest flow path for a given watershed, a parameter needed in hydrologic models for estimation of time to steady state (how long it will take for all water in the watershed to reach the outlet) or to compute the slope length factor for the original version of the Universal Soil Loss Equation (USLE, see Section 5.5.2).

Flow accumulation from massive DEMs Flow accumulation from massive DEMs (thousands of rows and columns) that cannot be handled by r.watershed can be computed by the r.terraflow module (Arge et al., 2003). This module is also useful for generating dispersal flow pattern over hillslopes because it uses the MFD flowrouting. Optionally, it computes topographic wetness index (available also in r.topidx, see Appendix A.2 for the equation).

```
g.region rast=elev_lid792_1m -p
r.terraflow elev_lid792_1m fill=elev_lidfilled_1m     \
            dir=dir_terrafl_1m swater=swatershed_1m \
            accum=accumterra_1m tci=tci_1m
d.erase
d.rast swatershed_1m
d.rast accumterra_1m
d.rast tci_1m
```

Note that the watersheds map computed by r.terraflow represents the contributing areas for sinks found in the elevation data so they are slightly different from watersheds computed by r.watershed or other modules that are defined for terrain after the sinks were filled (or passed through). The flow accumulation is already computed as upslope contributed area in square meters (or sq. feet).

When compared to streams digitized from airphotos and verified on ground we can see that the road represents an obstacle for automated flow routing. It diverts the flow accumulation derived by r.watershed or r.terraflow to the road's lowest point that is located several meters east from the culvert that drains water under the road. To see the problem, you can display the relevant data as follows:

```
d.rast ortho_2001_t792_1m

# get range of flow accumulation
r.info -r accumterra_1m
# selective display
d.rast accumterra_1m val=800-300000 -o
d.rast accum_5K cat=800-1000000 -o
d.vect streets_wake col=red
d.vect streams col=green
```

Carving a channel into a DEM   To modify the terrain so that the channel is located along the digitized stream in the Wake county stream map streams, use r.carve:

```
r.carve rast=elev_lid792_1m vect=streams width=2 depth=0.8 \
        out=elev_lidcarved_1m points=carved_pts
r.colors elev_lidcarved_1m rast=elev_lid792_1m
d.rast elev_lidcarved_1m

# extract streams from the carved DEM
r.watershed elev_lidcarved_1m thresh=50000 accum=accumc_5K1m
d.rast -o accumc_5K1m cat=1500-10000000
d.vect streams
```

The resulting flow accumulation from r.watershed applied to the carved DEM now passes correctly through the road and in the flat area beyond it; compare this result with accum_5K and accum_50K. Note that the lower (10m) resolution flow accumulation accum_50K passes the road correctly as opposed to the 1m resolution flow accumulation accum_5K due to the fact that in the 10m resolution DEM the road is smoothed out. However, carving does not help to improve the result of r.flow as it stops the flow on the edges of the carved channel.

Lake filling If the culvert under the road clogs with sediment and debris, the road serves as a dam and a lake is created above it. We can simulate it using a point near the culvert as a seed point for the r.lake module:

```
r.lake elev_lid792_1m wl=113.5 lake=flood1 xy=638728,220278
d.rast elev_lid792_1m
d.rast -o flood1

# display seed, overlay flow accumulation from r.watershed
echo "638728 220278 1"|v.in.ascii out=lakeseed fs=space
d.vect lakeseed
d.rast -o accum_5K cat=1000-1000000
d.vect streets_wake col=red

# use previous result as seed and increase water level
r.lake elev_lid792_1m wl=113.6 lake=flood2 seed=flood1
d.rast -o flood2
```

The second run puts the elevation over the lowest point on the road, flooding a large area downslope. You can create a sequence of lakes by gradually increasing elevation, to create a simplified flooding effect.

To summarize flowtracing, you can compare the pattern of flow accumulation produced by different modules in Figure 5.9. Based on the algorithms used in the modules these are general recommendations: use r.watershed to extract stream networks and watershed boundaries, r.flow for hillslope flow-patterns, erosion modeling, and flowline generation useful for simple surfaces, r.terraflow for flow accumulation over massive terrains and for applications requiring MFD (simplified floodplain mapping, wetness index).

Geomorphometry and landforms The module r.param.scale extracts terrain parameters at selected level of detail controlled by the size of moving window. It also derives a map of terrain features including peaks, ridges, passes, channels, pits and planes. As an example, we extract morphometric features from our elevation maps at different resolutions and levels of detail:

```
# setting region to 10m resolution DEM
g.region rast=elevation -p
d.erase
```

```
# analyze 90m neighborhood
r.param.scale elevation out=feature9c_10m size=9 param=feature
d.rast.leg feature9c_10m

# rural subregion at 1m resolution, 9m and 45m neighborhood
g.region rural_1m -p
r.param.scale elev_lid792_1m o=feature9c_1m  size=9  p=feature
r.param.scale elev_lid792_1m o=feature45c_1m size=45 p=feature
d.rast.leg feature9c_1m
d.rast.leg feature45c_1m

# display with shaded relief
r.shaded.relief elev_lid792_1m shadedmap=elev_lid792_1m_shade
d.his i=elev_lid792_1m_shade h=feature9c_1m
d.his i=elev_lid792_1m_shade h=feature45c_1m
```

The resulting maps partition the elevation surface into the areas that represent terrain features – you can see that the level of detail depends on the resolution and the size of the neighborhood. If you plan to use this module for a serious research or applications, we suggest that you read Wood (1996), which provides detailed explanations and equations used in the module.

Sun illumination and solar energy maps Many earth processes are influenced by the received solar energy. Incoming radiation is needed in environmental modeling as an input for evapotranspiration models and in urban planning it is important for designing buildings and parks. For solar illumination effects and potential radiation calculations, GRASS provides two modules: **r.sunmask** to calculate sun position and a cast shadow map, and **r.sun** to calculate solar radiation (irradiance, energy and cast shadow) maps.

The module **r.sunmask** uses the SOLPOS2 algorithm from NREL (National Renewable Energy Laboratory) which includes refraction in the atmosphere to calculate the position of sun in the sky for a given date, time, and a location on earth. You may use this feature for other purposes in case you need the sun position (**r.sunmask** provides a flag -s to calculate the sun position and then to exit). For example, we can compute the cast shadow map in our rural region for terrain and the planned facility on 22. December at 14h 25min. For this, we first calculate the solar parameters for the given date, time and location and use this as input for the cast shadow map calculation:

```
# set the region and add the planned building to the DEM
g.region rural_1m -p
r.mapcalc "elevfacility_1m=if(isnull(facility), \
          elev_lid792_1m,138.)"
r.colors elevfacility_1m col=elevation
d.rast elevfacility_1m
```

```
# compute the sun position
r.sunmask -s --v elevfacility_1m out=dummy year=2001 month=12 \
         day=22 hour=14 minute=25 sec=0 timezone=-5
 Calc. sun position..(using solpos (V.11 April 2001) from NREL)
 2001.12.22, daynum 356, time: 14:25:00(decimal time: 14.41667)
 long: -78.678856, lat: 35.736160, timezone: -5.000000
 Solar position: sun azimuth: 212.793472,
 sun angle above horz.(refraction corrected): 23.192472
 Sunrise time (without refraction): 07:26:30
 Sunset time (without refraction): 17:00:28
 No map calculation requested. Finished.
```

The sun position and date calculation results from r.sunmask can be used
as input for computation of solar radiation using r.sun (Hofierka and Súri,
2002; Súri and Hofierka, 2004), which requires the day-of-the-year and the
decimal local (solar) time for the given location as input. Optionally you can
incorporate a shadowing effect of terrain and buildings using the -s flag. In
mountainous and even hilly areas this can lead to very different results, so we
highly recommend to enable it. Note that calculating the shadowing effect of
relief can be computationally demanding:

```
# prepare input maps
r.slope.aspect elevfacility_1m asp=aspect_elevfacility_1m \
            slo=slope_elevfacility_1m
# calculate incidence angles including cast shadows
r.sun -s elevfacility_1m asp=aspect_elevfacility_1m \
         slo=slope_elevfacility_1m inc=incid_elevfacility_1m \
         day=356 time=14.416667

r.mapcalc "shadow_1m=if(incid_elevfacility_1m == 0, 1, null())"
r.colors shadow_1m col=rules
 > 1 grey
 > end

d.rast elevfacility_1m
d.rast -o shadow_1m
d.his i=aspect_elevfacility_1m h=shadow_1m
```

The shadow maps shows a curvature in the cast shadow due to the terrain.
You can write a shell script (see Chapter 9) that will compute the shadow
starting in the morning at each 30 minutes and then animate the results (see
Chapter 7.1.3) to see how the shadows move throughout the day. As another
example, we calculate radiation for our entire region at 30m resolution for day
356 (22. December):

```
g.region rast=elev_ned_30m -p
r.slope.aspect elev_ned_30m asp=asp_ned_30m slo=slp_ned_30m
r.sun -s elev_ned_30m asp=asp_ned_30m slo=slp_ned_30m lin=2.5\
 alb=0.2 beam=b356 dif=d356 refl=r356 insol_time=it356 day=356
```

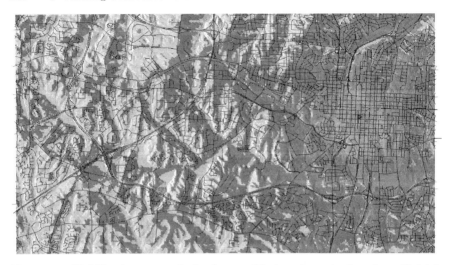

Fig. 5.10. Visibility impact analysis of a new 32 story tower in downtown Raleigh
The visibility map is overlayed over the elevation model, and it is displayed together
with the vector streets map

```
d.rast.leg d356
d.rast.leg it356
```

The radiation output maps b356, d356, and r356 represent direct (cloudless
direct beam radiation), diffuse, and reflected radiation for the given day (in
$Wh.m^{-2}.day^{-1}$), respectively. The sunshine duration is recorded in the map
it356 (in hours).

Line of sight The line of sight analysis creates a viewshed for a specific
point in an area based on the digital elevation model. The module r.los
generates a raster map output with the cells that are visible from a user-
specified observer location at a given altitude over the ground. The output map
cell values represent the vertical angle (in degree from the ground) required
to see those cells from the observer location. We introduce the usage of this
module with an example that maps the visibility of a new skyscraper that is
being built in downtown Raleigh.

The building is 32 story (165m) tall and it is being built in downtown area
at coordinates 593670E and 4926877N (NC SPM). The tower should enhance
the Raleigh skyline and there is a lot of interest in locations from where it will
be possible to see it and also what can be viewed from the top of the building.
To map the tower visbility, we use a "line of sight" method implemented in
the module r.los. We provide the elevation model as an input for analysis,
specify the coordinates and the height of the tower. The max parameter is
needed to define the maximum visibility distance (here 50km):

```
g.region rast=elev_ned_30m -p

# visibility from a new 32 story tower being built in downtown
echo "642212,224767,165"|v.in.ascii -z out=tower_165m fs=, z=3
r.los elev_ned_30m out=tower_165_los coord=642212,224767 \
     obs=165 max=50000
d.erase
d.his h=tower_165_los i=elevation_shade
d.vect streets_wake
d.vect tower_165m siz=10 col=orange icon=basic/marker
d.barscale at=0,0

# we can also do visibility from RedHat headquarters
echo "638898,224528,25" |v.in.ascii -z out=redhat_25m fs=, z=3
r.los elev_ned_30m out=redhat_25_los coord=638898,224528 \
     obs=25 max=50000
d.rast -o redhat_25_los
d.vect redhat_25m siz=10 col=red icon=basic/marker
```

We have neglected the influence of trees and existing buildings – those can be added by replacing the bare ground DEM with DSM derived from the first return lidar data or you can try the SRTM DSM. The results show that the tower will be widely visible up to about 5km, its visibility starts to diminish beyond that due to impact of topography (Figure 5.10). You can see the simulated views on the tower developers Web site.[7] The second example illustrates the same analysis for a smaller building located west of downtown at coordinates 638898E, 224528N, that is the RedHat headquarters. You can use **r.report** to compare the size of the area that can be viewed from each building (over 10,000ha from the tower as compared to 2,550ha from RedHat). See Section 6.5.5 for the extraction of the points of interest visible from top of the new skyscraper.

5.4.5 Landscape structure analysis

Quantitative analysis of the landscape structure or 2D patches represented by a raster map can be performed using the **r.li** set of commands (based on the former **r.le** suite, Baker and Cai, 1992; Baker, 2001): **r.li.setup**, **r.li.patchnum**, **r.li.richness**, **r.li.simpson**, and others. Patches may represent disturbances, natural resource areas, vegetation types, development or other landscape elements. The **r.li** suite provides options for controlling the shape (rectangle or circle), size, number, and distribution of sampling areas used to compute the landscape structure indexes. The outputs are single values that characterize the patches (e.g. mean patch size) or distribution of

---

[7] Raleigh Tower developers Web site, `http://www.rbcplazacondos.com/`, click on "views"

values (e.g. frequency distribution). These measures are raster maps or output tables, depending on the selected landscape index.

In general, landscape structure analysis can be very memory intensive. To minimize problems, the r.li suite is a client-server, multiprocess implementation which scales very well for larger maps. As an exercise, we perform analysis on a forest raster map representing their distribution in the year 1996:

```
g.region rast=landclass96 -p
r.mapcalc "forest1996=if(landclass96==5,1,null())"
d.erase
d.rast forest1996
r.li.setup
```

The r.li.setup module opens a graphical dialog which interactively queries the user to define the settings for the landscape structure analysis. With New a new configuration dialog is started. As configuration file name, we choose the name movwindow3 as it is independent from the analyzed raster maps. As Raster map to use to select areas we select the raster map to be analyzed forest1996. Next click on the Setup sampling frame button and select Whole maplayer as sampling frame and confirm with OK. Then click on Setup sampling areas to continue the dialog. We now define the sampling areas, in our case we use a moving window (the r.li.setup manual page shows several graphical illustrations of sampling areas). We confirm with OK and now click on Use keyboard to enter moving window dimension. The shape of the mowing window shall be rectangle and we need to provide the widths and height size. The resolution is given in cell units. The forest1996 map has a resolution of 28.5m × 28.5m; we choose 3 × 3 as moving window dimension. We store these settings with Save settings and return to the main window of r.li.setup. The new configuration should be listed there. We close this module now.

The configuration is now done. At this point, we can calculate a series of indices. We start with indices based on the patch number (here: in the moving window) using a 4 neighbour algorithm:

```
r.li.patchnum forest1996 conf=movwindow3 out=forest_p_num3
r.li.patchdensity forest1996 conf=movwindow3 out=forest_p_dens3
```

To better understand the results, we can convert the original forest map to vector boundaries and overlay them:

```
r.to.vect forest1996 out=forest1996 feature=area
d.rast.leg forest_p_num3
d.vect forest1996 type=boundary

d.rast.leg forest_p_dens3
d.vect forest1996 type=boundary
```

We can define a different moving window size for these or other indices by following above described procedure using r.li.setup. We can compute an index based on mean patch size or patch edge density as follows:

```
r.li.mps forest1996 conf=movwindow7 out=forest_p_mps7
d.rast.leg forest_p_mps7
r.li.edgedensity forest1996 conf=movwindow7 \
                out=forest_p_edgedens7
d.rast.leg forest_p_edgedens7
```

A different index type is the set of diversity indices. For example, we can calculate Shannon's diversity index for a raster map:

```
r.li.shannon forest1996 conf=movwindow7 \
            out=forest_p_shannon7
d.rast.leg forest_p_shannon7
d.vect forest1996 type=boundary
```

Note that if you want to run another index with the same area configuration, you do not have to create another configuration file. You can also use the same area configuration file on another map. The **r.li.*** modules rescale it automatically. For instance, if you have selected a 5×5 sample area on 100×100 rows/cols raster map, and you use the same configuration file on a 200×200 rows/cols raster map, then the sample area is 10×10.

Besides moving window, other types of sampling areas can be defined. They can be distributed across the landscape in a random, systematic contiguous and non-contiguous, or stratified-random manner. It can be also centered over sites. The manual page of **r.li.setup** contains related figures which explain the sampling design.

5.5 Landscape process modeling

GRASS has a long tradition of providing tools to support landscape process modeling. It has been coupled with various external modeling tools and used for processing of input data, analysis of results and visualization. At the same time, several models have been fully integrated with GRASS as modules, or were implemented using map algebra and other GRASS tools. In the following sections, we will focus on hydrologic, groundwater and erosion modeling. We suggest that you start **grass63** with a new MAPSET, called, for example, **simulations** to run examples in this section, because we will be computing a large number of new raster maps.

5.5.1 Hydrologic and groundwater modeling

We have already shown how to compute wetness index as a function of upslope contributing area and slope using **r.terraflow** (see Section 5.4.4). Alternatively, we can compute it using the module **r.topidx** for our high resolution, rural terrain:

```
g.region rural_1m -p
r.topidx elev_lid792_1m out=wetness_1m
d.erase
d.rast wetness_1m
d.vect elev_lid792_cont_1
```

The module uses the MFD algorithm for flowtracing and the result is a raster map that represents topographic potential for wetness or spatial soil moisture pattern (see Appendix A.2 for the equation).

More complex, dynamic study of watershed hydrology in terms of predicting surface and subsurface flow and related phenomena can be performed using several current and older versions of hydrologic models integrated with GRASS, such as r.topmodel and r.sim.water (see example of its application in the erosion modeling section below). Use of these models requires some hydrologic background, especially familiarity with hydrologic terminology and access to input data that are not as widely available as basic GIS maps. Use of these models is beyond the scope of this book; however, it is important to note that to fully evaluate the impact of spatial distribution of land use on water flow, this type of models is needed. They capture spatial aspects of such important effects as the reduced velocity of water flow and higher infiltration in areas covered by dense vegetation, or increased risk of flooding due to development when vegetated area is replaced by an impervious surface.

Modeling of several processes that form the hydrologic cycle has been integrated in HydroFOSS (Cannata, 2006). The GRASS add-ons for modeling evaporation, vegetation intercept, snow melt, and runoff (h.evapo.PM, h.hydroFOSS.init, h.hydroFOSS.runoff, h.interception, h.snow) can be downloaded from the GRASS AddOns Wiki. HydroFOSS supports continuous time simulations, determining flow rates and conditions during both runoff and dry periods. It is coupled with the automatic inverse calibration model UCODE-2005 (Poeter et al., 2005). Comprehensive hydrological and geomorphological analyses and modeling are provided by JGrass: a Java based GIS built on top of GRASS combined with R and recently integrated with uDig. The sources can be found at the JGrass Web site, Italy.[8]

Groundwater modeling   The most recent addition to GRASS that should greatly enhance its potential for development of physics based models is a new library gpde_lib that is designed to support solution of partial differential equations (Gebbert, 2007). Two new modules r.gwflow and r3.gwflow for 2D and 3D groundwater modeling have been developed using the library (see Gebbert, 2007). We illustrate their functionality using hypothetical input data for piezometric head phead_10m, boundary condition status status_10m, inner water sink well_10m, hydraulic conductivity tensor hydcond_10m, and specific yield syield_10m for a confined aquifer with its top and bottom surfaces

---

[8] JGrass Web site, http://www.jgrass.org

defined by horizontal planes at elevation 30m and 65m, and an unconfined aquifer with top at 98m:

```
# set region and prepare input data
g.region rural_1m res=10 -pa
d.erase
r.mapcalc "phead_10m= if(row() < 3, 85, 70)"
d.rast phead_10m
r.mapcalc "status_10m=if(row() < 3, 2, 1)"
d.rast status_10m
r.mapcalc "well_10m=if(row()==28 && col()==26, -0.005, 0)"
d.rast well_10m


r.mapcalc "hydcond_10m=0.00021"
r.mapcalc "top_conf_10m=65.0"
r.mapcalc "top_unconf_10m=98.0"
r.mapcalc "bottom_10m=30.0"
r.mapcalc "poros_10m=0.12"
r.mapcalc "syield_10m=0.0002"
r.mapcalc "recharg_10m=0"


# run the model for a confined reservoir
r.gwflow -s solv=cg top=top_conf_10m bottom=bottom_10m \
   phead=phead_10m status=status_10m hc_x=hydcond_10m \
   hc_y=hydcond_10m q=well_10m s=syield_10m r=recharg_10m \
   out=gwres_conf_10m dt=8640000 type=confined \
   velocity=gwres_confvel_10m
d.rast.leg gwres_conf_10m


# run the model for unconfined reservoir
r.gwflow -s solv=cg top=top_unconf_10m bottom=bottom_10m \
   phead=phead_10m status=status_10m hc_x=hydcond_10m \
   hc_y=hydcond_10m q=well_10m s=poros_10m r=recharg_10m \
   out=gwres_uconf_10m dt=8640000 type=unconfined \
   velocity=gwres_uconfvel_10m
d.rast.leg gwres_uconf_10m
```

The output raster maps `gwres_conf_10m` and `gwres2_uconf_10m` represent the piezometric head after time defined by the parameter `dt`, in our case 100 days given in seconds. You can write a script to run the command with gradually increasing `dt` to obtain a series of raster maps that shows the piezometric head development over time and then view the resulting animation using `xganim` (see the next section for a practical animation example). The groundwater modules are under active development, so check the related manual page for the latest enhancements and more complex examples (see also Gebbert, 2007).

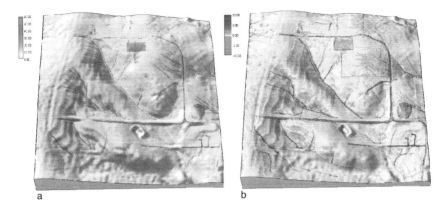

Fig. 5.11. Spatial pattern of a) sediment flow and b) erosion and deposition estimated for spatially variable cover using the USPED model

5.5.2 Erosion and deposition modeling

To compute simplified erosion risk maps, we can use the widely applied Universal Soil Loss Equation (USLE), modified for complex terrain (see Mitasova and Mitas, 2001). It is often referred to as USLE3D or the updated RUSLE3D. The detailed equations and units are in Appendix A.2, here we show only the basics needed for writing the related map algebra expressions. USLE estimates annual soil loss by the following multiplication of empirical factors:

$$A = RKLSCP \tag{5.3}$$

where A is soil loss, R is rainfall factor, K is soil erodibility, LS is a topographic factor, C is a cover factor, and P is a prevention measures factor (the last three factors are dimensionless). You can learn more about the recent developments in RUSLE at the associated Web site.[9] The LS factor can be modified for complex terrain by replacing the slope length, used in the original equation, by upslope area (Moore and Burch, 1986; Mitasova et al., 1996; Desmet and Govers, 1996; Moore and Wilson, 1992):

$$LS = (m + 1) \left(\frac{U}{22.1} \right)^m \left(\frac{\sin \beta}{0.09} \right)^n \tag{5.4}$$

where $U[m]$ is the upslope area per unit width (measure of water flow, m^2/m), β is the slope angle in degrees, 22.1m is the length and $0.09 = 9\% = 5.15°$ is the slope of the standard USLE plot, $m = 0.4$ and $n = 1.3$ are empirical constants described in more detail in the Appendix. The difference between the original LS factor based on slope length and its 3D modification is ex-

---

[9] RUSLE Web site, http://www.ars.usda.gov/Research/docs.htm?docid=5971

plained in more detail, in the relevant CERL report.[10] When computing the
3D *LS* factor, we multiply the flow accumulation computed by r.flow by the
resolution to get the upslope contributing area per cell width and calculate
the Equation 5.4 using r.mapcalc:

```
g.region rural_1m -p
r.slope.aspect elev_lid792_1m slo=slope_1m asp=aspect_1m
r.flow elev_lid792_1m dsout=flowacc_1m
r.mapcalc "lsfac3d_1m=1.4 * exp(flowacc_1m*1./22.1,0.4)\
                      * exp(sin(slope_1m)/0.09,1.2)"

# assign special color table and display result
r.info lsfac3d_1m -r
r.colors lsfac3d_1m col=rules <<EOF
 0 white
 3 yellow
 6 orange
 10 red
 50 magenta
 100 violet
EOF
d.rast.leg lsfac3d_1m
```

Note that multiplication by the cell size equal to 1.0 in the r.mapcalc expression for lsfac3d_1m is not necessary, we keep it to illustrate that this
is required if different resolution is used. Before displaying the lsfac3d_1m
raster map, we have assigned it a special color table to account for its skewed
distribution.

Alternatively, we can use flow accumulation computed by r.terraflow
divided by resolution (i.e., accumterra_1m/1., because here the flow accumulation output represents total upslope contributing area. The values of the
LS factor will be slightly different due to different flowrouting algorithm used
in r.terraflow and r.flow. The r.terraflow result will have lower values
on ridges due to dispersal flow and higher values in some concentrated areas
because it routes flow through depressions; overall it would be more suitable
for areas with prevailing dispersal flow, typical, for example, for vegetated
terrains.

The computation of slope length based RUSLE LS factor has been implemented in r.watershed, but its application is suitable only to lower resolution
data (10m and more) because it internally uses integer (CELL) values that
at higher resolution create artificial flats. The output soil loss is multiplied by
100. We can apply it for a regional scale erosion risk estimate as follows:

```
# compute length-based LS for larger area at lower resolution
g.region swwake_30m
```

---

[10] Mitasova et al., Terrain modeling and simulation,
http://skagit.meas.ncsu.edu/~helena/gmslab/reports/cerl99/rep99.html

```
r.watershed elev_ned_30m thresh=50000 length=lsfac_rws
r.mapcalc lsfac_rwsfp=lsfac_rws/100.
r.colors lsfac_rwsfp rast=lsfac3d_1m
d.rast.leg lsfac_rwsfp
```

The resulting LS values are smaller than in our previous example, due to underestimation of slope typical for this resolution and omission of concentrated flow.

The R factor is not spatially variable in our study area and we can use a constant $R = 270$ (4600 in SI units). For most areas, you can find the values of annual and single-storm R in any USLE handbook or a related textbook such as Haan et al. (1994). The soil erodibility factor K is already included in the sample data set as a raster map `soils_Kfactor`. You can check its values by running `d.what.rast` on the displayed map or by printing a table using `r.report soils_Kfactor`. The values range from 0.15 to 0.28 in US units, parking lot areas are assigned 0. If the R and K factors are not available, you can compute them using add-on modules[11] `r.usler`, `r.uslek` from annual precipitation data and soil texture, respectively. The C-factor maps `cfactorbare_1m` and `cfactorgrow_1m` are also included in the data set. They were computed by recoding the landcover raster map `landcover_1m` derived from aerial photography. The first map handles C-factor in the agricultural fields as mostly bare, while the second map considers vegetation cover typical for growing season (the recoding tables are provided in the `ncexternal/` subdirectory of our test data set).

Finally, we can calculate the erosion risk (soil detachment) for the given vegetation cover using `r.mapcalc`:

```
g.region rural_1m
r.mapcalc "soillossbare_1m=270. * soils_Kfactor * \
          lsfac3d_1m * cfactorbare_1m"
r.mapcalc "soillossgrow_1m=270. * soils_Kfactor * \
          lsfac3d_1m * cfactorgrow_1m"
r.colors soillossbare_1m col=rules <<EOF
 0 220 255 220
 5 yellow
 40 orange
 90 red
 250 magenta
 4000 violet
EOF
r.colors soillossgrow_1m rast=soillossbare_1m
d.erase
d.rast soillossbare_1m
d.rast soillossgrow_1m
r.univar soillossbare_1m
r.univar soillossgrow_1m
```

---

[11] GRASS AddOns Web site, http://grass.gdf-hannover.de/wiki/GRASS_AddOns

```
# compare with land use
d.his h_map=soillossbare_1m i_map=ortho_2001_t792_1m
```

The resulting maps represent average annual soil loss rate at the center of each grid cell for two scenarios: with agricultural fields mostly bare and with the fields covered by some vegetation throughout the year. The map soillossbare_1m shows severe erosion and potential gully formation in the fields with the average annual erosion rate for the entire area 11.64 ton/(acre.year) computed by r.univar. Keeping the fields vegetated reduces the average annual erosion to 2.20 ton/(acre.year), but the dirt roads, bare ground and vineyard can still produce significant amount of sediment. The orthophoto for this area ortho_2001_t792_1m shows that the fields are split into strips with rotating crops to ensure that at least part of the field area is always vegetated. Note that we have again used a special color table to highlight the spatial pattern of the erosion rates.

The USLE model assumes detachment limited regime of sediment transport that does not include deposition. The Unit Stream Power Based Erosion/Deposition model (USPED, Mitasova and Mitas, 2001) estimates a simplified case of erosion/deposition using the idea originally proposed by Moore and Burch (1986). It combines the RUSLE parameters and upslope contributing area per unit width U to estimate sediment flow T:

$$T \approx RKCPU^m(sin\beta)^n, \tag{5.5}$$

where the exponents m, n control the relative influence of water and slope terms and reflect the impact of different types of flow (see more details in the Appendix A.2). The net erosion/deposition D [ton/(acre.year] is computed as a divergence of sediment flow T:

$$D = \nabla \cdot (T\mathbf{s_0}) = \frac{d(T\cos\alpha)}{dx} + \frac{d(T\sin\alpha)}{dy} \tag{5.6}$$

where α in degrees is the aspect. As an example, we compute the sediment flow map sedflow using the parameters $m = n = 1$ and then the net erosion/deposition using the equation 5.6 (Figure 5.11):

```
# compute sediment flow and its components in x, y directions
r.mapcalc "sedflow_1m=270.*soils_Kfactor*cfactorgrow_1m* \
          flowacc_1m*sin(slope_1m)"
r.colors sedflow_1m rast=soillossbare_1m
d.rast sedflow_1m
r.mapcalc "qsx=sedflow_1m * cos(aspect_1m)"
r.mapcalc "qsy=sedflow_1m * sin(aspect_1m)"

# compute change of sediment flow in the x and y directions
# and then in the direction of flow using divergence
r.slope.aspect qsx dx=qsx_dx
r.slope.aspect qsy dy=qsy_dy
```

```
r.mapcalc "erdep=qsx_dx + qsy_dy"
r.info -r erdep
r.colors erdep col=rules <<EOF
 -15000 100 0 100    #dark magenta
 -100 magenta
 -10 red
 -1 orange
 -0.1 yellow
 0 200 255 200      #light green
 0.1 cyan
 1 aqua
 10 blue
 100 0 0 100        #dark blue
 18000 black
EOF
d.rast.leg erdep

# compute summary statistics
r.recode erdep out=erdep_class <<EOF
 -1200:-50:-4:-4
 -50:-5:-3:-3
 -5:-1:-2:-2
 -1:-0.1:-1:-1
 -0.1:0.1:0:0
 0.1:1:1:1
 1:5:2:2
 5:50:3:3
 50:1500:4:4
EOF
# add labels (see report table below)
r.support erdep_class

r.sum erdep
 SUM = -2374.473814
r.report erdep_class unit=p,h,a
 [...]
```

#	description	%	hectares	acres
-4	severe erosion . . .	0.19	0.101300	0.25031
-3	high erosion	1.34	0.701600	1.73365
-2	moderate erosion . .	3.89	2.042600	5.04726
-1	low erosion	19.74	10.366000	25.61438
0	stable	61.32	32.192000	79.54643
1	low deposition . . .	8.40	4.407600	10.89118
2	moderate deposition	2.49	1.307500	3.23083
3	high deposition . .	1.29	0.676900	1.67262
4	severe deposition .	0.24	0.126100	0.31159
*	no data.	1.10	0.578400	1.42922

TOTAL	100.00	52.500000	129.7275

The resulting map shows a very rich pattern of erosion and deposition typical for areas with complex land cover and topography. Red-orange-yellow areas show erosion and blue shades represent deposition. The concentrated flow in the valley has the highest erosion rates, about one magnitude lower erosion is predicted in the agricultural fields. Not all of the eroded soil will be transported out of the fields as a substantial portion can be deposited directly in the field concave areas and at the border of the field where water is slowed down by grass (Figure 5.11). The model computes extremely high values in few cells due to extremes in slope and flow accumulation change that are often artifacts of the data or due to simplifications in the algorithm. Such cells should be assigned the highest realistic erosion or deposition values. The report shows that most of the area is stable (61%) or has a low rate of erosion or deposition. Caution should be used when interpreting the results from USPED, because the RUSLE parameters were developed for simple planar fields and detachment limited erosion, while here we are applying them to much more complex conditions.

Process-based water flow and erosion modeling   Shallow overland flow can be described by St. Venant equation for continuity of flow. The solver based on path sampling method is implemented in the module r.sim.water, see Mitas and Mitasova (1998); Mitasova et al. (2005c) for equations and detailed explanation. The module computes overland flow depth or discharge based on steady, spatially distributed rainfall excess (rainfall intensity minus infiltration rate in $[mm/hr]$), elevation surface gradient given as (dx, dy), surface roughness given by Manning's coefficient, and overland flow infiltration rate in $[mm/hr]$. The gradient allows us to include prescribed flow direction by combining (dx, dy) derived from a DEM with (dx, dy) derived from line data representing channels, ditches or pipes that are not captured by the DEM. We illustrate its functionality in our rural area. For the sake of simplicity, we consider uniform land use, soils and rainfall, and no infiltration rate for flowing water:

```
# compute input raster maps for uniform land cover
# and uniform rainfall excess
g.region rural_1m res=2 -p
r.mapcalc man05=0.05
r.mapcalc infil0=0.
r.mapcalc rain50mmhr=50.

# calculate partial derivatives
v.surf.rst -d input=elev_lid792_bepts layer=0 \
        elev=elev_lid792_2m slope=dx_2m aspect=dy_2m \
        ten=15 smooth=1.5 segmax=25 npmin=100

# run the model
r.sim.water -t elevin=elev_lid792_2m dxin=dx_2m dyin=dy_2m \
        rain=rain50mmhr infil=infil0 manin=man05 \
```

```
                depth=wdp_2m disch=disch_2m nwalk=400000 \
                niter=500 outit=20 hmax=0.2 halpha=8.0 hbeta=1.0
d.rast.leg wdp_2m.0498
d.rast.leg disch_2m.0498

# animate the time series (requires Motif)
# click circle arrow to run a loop, snail to slow it down
xganim view1="wdp_2m*" view2="disch_2m*"
```

We have used v.surf.rst to compute the gradient (partial derivatives) and at
the same time, smooth the elevation surface. You can also use r.slope.aspect
and the provided DEMs to derive (dx, dy). We set the number of walkers
nwalk to approximately 2-3 per cell (it will run faster than with the default 2
million walkers). The results can be displayed by xganim to see the evolution
of overland flow represented as water depth in $[m]$ and discharge in $[m^3/s]$.
You can see that there are several depressions with ponding water, the largest
is upstream from the road which behaves as a dam. To simulate water flow
below the road, we can combine the gradient derived from the DEM with
the gradient derived from the stream network as we show in the following
example:

```
# compute direction (aspect) of the given streams
v.to.rast streams out=streams_dir_2m use=dir

# compute stream dx,dy using direction and slope=2deg
r.mapcalc "dxstr_2m=tan(2.)*cos(streams_dir_2m)"
r.mapcalc "dystr_2m=tan(2.)*sin(streams_dir_2m)"

# compute combined DEM and stream dx,dy
r.mapcalc "dxdemstr_2m=if(isnull(dxstr_2m), dx_2m, dxstr_2m)"
r.mapcalc "dydemstr_2m=if(isnull(dystr_2m), dy_2m, dystr_2m)"

# run the model
r.sim.water -t elevin=elev_lid792_2m dxin=dxdemstr_2m \
      dyin=dydemstr_2m rain=rain50mmhr inf=infil0 man=man05 \
      depth=wdpstr_2m disch=dischstr_2m nwalk=400000 \
      niter=500 outit=20 hmax=0.2 halpha=8.0 hbeta=1.0
d.rast.leg dischstr_2m.0498
xganim view1="dischstr_2m*" view2="disch_2m*"
```

Predefined gradient of water flow allows us to simulate water flowing "through"
the road, reducing the extent of flooding on the northern side of the road (for
illustration, see photos from storm events in this area here[12]). You can further
enhance the simulation by adding the impact of spatially variable landuse and
soil properties reflected in the Manning's and rainfall excess maps, created by
recording the landcover and soils maps.

The resulting water depth maps can be used as input for modeling sediment transport and net erosion/deposition using the module r.sim.sediment.

---

[12] Storm events photos,
  http://skagit.meas.ncsu.edu/~helena/wrriwork/lakewh/photos.html

This module uses a generalization of sediment transport equations used in the WEPP model (Mitas and Mitasova, 1998) and we will illustrate its functionality using simplified uniform soil and land cover conditions given by the transport capacity coefficient[s], detachment capacity coefficient [s/m], Manning's coefficient, and critical shear stress in [Pa]. We run the simulation only for the upper part of the watershed to avoid simulating the flooded areas and we put it in the background so that we can display the transport capacity map and transport capacity limited (TCL) erosion/deposition map, as these two are computed immediately:

```
# set region to the upper part of the watershed
g.region s=s+290

# compute input transport capacity and detachment coef. maps
r.mapcalc tranin=0.001
r.mapcalc detin=0.001

# compute input critical shear stress
r.mapcalc tauin=0.01

# run the model, use last depth from previous run
g.copy rast=wdp_2m.00498,wdp_2m
r.sim.sediment elevin=elev_lid792_2m dxin=dx_2m \
      dyin=dy_2m wdepth=wdp_2m detin=detin tranin=tranin \
      tauin=tauin manin=man05 nwalk=600000 niter=600 \
      tc=tcapacity et=erdepmax flux=sedflow erdep=erdepsimwe

# display these results after few seconds:
d.rast tcapacity
d.rast erdepmax

# display the final results
d.rast sedflow
d.rast erdepsimwe
```

The transport capacity map tcapacity in $[kg/ms]$ has spatial pattern similar to RUSLE3D while the TCL erosion/deposition map erdepmax in $[kg/m^2s]$ will be close to the USPED result for uniform conditions. Note that to use the same color table as for the USPED results, you will have to multiply the r.sim.sediment result by 10000 (for more accurate conversion and comparison convert the units in either of the models). The sediment flow and net erosion/deposition are results of sediment routing by water flow given by wdp_2m. The resulting spatial pattern depends on the given parameters, ranging between erosion almost everywhere (if tranin >> detin) to the result simular to USPED (transport capacity limted erosion/deposition with tranin <= detin). You can try to rerun the model by replacing detin=0.001 with detin=0.0001 to see the effect (you should get much smaller area with deposition). If the results are too noisy, you need to use larger number of walkers given by `nwalk`.

5.5.3 Final note on raster-based modeling and analysis

In addition to water and sediment flow modeling tools, GRASS also provides a set of modules for computing maps representing rates of spread `r.ros` and spread time `r.spread` designed to simulate spread of wildfires. You can learn more about these modules from the manual and try them out using a data set provided with the modules. Further examples of raster modules, including time series analysis, map algebra, filtering, and color management can be found in the Chapter 8 on Image Processing.

Several interesting modules are being developed as add-ons, for example, a landscape evolution module `r.landscape.evol` (Ullah et al., 2007), set of energy balance modules `r.eb.*`, biomass growth module `r.biomass`, and others. Check the GRASS AddOns Web page for these modules and possible new additions.

5.6 Working with voxel data

Although the support for 3D raster data is still rather limited, several new tools were added for GRASS 6. You can now perform various conversions between the 2D and 3D raster representation, for example convert a 3D raster into a series of 2D raster maps that can be analyzed using the standard 2D raster tools, or convert a set of 2D maps into a 3D raster. It is also possible to create a 3D raster by binning vector point data (assigning points to 3D raster cells). You can visualize the 3D rasters in `nviz` or export the result in VTK format for a more sophisticated visualization using `paraview` (see also Section 7.4). Here, we will use 2D horizontal slices to browse through the volumes.

In the first example, we create a 3D raster from multiple return lidar data by binning the points into the volume with 2m deep horizontal layers. We then convert the volume to a set of 2D raster maps, each representing a set of points captured at the respective elevation level:

```
# create volume from multiple return points (3D)
v.info elev_lidrural_mrpts
 [...]
 |     B:        102.22768553    T:        153.0658283         |
 [...]
g.region rural_1m res3=3 t=154 b=102 tbres=2 -ap3
d.erase
d.vect elev_lidrural_mrpts
v.to.rast3 elev_lidrural_mrpts out=elev_lid_mrvol col=Return
r3.stats elev_lid_mrvol
```

```
#convert the volume to series of 2D maps
r3.info elev_lid_mrvol
r3.to.rast elev_lid_mrvol out=elev_lid_mrlev
d.slide.show elev_lid_mrlev across=6 down=5
xganim view1="elev_lid_mrlev_*"
d.erase -f
```

We have displayed the volume as a series of horizontal planes with values representing the lidar return number allowing us to visually analyze returns available at different elevations.

In the next example, we create a volume map with values representing a soil property (in our case K-factor) by filling the volume below elevation surface `elev_lid792_1m` with the values of K-factor from the raster map `soils_Kfactor`:

```
g.region rural_1m
r.univar elevation
 [...]
 minimum: 103.973
 maximum: 131.708
 [...]
g.region rural_1m res3=3 t=132 b=102 tbres=2 -ap3
r.to.rast3elev -l soils_Kfactor elev=elev_lid792_1m \
               out=soils_Kvol
r3.info soils_Kvol

# convert to horiz. slices and display
r3.to.rast soils_Kvol out=soils_Kvolslice
d.slide.show soils_Kvolslice
xganim view1="soils_Kvolslice*"
d.erase -f
```

Also, you can create a 2D raster as an intersection between a volume and a plane or a complex surface:

```
# create a cutting plane and a cutting surface
r.plane name=cutplane dip=4 azimuth=120 easting=638600 \
        northing=220350 elevation=115 type=float
d.rast cutplane
r.mapcalc "cutsurf=if(elev_lid792_1m>115, elev_lid792_1m-8, \
           elev_lid792_1m-3)"
d.rast cutsurf
# increase 'z-exag' for better visibilit
# increase 'z-exag' for better visibilityy
nviz elev_lid792_1m,cutplane,cutsurf

# create 2D rasters as intersections with the volume
r3.cross.rast soils_Kvol elev=cutplane output=soils_Kcutpl
r3.cross.rast soils_Kvol elev=cutsurf output=soils_Kcutsurf
```

```
d.rast soils_Kcutpl
d.rast soils_Kcutsurf
```

The cross-section with the plane `soils_Kcutpl` covers only a subarea of the current region because part of the plane is above the elevation surface where the volume is filled with NULLs.

Voxel calculator and summary statistics Creation of new volume maps, reclassification, 3D analysis and modeling can be performed using the 3D version of the map algebra module `r3.mapcalc`. The module is similar to `r.mapcalc`, therefore, we include only a small example where we reclassify the soils K-factor volume map into two classes based on the K-factor value and the depth and then compute the volume for each of the two classes:

```
g.region -p3
r3.univar soils_Kvol
r3.mapcalc "soilKlow_vol=if(soils_Kvol<0.24 && depth()<10,1,2)"
r3.stats soilKlow_vol nsteps=2
num | min<=value | value<max | volume      | perc     | cell count
1     1.00000     1.500000    5621276.395 38.23998    311847
2     1.50000     2.000000    4646029.185 31.60564    257744
3          *            *     4432694.421 30.15438    245909
```

Volume analysis often requires a mask (for example, defined by bathymetry if a temperature in lake volume is modeled). You can create a volume mask based on the values in the 3D raster map `soils_Kvol` as follows:

```
# set volume MASK and see its effect
r3.mask soils_Kvol maskvalues=0.25-0.33
r3.to.rast soils_Kvol out=soils_Kvolslice_mask
d.slide.show soils_Kvolslice_mask
d.slide.show soils_Kvolslice
# remove volume MASK
g.remove rast3d=G3D_MASK
```

In the next chapter, we describe how to create volumes from 3D vector points using trivariate interpolation. Then we will get back to volumes in the visualization chapter and show how to view the volume data in 3D space. Voxel support is still relatively new in GRASS and new capabilities are added, so check the manuals for latest additions.

6

Working with vector data

The vector data model is used for representation of geographic phenomena as geometric objects composed of points, lines and areas. Point data represent either a discrete feature at a given scale, such as a city, an archaeological site or a hospital, or they are discrete samples of continuous fields such as data from climatic stations, measured elevation points, or bore-hole data. Lines are used for roads, railroads, streams, or utility networks, while areas can represent soil types, land use categories, lakes, or zoning in urban areas. Vector data are stored using their coordinates. In GRASS, the vector data model includes the description of topology.

GRASS provides tools for management and analysis of vector maps including the attributes that can be stored in a database management system. GRASS vector map operations are always performed on the full map. If this is not desired, the input map has to be clipped to the current region or a polygon beforehand as we explain in the Section 6.5. Please refer to Sections 4.2 (data model) and 4.2.2 (data exchange) for the vector model description and import/export of data.

6.1 Map viewing and metadata management

In this section, we describe how to display different types of vector maps and manage their metadata. We again use our sample data set, so you should have GRASS 6 opened with the LOCATION nc_spm (see Section 3.1.4).

6.1.1 Displaying vector maps

As we have already explained, the most efficient approach to display GRASS raster and vector data is through GUI, so check for the latest release, as the GUI is being updated frequently. To save space, we continue using the more stable display in the GRASS monitor window, using the command d.vect.

For example, to display the streams and overpasses in our region, run (see Section 3.1.5 for display management):

```
g.region region=swwake_10m -p
d.mon x0
d.erase
d.vect streams col=blue
d.vect streets_wake

# display symbols
d.vect overpasses icon=extra/bridge size=15 fcol=red
```

The d.vect command displays the selected vector map in the current GRASS monitor; in case that another map is already present, it will be overlayed over the existing map. If you want to display only selected vector objects in a map, you can use the cats parameter. The selection is done by category number (vector ID), given as a single value or a comma-separated list. See Section 4.2.1 for details about the vector model. More sophisticated selections can be done with SQL statements using the where parameter which we explain later in this chapter. Vector areas are automatically shown as filled polygons with centroids. If you don't define the color, the vector map will be displayed in black/grey, so you need to have a non-black background or raster map in your monitor to see it. Optionally, the border and polygon fill colors can be changed from default grey to another color. To individually colorize the polygons, there are two options. One is the random colorization, e.g. a colored soils vector map is shown with:

```
# display areas with centroids (sized 2 pixels)
d.vect -c soils_wake size=2

# display areas without borders and centroids
d.vect -c soils_wake type=area
```

Alternatively, an existing column containing RGB definitions can be specified. The column is a variable character field (varchar) type containing RRR:GGG:BBB values (e.g., 223:45:237 for purple).

To zoom or pan within the map, you can use d.zoom The zoom module is controlled with the mouse buttons, the context menu is shown in the terminal window see also (Section 3.1.5).

To display the vector map attributes, use d.vect with the display parameter, for example:

```
d.vect -c census_wake2000 disp=shape,attr attrcol=FIPSSTCO \
       siz=5 lcol=black
```

Font size, colors, label marker etc., can be customized accordingly, in the example above, we have set the size of the centroid symbol to 5 and the color of the label to black. Type d.vect help to see all available options for customization. You can use the script d.slide.show with flag -v to display all or selected vector maps:

```
d.slide.show -v mapsets=PERMANENT
```

Optionally, you can define a name **prefix** to see only selected maps matching the **prefix** pattern. To learn how to view vector maps in 3D using the **nviz** module, see Chapter 7.

Vector map legends Thematic vector maps can be created with d.vect.thematic. Various theme types are supported and the legend can be added in a second GRASS monitor. To illustrate the usage, we display the total capacity of schools in the south-west Wake County as graduated points overlayed over the census block map colored according to the number of households:

```
g.region swwake_30m -p
d.erase
d.vect.thematic -l censusblk_swwake column=HOUSEHOLDS \
            nint=6 color=yellow-cyan
d.vect.thematic -l schools_wake column=CAPACITYTO type=point \
            size=10 nint=6 themetype=graduated_points
```

We have used a graduated color scheme from yellow to cyan for the census block map and graduated icon sizes for school point data. Graduated line widths can be used for lines and boundaries with associated attributes. We have defined the numerical attribute to be mapped using parameter column and the number of intervals for color scheme or symbol size by the parameter nint.

Vector map charts To display charts based on vector data attributes, use d.vect.chart. The charts are positioned with their lower edge on line centers for vector lines, area centroids for vector areas, and at points for vector points by default. We can use a chart to display monthly precipitation (30 year normals) at the two stations located in the south-west Wake county as follows:

```
# set the region and display DEM, roads and lakes
g.region swwake_10m -p
d.erase
d.rast elevation
d.vect roadsmajor
d.vect lakes type=area fcol=cyan col=cyan

# display chart
d.vect.chart -c precip_30ynormals ctype=bar size=80 scale=0.6 \
    column=jan,feb,mar,apr,may,jun,jul,aug,sep,oct,nov,dec \
    color=cyan,cyan,yellow,yellow,yellow,green,green,green,\
    blue,blue,blue,cyan
```

We have used the flag -c to position the center of the chart at the given points, increased its size and reduced its relative height using the parameters size and scale and defined different colors for winter, spring, summer and fall months. You can observe slight differences between the distribution of 30 year average monthly precipitation for these two meteorological stations.

6.1.2 Vector map metadata maintenance

Metadata play an important and often underestimated role in GIS data processing. They are the key factor for data consistency and quality control, especially in large GIS projects. Metadata describe the map's origin (data source), the producer, the map scale and other geographic references as well as the time of production/modification and information on data accuracy. US government data usually include a comprehensive, standardized metadata file.

As we have already explained in Section 3.1.4, we can display general information about a vector map, such as map title, creation date, scale, number of categories, lines and areas, and boundary coordinates, with the command v.info. To display basic metadata for a point, line or polygon vector map, run:

```
v.info schools_wake
v.info streets_wake
v.info census_wake2000
```

The output includes the map title, production date, creator, vector level (topology present or not), number of categories, points, lines, boundaries, centroids, areas, islands (areas in areas), faces, kernels, projection, map boundary coordinates, map scale, and further comments. The v.info command can also display column types and names for a map. For example, the NC soils_general map has the following columns in the attribute table:

```
v.info -c soils_general
```

The vector map history (information on how the map was generated) can be retrieved as follows:

```
v.info -h soils_general
```

Most metadata for vector maps can be managed with v.support in the current MAPSET:

```
# update scale information to 1:24000
v.support myvectmap scale=24000

# update organization
v.support myvectmap organization="OSGeo labs"
v.info myvectmap
```

6.2 Vector map attribute management and SQL support

An attribute table, linked to the vector geometry, is usually generated for each newly created vector map. The table must contain a column `cat` to store the category numbers (the vector IDs) which connect the individual vector objects to one or more attributes. Hence, each table row corresponds to a category number. Several vector objects can be assigned to the same category number (table row). Category numbers are maintained with `v.category`. For more details, please read Section 4.2.1 or see the integrated manual:

```
g.manual databaseintro
```

To quickly see the contents of a vector map table, run `v.db.select`, for example:

```
v.db.select roadsmajor
```

Attributes can be modified with the module `v.db.update` with a SQL condition, or interactively through a database browser connected to the used DBMS. Here are some recommendations for tools that can be used to manage the attribute tables:

- DBF: OpenOffice.org Base (`http://www.openoffice.org`);
- MySQL: MySQL Administrator and MySQL Query Browser (`http://www.mysql.com/products/tools/`;
- PostgreSQL: pgadmin3 (`http://www.pgadmin.org/`);
- SQLite: sqlitebrowser (`http://sqlitebrowser.sf.net`); OpenOffice.org Base with sqliteodbc driver (see Appendix A.3 for setup details).

GRASS can be connected to various RDBMS and embedded databases. It supports SQL (Structured Query Language) which is a computer language used to create, retrieve, update, and delete data from relational database management systems. GRASS unifies the different drivers in an abstraction layer called DBMI – database management interface – to assist the user. SQL queries are directly passed to the underlying database system. The set of supported SQL commands depends on the RDMBS and selected driver. Table 6.1 shows the available drivers in GRASS 6.2 and later.

DBMI driver	Database site
dbf	DBF files, `http://shapelib.maptools.org/dbf_api.html`
sqlite	SQLite database, `http://sqlite.org`
pg	PostgreSQL ORDBMS, `http://postgresql.org`
mysql	MySQL RDBMS, `http://mysql.org`
mesql	MySQL embedded database, `http://mysql.org`
odbc	unixODBC (with various DBMS drivers), `http://unixodbc.org`

Table 6.1. List of available database drivers in GRASS (list may vary among various binary distributions of GRASS)

6.2.1 SQL support in GRASS 6

Usually an attribute table name is identical to the connected vector map name. Due to the SQL language definition, there are some constraints for the selection of a vector map name. SQL does not support "." (dots) in table names. Supported table name characters are limited to [A-Za-z][A-Za-z0-9_]*. A table name must start with a character, not a number. Text-string matching requires the text part to be 'single quoted'. When run from the command line, multiple queries should be contained in "double quotes", e.g.

```
g.region vect=schools_wake -p
d.erase

# show all schools in Wake County
d.vect schools_wake col=red icon=basic/circle siz=5

# show a subset of all elementary schools in Raleigh
d.vect schools_wake where="ADDRCITY='Raleigh' and GLEVEL='E'"
```

Before looking into the details of the various SQL drivers in GRASS, a general recommendation: to avoid the need of quoting column names in certain SQL backends, it is a good idea to avoid using capital letters for column names.

The default DBMS connection settings in MAPSET are managed with db.connect. By default, the DBF driver is used. Note that this command does not check for valid settings. In case you define incorrect parameters, it will be discovered only once you use the connection. For this reason, we always use db.tables in the following section to test the settings.

DBF driver

If a DBF table has to be created manually, db.execute can be used or even a spreadsheet software. Also db.copy can be used to transfer between different DBMS engines. The DBF driver supports only a few SQL statements since the DBF tables are intended for simple table storage. DBF column names are limited to 10 characters (as defined in the DBF specifications).

As an example, we create a new table with column cat for integer numbers and stype for character strings (additional supported type is double precision):

```
# define MAPSET DB connection as DBF (which is the default)
# use single quotes to avoid variable expansion in the shell
db.connect driver=dbf \
   database='$GISDBASE/$LOCATION_NAME/$MAPSET/dbf/'
# print current connection
db.connect -p
```

```
# copy map from PERMANENT: converts (or keeps in) table to DBF
g.copy vect=roadsmajor,myroadsmajor
# show attribute connection
v.db.connect -p myroadsmajor

# show the DBF tables that can be modified (in current MAPSET)
db.tables -p
```

The current GRASS MAPSET is now configured to use the DBF driver with the defined connection for attribute storage. The manual page is accessible via **g.manual grass-dbf**.

SQLite driver

The SQLite driver is much more powerful than the DBF driver and it is very simple to maintain (in fact, it is a file based SQL DBMS which does not require any configuration nor a server). Instead of using a server based system, SQLite databases are local files which are usually stored in the MAPSET. It is a good idea to create a new MAPSET when using the SQLite driver in order to manage all vector maps within this MAPSET with SQLite connection. We suggest to always choose **sqlite.db** as database name in a MAPSET. After creating the new MAPSET using **g.mapset**, the DBMS settings are defined as follows:

```
# use single quotes to avoid variable expansion in the shell
db.connect driver=sqlite \
        database='$GISDBASE/$LOCATION_NAME/$MAPSET/sqlite.db'
db.connect -p

# copy map from PERMANENT, this converts table to SQLite
g.copy vect=roadsmajor,myroadsmajor
# show attribute connection
v.db.connect -p myroadsmajor

# show SQLite tables that can be modified (in current MAPSET)
db.tables -p
```

The SQLite database is created automatically when used the first time. If you now copy an existing vector map with attribute table from another MAPSET using **g.copy**, the attribute table will be automatically converted to SQLite and imported into the **sqlite.db** file as defined above. The advantage of this driver is that it offers rather full SQL support[1] while no administration is needed since the database is a local file. The manual page is accessible via **g.manual grass-sqlite**.

---

[1] SQL As Understood By SQLite, http://www.sqlite.org/lang_expr.html

PostgreSQL driver

The PostgreSQL driver can be used for server based storage of attributes. Administrator privileges and password may be required for database creation and access, please refer to the PostgreSQL documentation for details. We use similar example as in previous cases (the PostgreSQL commands can be used within or outside of a GRASS session):

```
# create a new database "nc_usa" using PostgreSQL command
createdb -h localhost nc_usa

# enter PostgreSQL to create a user ('nc_usa=#' is the prompt)
psql -h localhost nc_usa
nc_usa=# CREATE USER grassuser ENCRYPTED PASSWORD 'my12sec!';
nc_usa=# \q

# define GRASS connection
db.connect driver=pg database="host=localhost,dbname=nc_usa"
# db.login allows to enter the password from above
db.login user=grassuser
db.connect -p

# copy map from PERMANENT, this converts table to PostgreSQL
g.copy vect=roadsmajor,myroadsmajor
# show attribute connection
v.db.connect -p myroadsmajor

# show user modifiable PostgreSQL tables (in current MAPSET)
# (e.g. public.myroadsmajor)
db.tables -p
```

The current GRASS mapset is now configured to use the PostgreSQL driver with the defined connection for attribute storage. The manual page is accessible via **g.manual grass-pg**.

Database schemas are currently supported only by PostgreSQL connections. Schemas enable database objects to be grouped together in distinct namespaces within the same database to achieve cross-database connectivity (e.g. between different databases in a PostgreSQL database cluster). The default schema is the "public" schema which is used by GRASS. The default schema can be set with **db.connect**. Note that the default schema will be used by all **db.*** modules. The **db.tables** command returns table names as "schema.table" if schemas are available in the database.

MySQL driver

An alternative DBMS for server based storage of attributes is MySQL. Both standard and MySQL embedded database are supported. Permissions and a password may be needed to modify tables, please refer to the MySQL documentation for details. We use the example with roads again:

```
mysql -h localhost
# create new database "nc_usa" within MySQL ('mysql>' is prompt)
mysql> CREATE DATABASE nc_usa;
mysql> CREATE USER 'grassuser'@'localhost';
mysql> GRANT ALL PRIVILEGES ON *.* TO 'grassuser'@'localhost';
mysql> SET PASSWORD FOR 'grassuser'@'localhost' =
        PASSWORD('my12sec!');
mysql> quit;

# define GRASS connection
db.connect driver=mysql database="host=localhost,dbname=nc_usa"
# db.login allows to enter the password
db.login user=grassuser
db.connect -p

# copy map from PERMANENT, this converts table to MySQL
g.copy vect=roadsmajor,myroadsmajor
# show attribute connection
v.db.connect -p myroadsmajor

# show available MySQL tables
db.tables -p
```

The current GRASS mapset is now configured to use the MySQL driver with the defined connection for attribute storage. The manual page is accessible via **g.manual grass-mysql**.

unixODBC driver

The unixODBC driver permits to connect GRASS to various DBMS including database/spreadsheet software, especially from the MS-Windows world. The settings have to be defined with **ODBCConfig** or a text editor modifying the config file. Each ODBC database connection is given a name which is called "DSN" (data source name). The DSN must be defined in **$HOME/.odbc.ini** (at individual user level) or in **/etc/odbc.ini** for (for all users). Please refer to the unixODBC documentation for details. The DSN is then used in GRASS to establish the connection:

```
# define GRASS connection
db.connect driver=odbc database=myodbcdsn
# db.login allows to enter the password
db.login user=myname
db.connect -p

# copy map from PERMANENT, this converts table to ODBC DBMS
g.copy vect=roadsmajor,myroadsmajor
# show attribute connection
v.db.connect -p myroadsmajor
# show available ODBC linked tables
db.tables -p
```

The current GRASS mapset is now configured to use the unixODBC driver with the defined connection for attribute storage. The manual page is accessible via g.manual grass-odbc.

Converting CSV files and spreadsheet tables to SQL

The import of external tables without geometry is easiest done with db.in.ogr. It just requires the CSV file (comma separated) and an output name, the MAPSET DBMI settings are used for database storage.

Alternatively, using OpenOffice.org "Base", you can easily convert a spreadsheet table into a true SQL database (PostgreSQL, MySQL etc). The procedure is explained in the GRASS WIKI.[2] You simply select the table contents and then paste them into the OpenOffice.org DBMS driver in a graphical dialog.

You can also convert tables on command line with OGR. To convert an external attribute table to a DBMS format supported by GRASS (e.g. SQLite), you can use ogr2ogr. This tool also works if no geometry is present. We can convert, for example, the CSV table wake_soil_groups.csv (Wake county hydrologic soils groups) into SQLite format and add this new table into our existing SQLite database file sqlite.db which is stored in our mapset (this also works likewise for other DBMS drivers):

```
# convert CSV table into SQLite format and add as table in
# sqlite.db (adapt path to your settings, input file needs
# '.csv' extension)
ogr2ogr -update -f SQLite \
  $HOME/grassdata/nc_spm/sqlite/sqlite.db wake_soil_groups.csv

# or simply
db.in.ogr wake_soil_groups.csv out=wake_soil_groups
```

This conversion works for any OGR supported vector format. When listing the available tables, this new table appears, too:

```
db.tables -p
```

The wake_soil_groups table can be now joined to the attribute table of the soils_wake map (DSL_NAME column) (see Section 6.2.1).

Map-table connections

There are several ways to find out if and how a vector map is connected to one or many tables:

---

[2] OpenOffice.org with SQL Databases,
   http://grass.gdf-hannover.de/wiki/Openoffice.org_with_SQL_Databases

```
# show table connection(s) of a map
v.db.connect -p schools_wake
# show attribute column names and types of a map
v.info -c schools_wake

# show available tables in current mapset
db.tables -p
# describe details of a table
db.describe mysoils
# describe it in shortened form
db.describe -c mysoils
```

The **v.db.connect** can also be used to (re)define a link between a vector map and a table that can be stored locally, or on a network server. By default, the MAPSET-wide definition is used as defined by **db.connect** but maps can be individually linked to different backends. We will show additional examples later on with **v.db.connect**.

Attribute table maintenance and access control

Table maintenance can be done with SQL commands and **db.execute**. In the common case that a table is connected to a vector map, there are special vector commands which simplify maintenance tasks:

- to add a new column, use **v.db.addcol**;
- to remove a column from a table, use **v.db.dropcol**;
- to rename a column, use **v.db.renamecol**;
- to add a new table to a vector map, use **v.db.addtable**;
- to delete an entire table, run **v.db.droptable** (you have to activate the -f flag to really remove the table connected to the map);
- to update a column, use **v.db.update**.

To learn to change a column type, refer to Section 6.2.2.

Access control Sometimes it is desired to grant read-only access to (attribute) data. GRASS and the DBMS backends offer various solutions to maintain access control:

- access to map: storing the map in a separate MAPSET (see Section 3.1.6 gives read-only access;
- access to attribute table:
  - DBF: modification of file permissions at operating system level;
  - SQL-DBMS: GRANT and REVOKE SQL commands to manage access on database user level;
- complete map storage in spatial SQL database (PostGIS, MySQL, Oracle etc.) including geometry: using **v.external** an external map can be virtually linked into the current mapset in read-only mode.

Attribute table joins

Attribute values are often stored as acronyms of longer words to save space. Sometimes it is desired to add explanations into additional columns which shall be merged from a separate legend table. This is done by a SQL join clause which combines columns of one table to that of another to create a single table. In this procedure, the values in a column are matched with those of a column in another table and further column contents are transferred. In GRASS, the v.db.join command performs simple table joins. As an example, we want to merge the legend of points of interest extracted from the Geonames database[3] into the map table. The original table looks like this:

```
v.db.select geonames_wake
 cat|GEONAMEID|NAME|ASCIINAME|ALTERNATEN|FEATURECLA|FEATURECOD...
24|4498303|West Raleigh|West Raleigh||P|PPL...
25|4487042|Raleigh|Raleigh|Raleigh,...|P|PPL...
32|4459467|Cary|Cary||P|PPL...
 [...]
```

We can get the legend file geonames_features.csv from the ncexternal/ directory and import it. This example requires to use a SQL driver different from DBF:

```
db.in.ogr geonames_features.csv out=geonames_features
db.tables -p
db.describe -c geonames_features

db.select geonames_features
 FEATURECLASS|FEATSHORTDESCRIPTION|FEATLONGDESCRIPTION...
 A|ADM1|first-order administrative division|a primary...
 A|ADM2|second-order administrative division|a subdivision...
 [...]
```

For the join itself, we want to use the FEATURECOD column of the map to be joined to the column FEATSHORTDESCRIPTION of the imported table geonames_features. To be able to modify the map table, we first copy the map into our current MAPSET:

```
g.copy vect=geonames_wake,mygeonames_wake

# note: DBF driver not supported for join
v.db.join mygeonames_wake col=FEATURECOD \
         otab=geonames_features ocol=FEATSHORTDESCRIPTION

# query new column
v.db.select mygeonames_wake \
         col=NAME,FEATLONGDESCRIPTION,POPULATION
```

---

[3] Geonames Web site, http://www.geonames.org

```
NAME|FEATLONGDESCRIPTION|POPULATION
West Raleigh|populated place|338759
Raleigh|populated place|276093
Cary|populated place|103945
[...]
```

The table now contains the legend columns and we can display the points of interest along with the text description:

```
g.region vect=mygeonames_wake -p
d.erase
d.vect streets_wake
d.vect -c mygeonames_wake icon=basic/pushpin
d.vect mygeonames_wake disp=attr attrcol=FEATLONGDESCRIPTION
```

This shows the map with textual labels instead of the abbreviations. If you prefer to perform joins in a graphical way, for example, OpenOffice.org "Base" supports this for various SQL backends (see also Section 6.2).

6.2.2 Sample SQL queries and attribute modifications

In this subsection, we show a series of SQL statements to illustrate SQL usage for GIS related tasks. Note that we have written all SQL keywords in capital letters for better readability, but there is no need to do so. The examples are based on the North Carolina data set. A good tutorial to learn SQL queries is the PostgreSQL documentation (note that MySQL and SQLite dialects may slightly differ, but our examples below should work everywhere, many of them even with the DBF driver).[4]

SQL selection examples In this example, we select all attributes from a table where "COUNTY" column values are equal to 'WAKE'. The **db.*** commands require to have the map in the current mapset if the DBF or the SQLite drivers are used. For this, we first copy the map from the PERMA-NENT MAPSET into our current MAPSET. Note that the **v.db.*** commands are able to search in any mapset since the map-DBMS connection is used to find the attached table.

```
g.copy vect=boundary_municp,mybnd_mun
echo "SELECT * FROM mybnd_mun WHERE COUNTY = 'WAKE'"|db.select

# combine with second condition
echo "SELECT * FROM mybnd_mun WHERE COUNTY = 'WAKE' AND \
     CENSUSTYPE <> 'Village'" | db.select
```

---

[4] PostgreSQL documentation, http://www.postgresql.org/docs/

SQL subquery expressions example   In this example, we want to select vector objects from a list of given items (note: does not work for DBF driver):

```
v.db.select schools_wake where="ADDRCITY IN ('Apex', 'Cary')"
```

SQL pattern matching example   In this example, we want to select vector objects with attributes that matching certain text patterns:

```
# match exactly number of characters (here: 2)
# (note: does not work for DBF driver)
v.db.select geology where="GEO_NAME LIKE 'Za__'"

# define wildcard (any length)
v.db.select geology where="GEO_NAME LIKE 'Z%'"
```

Delete vectors by attribute selection example   Sometimes maps contain unwanted vector objects. In addition to editing of the geometry with a digitizer, these vectors can also be selected by attributes through a SQL statement and "deleted" from the map by reverse selection to a new map. We want to reduce the schools map to only the small schools:

```
# check what to delete (find all big schools)
v.db.select schools_wake where="CAPACITYTO > 300"

# perform reverse selection, save to new map
v.extract -r schools_wake out=small_schools_wake \
        where="CAPACITYTO > 300"
v.db.select small_schools_wake
```

With the first command we check for the correctness of the SQL statement, the resulting attribute rows should be the deletion candidates. Then we execute a reverse selection with the -r flag which will extract all vectors except for those matching the **where** statement. The new map contains all remaining vector objects, here, the small schools.

Null handling example   This example illustrates NULL handling in SQL. First, we check whether the map mylakes has any NULLs in the column named FTYPE. Then we selectively display lakes without (blue) and with NULL (red) to find out which type is undefined. You will see that the lakes missing FTYPE attribute are wetlands along streams so we will replace NULL with the landuse type WETLAND:

```
# copy the map into your MAPSET and check for NULL
g.copy vect=lakes,mylakes
v.db.select mylakes
v.db.select mylakes where="FTYPE IS NULL"

# display the lakes, show undefined FTYPE lakes in red
```

```
g.region swwake_10m
d.erase
d.vect mylakes where="FTYPE NOT NULL" type=area col=blue
d.vect mylakes where="FTYPE IS NULL" type=area col=red

# replace NULL with FTYPE WETLAND
v.db.update mylakes col=FTYPE value=WETLAND \
          where="FTYPE IS NULL"
v.db.select mylakes
```

Column type converting example (type casts) A column type cast is a conversion from one SQL data type to another. Such conversions are possible under certain circumstances, if the content of a column is also representable in a different column type. For example, numeric values can be sometimes stored in a character type column. To work with such numbers, we can add a new numeric column and convert the character numbers to true numbers. In the map `geodetic_pts`, the point elevation is stored in a character type column called `Z_VALUE` which we want to type cast into a new double precision column `zval` (note: `CAST()` is not supported by DBF driver, so you need to use SQLite, PostgreSQL or MySQL to run this example):

```
v.info -c geodetic_pts
# copy map into current mapset
g.copy vect=geodetic_pts,mygeodetic_pts
v.db.addcol mygeodetic_pts col="zval double precision"

# the 'z_value' col contains 'N/A' strings, not to be converted
v.db.update mygeodetic_pts col=zval \
          qcol="CAST(z_value AS double precision)" \
          where="z_value <> 'N/A'"
v.info -c mygeodetic_pts
v.db.select mygeodetic_pts col=Z_VALUE,zval

# fix 0 in 'zval' to NULL (orig. 'N/A' entries in 'Z_VALUE')
echo "UPDATE mygeodetic_pts SET zval=NULL WHERE zval=0" \
     | db.execute
v.db.select mygeodetic_pts col=Z_VALUE,zval
```

The new column `zval` now contains double precision elevation values needed to perform numerical operations or interpolation.

SQL updates with character substitution Sometimes, the field values of character columns are not formatted in the desired way. For example, the column describing GRASS vector colors needs to be formatted as `RRR:GGG:BBB`. We assume to have a map with colors coded slightly differently, in our example as `RRR-GGG-BBB`. With SQL functions it is possible to create a new column and update it to the values of the adjacent column including character substitution on the fly (not supported by DBF driver):

```
v.db.select mymap col=RGB_COLOR
 RGB_COLOR
 230-000-077
 255-000-000
 204-077-242
 [...]
```

We add a new column with length of 11 characters and update it to the modified color values (we name it the standard GRASS color column name):

```
v.db.addcol mymap col="GRASSRGB varchar(11)"
# substr() extracts numbers, '||' operator appends strings
echo "UPDATE mymap SET GRASSRGB = (substr(RGB_COLOR,1,3)\
      ||':'||substr(RGB_COLOR,5,3)||':'||\
      substr(RGB_COLOR,9,3))" | db.execute

v.db.select mymap col=RGB_COLOR,GRASSRGB
 RGB_COLOR|GRASSRGB
 230-000-077|230:000:077
 255-000-000|255:000:000
 204-077-242|204:077:242
 [...]
```

The new column **GRASSRGB** is now understood by **d.vect -a**.

Complex SQL expression examples   In this example, we perform on the fly computation while updating a column to a new value. We bulk-define the maximum speed for the entire road network, then increase it selectively for multi-lane roads:

```
# copy map into current mapset for editing
g.copy vect=roadsmajor,myroadsmajor

# add a column and predefine its values with 55mph
v.db.addcol myroadsmajor col="speedmax double precision"
v.db.update myroadsmajor col=speedmax value=55
v.db.select myroadsmajor

# increase speedmax to 70mph for multilane roads

# a) example for SQL statement and db.execute
echo "UPDATE myroadsmajor SET speedmax=speedmax+15 \
      WHERE MULTILANE='yes' " | db.execute

# b) alternatively, use v.db.update
v.db.update myroadsmajor col=speedmax where="MULTILANE='yes'" \
            value="speedmax+15"
v.db.select myroadsmajor
```

SQL expressions with on the fly computation can also be used directly in a command. In the following example, we want to highlight the stations with annual precipitation greater than 50in using blue symbol while all the other stations are shown in grey. The data are provided in *mm* so we will do on the fly conversion to inches:

```
g.region nc_500m -p
v.info -c precip_30ynormals
d.erase
d.vect precip_30ynormals disp=shape icon=basic/box
d.vect precip_30ynormals disp=shape where="annual*0.03937>50" \
       icon=basic/circle fcol=blue
```

You can see that the stations with high precipitation totals are on the coast and in the mountains while the piedmont area gets less rain.

6.2.3 Map reclassification

Vector maps can be reclassified in a similar way as raster maps. The module **v.reclass** reclassifies vectors according to results of SQL queries or to a value in an attribute table column. Alternatively, a reclass rule file can be specified. The module is used in a similar way as **r.reclass**. As an example, we reclassify the counties of the points of interest map **geonames_NC** into several categories according to population. To understand the precise wording of the attributes, we first look at the attribute table (the command prints it to the terminal):

```
# fetch all counties
v.db.select geonames_NC \
            where="POPULATION<>0 and FEATURECOD='ADM2'"
```

An ASCII file **countypop.cls** containing the reclass rules can be written with any text editor. It contains:

```
cat 1
WHERE FEATURECOD='ADM2' AND POPULATION=0
cat 2
WHERE FEATURECOD='ADM2' AND POPULATION>0 AND POPULATION<1000
cat 3
WHERE FEATURECOD='ADM2' AND POPULATION>=1000 AND POPULATION<10000
cat 4
WHERE FEATURECOD='ADM2' AND POPULATION>=10000 AND POPULATION<100000
cat 5
WHERE FEATURECOD='ADM2' AND POPULATION>=100000 AND POPULATION<500000
cat 6
WHERE FEATURECOD='ADM2' AND POPULATION>=500000
```

This rules file we save under **countypop.cls** is applied to the map to generate a new reclassified vector map.

```
v.reclass geonames_NC rules=countypop.cls out=geonames_NC_recl
v.category geonames_NC_recl op=report
 Layer: 1
 type          count         min          max
 point          104            1            6
 line             0            0            0
 [...]
```

In a second pass, we have to add a new attribute table to the new map
geonames_NC_recl. We generate a single column to store a text label for each
category number (ID):

```
# add new table with one column
v.db.addtable geonames_NC_recl col="popclass varchar(50)"

# insert values into table
v.db.update geonames_NC_recl col=popclass value="unknown"   \
            where="cat=1"
v.db.update geonames_NC_recl col=popclass value="very low" \
            where="cat=2"
v.db.update geonames_NC_recl col=popclass value="low"        \
            where="cat=3"
v.db.update geonames_NC_recl col=popclass value="medium"    \
            where="cat=4"
v.db.update geonames_NC_recl col=popclass value="high"       \
            where="cat=5"
v.db.update geonames_NC_recl col=popclass value="very high"\
            where="cat=6"
```

Finally, we can verify the result and display the reclassified map:

```
# verify
v.db.select geonames_NC_recl
v.info geonames_NC_recl
d.erase
d.vect nc_state type=area
d.vect -c geonames_NC_recl where="popclass<>'unknown'"
```

With **v.what.vect** (see Section 6.4.1), this classification could be transferred
into the map **boundary_county** into a new column. In case of area maps, it
may be necessary to dissolve the boundaries between areas with an identical
attribute. See Section 6.5.3 for details.

6.2.4 Vector map with multiple attribute tables: layers

Features in a vector map may represent more than a single type of information,
for example, a road can also be a field boundary. GRASS vector format makes
it possible to create several vector map "layers" by linking the map geometry
to more than one external attribute table. The category number (vector ID)

is used to link each object to attribute table rows. All category numbers are stored both in the vector geometry file as well as in the "cat" column (integer type) in each attribute table. The category number is used to look up an attribute assigned to a vector object and vice versa. At user level, category numbers can be assigned to vector objects with the `v.category` command or generated with `v.to.db`.

In order to assign multiple attributes in different tables to vector objects, each map can be assigned multiple sets of category numbers. This is achieved by assigning more than one layer to the vector map, managed with the `v.db.connect` command. The layer number determines which table will be used for attribute queries. For example, a cadastral vector area map can include a layer 1 with an attribute table containing landuse descriptions maintained by department A while layer 2 is assigned to an attribute table containing owner descriptions maintained by department B. Each set of category numbers starts with 1; they do not have to be continuous.

6.3 Digitizing vector data

A paper map can be converted to digital form by manual digitizing. In general, there are two ways to interactively digitize a map:

- using a digitizing board, or
- digitizing heads-up (on screen).

In the first case, the map is placed on the digitizer board, which provides a special digitizing mouse. The corners are selected by a mouse click and their respective coordinates are entered using the keyboard. This process is called "registering a map". Then the lines and points on the map are digitized using a mouse. The advantage of this method is that the user always sees the entire map. However, the high cost of the equipment and the possibility that the map could be shifted during the digitizing, if it is not properly mounted, are significant disadvantages. Furthermore, the paper map must be free of distortions to prevent displacements.

On-screen digitizing requires a scanned and geocoded raster map that is displayed in the GRASS monitor. All features will be digitized using the mouse. It is not necessary to register such a map as it is already geocoded. The advantage of this method is the possibility to zoom in and thus achieve an improved accuracy. Apart from an access to a scanner, no additional equipment is needed. The major disadvantage is the more difficult orientation on the map. The following section deals only with heads-up digitizing using mouse.

6.3.1 General principles for digitizing topological data

To explain the digitizer module, we consider an example of vectorizing features from a scanned topographic map. We assume that the map was scanned, accurately geocoded and imported into GRASS in raster format (see Section 4.1.4

on geocoding scanned maps). Although it may be possible to automate the vectorization of a simple raster map using r.to.vect (see Section 5.3.1), problems often arise from overlapping lines, dots, map signatures etc. and manual digitizing is necessary.

There are few general recommendations for digitizing map features which can minimize potential accuracy problems. The recommendations are mostly based on the fact that to make the map readable some features are exaggerated at a given scale compared to their actual size:

- Line features should be digitized along their center-line, e.g. along the center of a road;
- Area features should be digitized by following the center-line of area boundary lines. An area centroid should be placed in the approximate center of the area;
- Point features should be digitized at the center of the object, e.g. a point in the center of a map symbol representing the point feature or at the reference point of such a symbol;
- The points defining the line or polygon boundary should be selected at a density that is sufficient for preserving the geometry of the digitized features.

General rules for digitizing in topological GIS   When working in a topological GIS such as GRASS, following certain rules is recommended, in order to benefit from the topological features of the software. The following rules apply to the vector data (from *GRASS 6 Programmer's Tutorial*, GRASS Development Team, 2006):

- Arcs (vector line primitives) should not cross each other (i.e., arcs which would cross must be split at their intersection to form distinct arcs);
- Arcs which share nodes must end at exactly the same points (i.e., must be snapped together using the *snapping* function of the digitizing module);
- Common boundaries should appear only once (i.e., should not be digitized twice);
- Areas must be explicitly closed. This means that it must be possible to complete each area by following one or more boundaries that are connected by common nodes, and that such tracings result in closed areas. Areas need a centroid to become valid areas;
- It is recommended that area features and linear features be placed in separate vector maps. However, if area features and linear features must appear in one map, common boundaries should be digitized only once. An area edge that is also a line (e.g., a road which is also a field boundary), should be digitized as an area edge to complete the area. The area feature should be labeled as an area. Additionally, the common boundary arc itself (i.e., the area edge which is also a line) should be labeled as a line to identify it as a linear feature.

Now we explain the digitizing process in detail.

6.3.2 Interactive digitizing in GRASS

Interactive digitizing is done by v.digit, we will use it to digitize unpaved roads and other features based on the provided orthophoto as follows (Figure 6.1):

```
g.region rast=ortho_2001_t792_1m
v.digit -n newmap bg="d.rast ortho_2001_t792_1m; \
          d.vect streets_wake col=red"
```

The -n flag is needed if you are going to create a new map. We have defined the background map(s) using the bgcmd parameter. Alternatively, they can be defined in the menu system (Open settings icon) of v.digit. Figure 6.1 shows the digitizer window with attribute form and GRASS monitor. After selecting one or more background maps, you can close the Settings window. By clicking the Redraw icon, everything will be shown. Usually an attribute table is desired for the new vector map. In the Open settings dialog, Table tab, columns can be defined. The cat column is used as ID column, other columns can be added (Add new column). For this, the column name and type have to be specified. The table is then created with the Create table button. We are now ready to start the digitizing procedure.

Digitizing vector points, lines and areas   Once a vector is digitized, v.digit will query by default for a new record in the attribute table via a form. If no attribute table is desired, deactivate the Insert new record into table in the main digitizer window. The digitizing is started by choosing the icon for the appropriate object type (point, line, boundary, or centroid). For now,

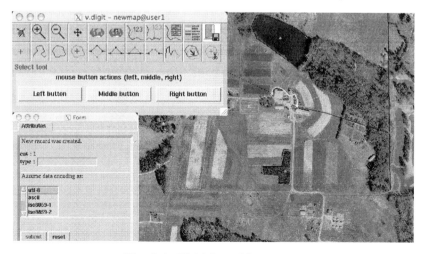

Fig. 6.1. Digitizing with v.digit

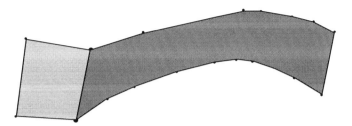

Fig. 6.2. Digitizing common area boundaries in a topological GIS

we assume that you want to digitize a line. Now switch over to the GRASS monitor, and start to draw the line using the mouse by clicking on the points representing the line with the left mouse button. Generally, we recommend making an extensive use of the zoom function which can improve the digitizing accuracy significantly. The pan (panning) function allows us to move the map into any direction without changing the zoom level.

The mouse menu is shown in the digitizer window, the buttons provide different functions such as drawing nodes and lines, removing the latest drawn node and finishing/omitting a line. This requires a bit of experience but you will quickly feel familiar with the concept. This way, you can digitize line by line (or just vector points). Topologically correct vectors have a different color (green) than incorrect ones (red).

When intersecting lines or connecting to lines, a node has to be inserted at the intersection. This is done by breaking the existing line which you want to cross or connect to. To break a line, click on the Split line icon and select the position on the line where to insert a node. You can now snap other lines to this node which is explained next. When working with polygons, it is important to digitize common boundaries of adjacent areas only as a single line (see Figure 6.2). GRASS will automatically assign the common boundary to both areas. Never digitize a common area edge as two parallel lines!

Note that the digitizing tool in QGIS comes with a similar look and feel as the GRASS' v.digit.

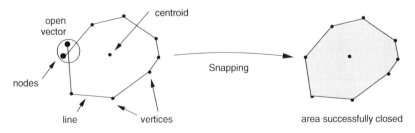

Fig. 6.3. The node snapping function in GIS

Snapping of nodes   Node snapping is necessary when digitizing a line which consists of several parts, or when closing a vector area. As mentioned above, closing areas is mandatory because only then is GRASS able to establish vector topology. A great help is the built-in snapping function. When two nodes are close to each other (depending on the current snapping threshold), they will be moved into the same node and reduced to one node (see Figure 6.3). The Move vertex icon provides this functionality. The snap radius is defined in the Open settings window, Settings tab. You can select between screen pixel threshold and map units threshold. The snapping threshold should be chosen appropriately for the map scale.

In general, reasonable values for snapping thresholds depend on the map scale, such as:

- 1:5000   - 1:10000: Snap distance 1-2m on ground
- 1:10000 - 1:25000: Snap distance 2-5m on ground
- 1:25000 - 1:50000: Snap distance 5-10m on ground

Common digitizing problems and solutions   A common digitizing problem is that area boundaries are not closed or lines intended to be connected ("polylines") are broken. Lines too short and not reaching another traverse line are called "undershoot", lines too long which cross another line without having a node at the intersection are called "overshoot" (see Figure 6.4). To fix these problems, you should use the snapping function. Note that snapping is also implemented in additional vector modules such as `v.clean` and `v.edit`. Polylines can be created with `v.build.polylines` and, for all or selected vectors with `v.edit` (`connect` tool).

Post-Digitizing issues   Maps containing intersecting vectors without nodes at the line intersections are called "spaghetti maps" which are topologically incorrect. To resolve this problem, the module `v.clean` can be used (see Figure 6.5 illustrating the functionality) which inserts nodes at the line intersections. The result is written to a new map. Optionally, erroneous vec-

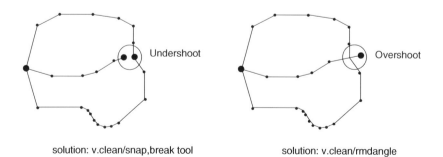

solution: v.clean/snap,break tool   solution: v.clean/rmdangle

Fig. 6.4. "Overshoots" and "undershoots" in vector maps

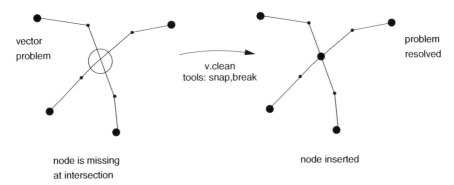

Fig. 6.5. Correction of "spaghetti digitizing"

tor objects can be stored into a separate error map for inspection. Additional functionality is "pruning" which can be used to remove excessive nodes from a map. Use it with care as you can oversimplify complex areas/polygons to squares or triangles if you remove too many nodes.

GRASS also provides a tool for non-interactive vector editing v.edit that can be used from command line. It provides a broad range of options for feature selection and modification. We use it in subsections below, for example for vector feature moving and reverse deletion in the Section 6.5.3. A new generalization tool is under development.

6.4 Vector map queries and statistics

In this section, we explain various possibilities to query vector maps, including interactive and automated attribute and geometry reports as well as cross-map queries (vector-vector and vector-raster maps).

6.4.1 Map queries

To get a list of the vector map attributes (category numbers and attributes), optionally with length or area sizes, use v.report:

```
# display ZIP code area sizes in hectares
v.report zipcodes_wake option=area units=hectares
```

The command allows us to query line lengths and area sizes of vector features when run with the parameter units. The map type, either line or area, has to be provided with parameter option.

You can interactively query vector data in a GRASS monitor with mouse. The module d.what.vect allows you to query vector objects in one or more maps. As an example, we query the zipcode map and the census map:

```
g.region vect=zipcodes_wake -p
d.erase
d.vect -c zipcodes_wake
d.vect census_wake2000 type=boundary
d.what.vect map=zipcodes_wake,census_wake2000

d.vect schools_wake
d.what.vect -x schools_wake
```

The mouse context menu is shown in the terminal window. The query form shows two tabs in this case, one for each queried map. The result can be shown in the GRASS terminal if the command is used with the flag -x.

The module **v.what** noninteractively retrieves area information from polygons queried by a list of coordinates, and optionally the attributes, when used with the flag -a. As an example, we query two points in the censusblk_swwake map:

```
# query census data at two given points
v.what -a censusblk_swwake \
        east_north=636982.5,218057.8,638012.2,224919.1
```

The module reports areas for the polygons which include the given coordinate pairs and the area attributes. To easily get coordinates from screen, use **d.where**.

A more sophisticated way to query vector maps is to upload the results to the attribute table of the base map. The command **v.what.vect** allows us to upload attribute values from a vector map queried at points given in a base vector point map into the attribute table of this base map. The base map has to be in the current MAPSET to be modifiable. In our example, we first copy the original map overpasses to our current mapset, then add a column to its attribute table and fill this column with values queried from the geology map:

```
# create own copy, add column, fill with geology
g.copy vect=overpasses,myoverpasses
v.db.addcol myoverpasses column="geology varchar(10)"
v.what.vect myoverpasses qvect=geology column=geology \
        qcolumn=GEO_NAME

# display overpasses loacted in areas in green
d.vect streets_wake col=grey
d.vect myoverpasses icon=extra/bridge size=10
d.vect myoverpasses disp=shape where="geology='CZig'" \
      col=green siz=10 icon=extra/bridge
v.db.select overpasses
```

The new map myoverpasses now contains a column geology which was populated with values from the map geology using its column GEO_NAME.

Hybrid queries transfer raster cell values to a vector map (e.g., to update a polygon attribute) with **v.what.rast**. For example, we can transfer the elevations from a DEM to the points given by myschools_wake map:

```
g.copy vect=schools_wake,myschools_wake
v.db.addcol myschools_wake column="elevation double precision"
# set region to raster map for query
g.region rast=elevation -p
v.what.rast myschools_wake rast=elevation column=elevation

# verification
v.info -c myschools_wake
v.db.select myschools_wake column=NAMELONG,elevation
```

The height above sea level for each school is now stored in the new column `elevation` of map `myschools_wake`. Note that several schools are outside of the `elevation` map extent and therefore without the elevation value.

6.4.2 Raster map statistics based on vector objects

To calculate basic univariate statistics from a raster map, individually for vector polygons, the command `v.rast.stats` can be used. Internally, the vector map will be rasterized according to the raster map resolution. Then univariate statistics are calculated per vector category number using the raster map and the results automatically uploaded to the vector map attribute table. New columns are generated in the attribute table if not already present.

Since the vector map table will be modified, it needs to be in the current MAPSET. In our example, we calculate and upload elevation statistics to the `zipcodes_wake` map:

```
g.copy vect=zipcodes_wake,myzipcodes_wake
g.region vect=zipcodes_wake res=500 -p
v.rast.stats vect=myzipcodes_wake rast=elev_state_500m \
            colprefix=elev
v.info -c myzipcodes_wake

# display zipcodes with given elevation properties
d.vect myzipcodes_wake disp=shape type=boundary
d.vect myzipcodes_wake disp=shape type=area \
       where="elev_mean<100" fcol=green
d.vect myzipcodes_wake disp=shape type=area \
       where="elev_range>70" fcol=brown
```

Nine new columns were added to the attribute table with the given prefix `elev` (number of cells, minimum, maximum, range, mean, standard deviation, variance, coefficient of variation, sum of values). The relevant values were computed from the raster map based on the vector polygon areas. The new attributes were then used to display the zipcodes with mean elevation less than 100m (green) and zipcodes that have quite varied topography with the elevation range greater than 70m (brown).

If you have a vector points map and you want to compare values of its attribute with values from a raster map at these points, you can use `v.sample`.

As an example, we compare geodetic control locations which were obtained through tower triangulation surveys, traverse surveys and Global Positioning System (GPS) surveys to an elevation raster map to evaluate the differences. The module v.sample is used to retrieve heights from the raster map elevation at the positions of the geodetic points and to calculate the height differences. The geodetic point map attribute table is extended and updated accordingly. We use the vector map geodetic_swwake_pts which is a subset of the map geodetic_pts (but with numeric zval column type, see Section 6.2.2 for how this was done):

```
g.region swwake_10m -p
d.erase
d.rast elevation
d.vect streets_wake
d.vect geodetic_swwake_pts fcol=cyan icon=basic/circle

v.sample geodetic_swwake_pts col=zval rast=elevation \
      out=elev_geod_diffs
v.db.select elev_geod_diffs where="pnt_val <> 0"
 cat|pnt_val|rast_val|diff
 26709|139.42|139.0857|-0.334307
 26713|102|103.1191|1.119095
 26751|100.53|100.5164|-0.013559
 [...]

d.vect elev_geod_diffs where="diff<-1. AND pnt_val <> 0" \
      size=10 fcol=blue icon=basic/circle
d.vect elev_geod_diffs where="diff<-10 AND pnt_val <> 0" \
      size=12 fcol=red icon=basic/circle
```

The resulting vector map includes the elevation values pnt_val from the input vector point map geodetic_swwake_pts, elevation rast_val sampled from the DEM elevation and their differences diff. By running v.db.select for all data points that had non-zero elevation at the given points, you can see that the differences are quite large. We have used d.vect with SQL statement to display the points with differences larger than 1m in blue and 10m in red. The three red points are apparently on top of buildings the large differences in the blue points are mostly caused by unsufficient vertical resolution (1m), low accuracy or location on structures not represented in the DEM. Additional error (around 0.2m) is due to different vertical datums (NGVD29 for the points and NAVD88 for the DEM). Apparently, this set of geodetic points cannot be used for evaluation of the DEM accuracy, accurate elevation points can be obtained from the NC Geodetic Survey (http://www.ncgs.state.nc.us/).

6.4.3 Point vector map statistics

Simple statistical analysis of vector points can be performed with several GRASS modules. For more sophisticated spatial statistics and geostatistics tools, it is recommended to take the advantage of the bridge between GRASS and R, as described in Chapter 10, as well as by Bivand (2000, 2007), or a link with gstat described in the same chapter.

Basic univariate statistics can be computed using v.univar. For example, for the elevation differences in geodetic points that we have generated above, we get:

```
v.univar elev_geod_diffs type=point col=diff where="pnt_val<>0"
  number of features with non NULL attribute: 217
  number of missing attributes: 46
  number of NULL attributes: 0
  minimum: -74.2899
  maximum: 1.54707
  range: 75.837
  mean: -1.58393
  mean of absolute values: 1.71452
  population standard deviation: 6.94657
  population variance: 48.2548
  population coefficient of variation: 4.38566
  sample standard deviation: 6.96263
  sample variance: 48.4782
  1st quartile: -0.984299
  median (odd number of cells): -0.375536
  3rd quartile: -0.12679
  90th percentile: 0.177383
```

The negative mean and median values indicate that the geodetic points have elevations slightly shifted above the DEM surfaces, the median value is close to the difference between the vertical datums. As we have explained above, the large minimum value is due to the given point location on top of a building. Please refer to Appendix A.1 for a list of basic statistical equations used to compute these measures.

Further statistical analysis can be performed using additional GRASS vector points commands, such as v.normal which supports computation of up to 15 different normality tests for a selected point attribute. The module v.qcount provides a test for complete spatial randomness using a quadrat method. Note that the commands do not include SQL support.

6.5 Geometry operations

In this section, we explain in detail how geometry operations are performed in GRASS. Since GRASS is a topological GIS, some special rules apply and, at the same time, capabilities to analyze spatial relationships specific to topological GIS are provided.

6.5.1 Topological operations

Topological operations are used to verify and enforce data integrity. At the expense of speed, data quality can be ensured and topologically broken data sets can be repaired. The topological vector model is explained in greater detail in Section 4.2.1. It is fundamentally different from the "Simple Features" model, which does not support topological operations. Digitizing data in the Simple Features model is error-prone when editing polygons or joining and connecting segments at nodes. In the topological GRASS vector model, erroneous vectors such as gaps/slivers or overshoots are immediately identified and related features remain intact.

Topological description and map cleanup   You can retrieve topological statistics for all vector types in a map with v.info (use the -t to get it in script style):

```
v.info -t soils_wake
 nodes=137682
 points=0
 lines=0
 boundaries=136379
 centroids=47156
 [...]
```

By default, GRASS 6 always builds the topology after vector map import, creation or modification. If it needs to be re-built, the v.build module is used. It optionally allows to copy the erroneous vector objects into a separate map for inspection. Topology can be visually inspected by d.vect:

```
g.region rural_1m -p
d.erase
# show topology
d.vect streets_wake disp=shape,topo
```

The map display takes some time since we are looking at small subset of a large and complex vector map. The d.vect command shows the line category numbers and nodes. Topological errors can be corrected either manually within v.digit (see Section 6.3.2) or, to some extent, automatically with v.clean or v.edit.

Vector line directions   Vector directions are defined by the original digitizing direction (e.g., from first node to last node on a line). The d.vect module has an option to display the line direction with a small arrow:

```
g.region rural_1m -p
d.erase
# show vector native directions
d.vect streets_wake disp=shape,dir
d.vect streams disp=shape,dir col=blue
```

The example shows the digitizing directions for streams and road network.

Line split/join, node insertion and segment extraction   Vector lines
which are unnecessarily split into segments can be joined into polylines with
v.build.polylines. To do the opposite, long lines can be split at given distances with v.split and v.segment. While v.split uses a maximum distance
to insert new nodes, the v.segment command permits to explicitly define the
positions where lines should be broken into segments. The module v.segment
creates new points or segments along the line (i.e., the orthogonal side offset
is zero by default; an offset leads to a point apart from the original line or
to a parallel segment). Nodes can be displayed using the module d.vect with
parameter display=shape,topo. As an example, we insert 1km nodes into the
railroads map:

```
g.region swwake_10m
d.erase
v.split railroads out=myrailroads_split length=1000
d.vect myrailroads_split disp=shape,topo
```

To illustrate the use of v.segment, we extract a segment from the 'NC-54' road
which is being road repaired. The base map roadsmajor contains, however,
some labeling errors which we have to fix first:

```
# road repair along 'NC-54'
g.region vect=roadsmajor -p
d.erase

# show line, road name, direction (to find the initial node)
d.vect -c roadsmajor disp=shape,attr,dir attrcol=ROAD_NAME

# two segments of 'NC-54' are unlabeled
d.erase
d.vect roadsmajor
d.vect roadsmajor where="ROAD_NAME='NC-54'" col=green
d.vect roadsmajor disp=attr attrcol=MAJORRDS_
# these segment IDs in MAJORRDS_ column are: 120 and 41
```

We observe that two segments are unlabeled. This needs to be fixed for the
later road selection by attribute:

```
# work on own map copy
g.copy vect=roadsmajor,myroadsmajor
v.db.update myroadsmajor col=ROAD_NAME value="NC-54" \
          where="MAJORRDS_=120"
v.db.update myroadsmajor col=ROAD_NAME value="NC-54" \
          where="MAJORRDS_=41"

d.vect myroadsmajor where="ROAD_NAME='NC-54'" col=green

# extract 'NC-54' road
v.extract myroadsmajor out=road_nc_54 where="ROAD_NAME='NC-54'"
```

```
g.region vect=road_nc_54
d.erase
d.vect -c road_nc_54 disp=shape,dir,cat
```

At this point we are able to extract the line segment between 1622m and 3470m measured from the beginning of line 221:

```
# Format: type segment_ID line_ID start_offset end_offset
echo "L 1 221 1622 3470 0" | v.segment road_nc_54 \
    out=road_nc_54_seg
d.vect road_nc_54_seg disp=shape,dir col=red width=2
```

The map **road_nc_54_seg** now includes the road segment under repair. To feed the start/end coordinates of this segment to a GPS based navigation system, they can be extracted with **v.to.db**:

```
v.to.db -p road_nc_54_seg option=start
v.to.db -p road_nc_54_seg option=end
```

Using **m.proj** the resulting NC State Plane Metric coordinates can be transformed into latitude-longitude coordinates or other projections:

```
# reproject the start node of the segment
v.to.db -p road_nc_54_seg option=start
 cat|x|y|z
 1|630984.675539017|226893.364917116|0
 [...]

echo "630984.68 226893.36" | m.proj -o
 78d45'48.341"W   35d47'42.387"N 0.000
```

Likewise you can reproject the end node (**option=end**) and use this for a GPS or other application.

Vertices/nodes extraction   Vertices or nodes (vertices at the end of a line) can be extracted with **v.to.points** using the -v or -n flag, respectively. In our example, we extract nodes from the railroads map modified earlier in this section into a separate vector point map:

```
g.region swwake_10m
d.erase
v.to.points -n myrailroads_split out=railroads_nodes
d.vect railroads
d.vect railroads_nodes fcol=green siz=10 icon=basic/marker
```

If the vertices on the lines are farther apart than what you need for your application (as it is sometimes the case when they are manually digitized) you can interpolate additional vertices along the line by setting the **dmax** parameter to the maximum distance between the points that you want to achieve. As an example, we interpolate additional points between the vertices along the lines given in the map `railroads` so that the points are at most 50m apart:

```
g.region swwake_10m -p
d.vect railroads

# output all vertices, then compute new ones at 50m max dist.
v.to.points -v railroads out=rail_vertices
v.to.points -vi railroads out=rail_vertpts50m dmax=50

# zoom into a small section of the railroad map
d.zoom
d.vect railroads_nodes fcol=green siz=20 icon=basic/marker
d.vect rail_vertices siz=10 col=red width=2
d.vect rail_vertpts50m siz=5 col=blue icon=basic/circle

# check attributes in layer 2 for the first 10 points
v.info rail_vertices
v.info -c rail_vertices layer=2
v.db.select rail_vertices layer=2 where="cat<10"
 cat|lcat|along
 1|1|0
 2|1|182.821656
 3|1|278.128115
 [...]
v.db.select rail_vertpts50m layer=2 where="cat<10"
 cat|lcat|along
 1|1|0              #original
 2|1|45.705414      #inserted
 3|1|91.410828      #inserted
 4|1|137.116242     #inserted
 5|1|182.821656     #original
 6|1|230.474886     #inserted
 7|1|278.128115     #original
 [...]
```

The output of `v.info` shows two `dblinks` because the output vector map includes two layers (see Section 6.2.4). Layer 1 holds the category and attributes of the input lines; in layer 2 each point has its unique category and attribute `lcat` for the category of the input line and `along` for the actual distance of the given point from line's start. When you compare the values of actual distances, you can see that the vertices are preserved and the distances between the inserted points are adjusted to fit the between the vertices. If no flag is given and no `dmax` is set, the module outputs points interpolated along the lines with approximately 100m distances without the original vertices (equivalent to running the command with -i flag and `dmax=100`) – we do not recommend these two options as they often produce pairs of points that are very close to each other. For a precise method to set milestones along vector lines, you may use a Linear Reference System (see Section 6.6.2) which separates logical nodes from physical nodes.

Analysis of adjacency and common boundaries   Adjacent polygons can be found by `v.to.db` (`sides` option). This is done by uploading the category numbers of the left/right polygons of shared boundaries to the attribute table linked to the vector geometry as a second layer (see Section 6.2.4). We show an example for the `boundary_municp` map. First, we add a second layer (the first layer will be copied as is to the new map) and attach a new table to this new second layer:

```
# add boundary categories into geometry as the layer 2
v.category boundary_municp out=mymunicp layer=2 type=boundary \
        option=add

# add a new table with left/right column to the layer 2
v.db.addtable mymunicp layer=2 col="left integer,right integer"
```

Next, we upload categories of left/right polygon the new table:

```
v.to.db mymunicp option=sides col=left,right layer=2 \
        type=boundary
```

To verify the result, we look at the table and the map and then display in green all polygons without shared boundary:

```
# select layer and set the region to display the map
v.db.select mymunicp layer=2
g.region swwake_30m -p
d.erase

# display polygon category numbers
d.vect mymunicp disp=shape,cat

# display left and right side category numbers
# and polygons without shared boundaries
d.vect mymunicp disp=attr layer=2 attrcol=left lcol=blue yref=top
d.vect mymunicp disp=attr layer=2 attrcol=right lcol=yellow \
        yref=bottom
d.vect mymunicp layer=2 where="left=-1 OR right=-1" col=green
```

You may need to zoom into areas with crowded polygons to read all category numbers.

Analysis of distances   Distances between vector objects are calculated with `v.distance`. For example, you can calculate distances of given points, stored as GRASS vector points, to the nearest vector line. We illustrate this functionality by computing the distances of schools to the closest road, given in a vector map. The individual distances can be just printed and/or can be stored as a new vector map (connecting lines):

```
g.region swwake_30m -p
d.erase
d.vect streets_wake col=grey
fcol=red icon=basic/marker siz=12
d.vect schools_wake col=red fcol=red icon=basic/marker siz=12

v.distance -p from=schools_wake to=streets_wake out=connectors\
         upload=dist column=dist
d.vect connectors col=blue width=2
```

The resulting **connectors** map could be now patched into the streets map to connect the schools to the street network. In general, these connectors do not correspond to the real school driveways, they are just the shortest connections between schools represented by points and the streets network. Note that for vector network map preparations, point to network connections can be done much easier with **v.net** (**connect** tool).

Vector geometry type conversion   The **v.type** module can be used to convert between different vector types, if such conversion is possible. For example, boundaries can be converted to lines and vice versa (see e.g., Subsection 6.7), likewise centroids can be changed from/to points. Vector line data can be transformed to vector points using **v.to.points**. You can transform only the point features (nodes) from your vector map, or all vertices that define the vector lines (see the example in Section 6.5).

Adding missing centroids   Vector lines, when converted to boundaries, require a centroid in each closed boundary to form topologically an area. Missing centroids can be added to areas with **v.centroids**. For area attributes, centroids are used as label points.

Random and spatially perturbed points   Randomly distributed vector points can be generated by **v.random**. Unlike the result of **r.random**, these vector points will include only the coordinates and no attributes. Those can be added with **v.db.addtable** and **v.db.update** or the query tools (see Section 6.4.1).

```
g.region rural_1m -p
d.erase
v.random -z out=randpts_3m n=500 zmin=1 zmax=100
v.info randpts_3m
d.vect randpts_3d siz=2
```

If needed, the spatial position of vector points can be randomly perturbed with **v.perturb** by adding a variable spatial deviation to the east and north coordinates using either a uniform or normal delta value. If the distribution is uniform, only one parameter, the maximum, is needed. For a normal distribution, two parameters, the mean and standard deviation, are required. Such perturbation can be useful to analyze the impact of horizontal error on interpolation or to render synthetic objects such as tree placements more natural (e.g., with the **v.trees3d** from the GRASS AddOns Wiki site). As an example, we spatially shift random sample points derived from a lidar-based DEM **elev_lid792_randpts** uniformly by 3m:

```
g.region rural_1m -p
d.erase
v.perturb elev_lid792_randpts out=elev_lid792_pertpts par=3
d.vect elev_lid792_randpts siz=2
d.vect elev_lid792_pertpts siz=2 col=red
v.db.select elev_lid792_randpts where="cat<10"
v.db.select elev_lid792_pertpts where="cat<10"
```

You can see that the new set of points has the same elevation values stored as attribute **value**, but their horizontal locations are shifted 3m.

6.5.2 Buffering

Buffers and circles (around points) can be generated with **v.buffer**. Half-side buffers are generated with **v.parallel**, or **v.segment**, with positive buffer value for the right side and negative for the left side in native vector direction. For example, to create a 400m buffer stored as vector area run:

```
g.region swwake_10m
d.erase
v.buffer schools_wake out=schools_buff_400m buff=400
d.vect schools_buff_400m
d.vect schools_wake icon=basic/marker col=red
```

You can see that the overlapping buffers were merged into single polygons; database query can be used whether these clustered schools are of the same type (e.g., elementary) or serve different student populations (e.g. middle and high schools).

The second example shows how to create parallel lines to the lines in the input map, in our case **railroads**:

```
# display railroad lines with line direction
d.vect railroads disp=shape,dir

# compute paralel lines to the right and left
v.parallel railroads out=myrailroads_parr dist=200
v.parallel railroads out=myrailroads_parl dist=-200
d.vect myrailroads_parr col=red
d.vect myrailroads_parl col=blue
```

You can see that there are few segments that were digitized in opposite direction that the segments before and after them, causing the parallel line to "jump" to the other side. You can select the offending segments and flip their direction using v.edit to avoid this problem.

6.5.3 Feature extraction and boundary dissolving

Vector objects can be extracted based on attribute selection using SQL rules with v.extract. We have already used this command in Subsection 6.2.2 and in 6.5.1 It also allows us to dissolve common boundaries, although the module v.dissolve does the same in an easier way. Objects can be selected and deselected by a mouse box using graphical extractor in the GRASS monitor d.extract. To select/delete vector objects by coordinates and a given distance, you can use v.edit.

To extract vector objects from a vector map, you can run v.extract with either the desired category(ies) listed by the parameter list or a where SQL statement. This will extract the selected vectors into a new map. Optionally, common boundaries can be dissolved with the -d flag. As an example, we extract the municipality *Cary* from the map boundary_municp into a new vector map:

```
# check column names first
v.info -c boundary_municp
v.extract boundary_municp out=cary where="TEXT_NAME = 'Cary'"
g.region vect=cary -p
d.erase
d.vect -c cary
```

The new vector map contains only the selected areas. You can dissolve common boundaries when adjacent polygons have the same category by using the command with the flag -d. However, this requires to also use the new parameter to assign a unique category number to the selected areas.

Alternatively, the v.dissolve command allows to dissolve common boundaries based on attributes found in the specified column. As an example, we dissolve the map of NC counties to a NC state map. Since no column with common value exists, we copy the map into our mapset, add a new column state and fill it with the value "NC" in all rows. The state column is then used for dissolving:

```
g.copy vect=boundary_county,mycounties
v.db.addcol mycounties col="state varchar(2)"
v.db.update mycounties col=state val="NC"
v.dissolve mycounties out=mync_state col=state

g.region vect=mync_state -p
d.erase
d.vect mync_state
d.vect mycounties type=boundary col=orange
```

Note that the resulting boundary extends approximately 36km into the Atlantic ocean. You can use SQL to delete the ocean area from the `mycounties` map before dissolving the boundaries (its category number is 4). In order to populate a database table from vector features such as geometric properties, use `v.to.db`. It calculates the values and uploads them to the attribute table.

Selective vector removal   As a way of selecting/deleting vector objects by coordinates and radius or box, you can use `v.edit`. As an example, we want to reduce the map `zipcodes_wake` to the area around Raleigh while selecting only complete polygons. Since the command extracts all vector objects in the given threshold (distance from center to selection box boundary), also incomplete areas will be saved into the resulting map. So we extract in a further step only the complete areas:

```
g.region vect=zipcodes_wake
d.erase
d.vect zipcodes_wake

g.copy vect=zipcodes_wake,myzipcodes_tmp
# take position for center of selection
d.where
# reverse deletion
v.edit -r myzipcodes_tmp coord=642730,224640 thresh=5000 \
    tool=delete

# extract only complete areas
v.extract myzipcodes_tmp out=myzipcodes type=area
d.vect myzipcodes fcol=red
v.db.select myzipcodes
```

Due to the reverse deletion, all polygons which do not match the query are deleted and the polygons of interest are kept.

6.5.4 Patching vector maps

New vector maps can be created by combining existing vector maps with `v.patch`. If the table structures are identical (same column names and types), also the attributes are transferred to the new table. Any vectors that are duplicated among the maps being patched together (e.g., border lines) will have to be edited or removed after v.patch is run. Such editing can be done automatically using `v.clean` (`tool=snap,break,rmdupl`).

  If several adjacent maps where combined (merging of tiles, e.g. with `v.patch`) then dissolving in order to remove tile boundaries requires an extra step since duplicated boundaries must be removed first:

```
# remove duplicated tile boundaries
v.clean tiles out=tiles_clean tool=snap,break,rmdupl thresh=0.1
```

```
# dissolve based on column attributes
v.dissolve tiles_clean output=tiles_dissolved col=mycolumn
```

6.5.5 Intersecting and clipping vector maps

Besides visual overlay of maps on the screen, you sometimes need to combine two vector maps and store the result into a new map, or perform spatial operations based on content of two maps. The module v.overlay provides this functionality for area maps, the v.select module for point and line maps. Intersection, union, merging, and clipping of vector maps is not a trivial task. For example, when intersecting, internally new vectors have to be generated, because for each new intersecting vector existing lines have to be broken and new nodes have to be inserted (compare Figure 6.5). Please note that in GRASS, all vector modification tools ignore the current geographic region settings and always operate on the full map. The map boundaries are extracted from the vector file headers. When intersecting maps, tables are joined accordingly.

Intersecting two area maps   Intersecting area (polygon) maps has two effects depending on the input data: either you want to use one map as a mask to cut out a spatial subsection from the second map or you want to intersect both maps to a single map. We illustrate the differences with two examples (which are somewhat synthetic):

To mask vector maps based on this method, you may digitize a mask area with v.digit or generate it with other tools (for a box, v.in.region is convenient). The mask area should be labeled with a category number (e.g. 1). The vector areas will determine the boundary of the final vector map. The module v.overlay produces a new map that contains all the vector data from the binput map that fall into the extent of the vector mask map given as map ainput. To perform a clip by polygon operation (vector mask), use the AND operator:

```
g.region vect=census_wake2000 -p

# create circular vector mask as "cookie cutter"
echo "642600|224640" | v.in.ascii out=mypoint
v.buffer mypoint out=mycircle buffer=10000

d.erase
d.vect -c census_wake2000 siz=2
d.vect -c mycircle

v.overlay ain=mycircle bin=census_wake2000 out=vover_and_mask \
        op=and
d.erase
d.vect -c vover_and_mask
```

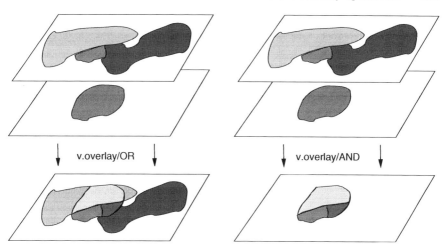

Fig. 6.6. Possible results of intersecting vector data using v.overlay with OR and AND operators

The resulting map `vover_and_mask` contains the subsection of the `census_wake2000` map restricted to the area covered by `mycircle`. We re-use the resulting map `vover_and_mask` to show intersection of two vector maps, each including several polygons (you can zoom into the circle area to better see the results):

```
v.overlay ain=vover_and_mask bin=boundary_municp \
        out=vover_and_map op=and
d.erase
d.vect -c vover_and_map siz=2
```

The resulting map `vover_and_map` contains the part of `boundary_municp` intersected with `census_wake2000` but restricted to the area covered by `vover_and_mask`. The areas with no-data (outside municipialities) become no-data in the resulting vector map.

Another operator is `OR` which performs union of two maps. Figure 6.6 illustrates the difference between the `AND` and the `OR` operator. We use the same module again with the same maps:

```
v.overlay ain=mycircle bin=zipcodes_wake out=vover_or op=or
v.overlay ain=mycircle bin=census_wake2000 out=vover_or op=or
d.erase
d.vect -c vover_or
```

The resulting map `vover_or` contains the map `census_wake2000` unified with the map `mycircle`. The polygons that are intersected with the circle boundary are split into two new polygons.

To copy features from either `ainput` or `binput` but not those from `ainput` overlayed by `binput`, use the XOR operator:

```
v.overlay ain=mycircle bin=censusblk_swwake out=vover_xor op=xor
d.erase
d.vect -c vover_xor siz=2
```

The resulting map `vover_xor` contains the map `censusblk_swwake` and
`mycircle`, except for the area covered by their intersection.

The fourth operator is NOT which copies features from `ainput` not overlayed
by features from `binput`:

```
v.overlay ain=mycircle bin=censusblk_swwake out=vover_not op=not
d.erase
d.vect -c vover_not siz=2
```

The resulting map `vover_not` contains the map with area covered by `mycircle`
except for the area covered by `censusblk_swwake`.

When performing intersect operations, we need to keep in mind that if we
plan to use the data assigned to the original polygons (e.g. population) we
need to ensure that the polygons that we need to use are preserved in their
original shape and size. A good example is the map `censusblk_swwake` that
was clipped from a larger map using the coordinates of the `swwake_10m` region
by `ogr2ogr` while preserving the entire polygons that fall at least partially
within this region. When using `v.overlay` you need to make sure that the
`ainput` map is sufficiently large to preserve the desired polygons.

Intersecting areas and lines/points: point in polygon   To select lines
or points which fall within an area, we an use the `v.select` module. As an
example, we calculate which points of interest can be seen from the new
skyscraper built in downtown Raleigh (see Section 5.4.4 for calculation of
the line of sight map `tower_165_los`):

```
g.region rast=tower_165_los -p
# create a binary viewshed map and convert it to vector map
r.mapcalc "tower_165_losbin=if(tower_165_los)"
r.to.vect -s tower_165_losbin out=tower_165_los feat=area

# first map must be point map
v.select ain=poi_names_wake bin=tower_165_los out=tower_poi
d.erase
d.his h=tower_165_los i=elevation_shade
d.vect streets_wake
d.vect tower_165m siz=10 col=orange icon=basic/marker
d.vect tower_poi icon=basic/circle fcol=yellow
d.vect tower_poi disp=attr attrcol=class
```

The `v.select` command intersects the vectorized line of sight area with the
points of interest and extracts the visible points to a new map. You could now
do another selection for directional view sectors to prepare the descriptions
on top of the skyscraper for visitors.

6.5.6 Transforming vector geometry and creating 3D vectors

Geocoding of unreferenced CAD drawings and other vector maps using ground control points can be done with v.transform. We have already explained the procedure in Section 4.2.3.

Another usage of this command is re-scaling. Sometimes, the vector geometry is stored with inconsistent units, for example heights may be stored in feet while the horizontal coordinates are in meters (as is sometimes the case with US GIS data). In the following example, we rescale the z coordinates of the lidar points stored in the elev_lidrural_mrptsft map from feet to meters:

```
v.transform elev_lidrural_mrptsft out=myelev_lidrural_mrpts \
        zscale=0.3048006096012192
# report map 3D spatial extension
v.info -g elev_lidrural_mrptsft
 [...]
 top=502.183472
 bottom=335.391998

v.info -g myelev_lidrural_mrpts
 [...]
 top=153.065828
 bottom=102.227686
```

The vertical range of the converted lidar points now matches the metric elevation model range. With v.sample differences can be sampled and calculated.

Extruding from 2D to 3D vectors   GRASS supports various methods to change 2D to 3D geometry in case that original true 3D vector data are not available. 2D vector polygons (e.g. house footprints) can be extruded to 3D block structures using v.extrude:

```
# set region to tile792 where we have both DEM and planimetry
g.region rast=el_D792_6m -p
```

Fig. 6.7. Buildings extruded from 2D and positioned on top of DEM

```
# extract buildings, based on original DXF data set
v.extract P079215 out=bldg_resid where="layer='BLDG_RESID_BL'"
v.extract P079215 out=bldg_cmcl where="layer='BLDG_COMMER_BL'"

# add height of the building
v.extrude -t bldg_resid out=bldg_resid_3d height=10 \
          elev=el_D792_6m
v.extrude -t bldg_cmcl  out=bldg_cmcl_3d height=15 \
          elev=el_D792_6m
```

You can visualize the resulting 3D vector buildings on top of digital terrain model using nviz, as we did for the cover of this book (see Figure 6.7, see also Chapter 7):

```
# set the region to area used on the book cover
g.region rural_1m -p
nviz elev_lid792_1m col=ortho_2001_t792_1m \
    vect=bldg_resid_3d,bldg_cmcl_3d
# use "Vector lines/3D Polygons" panel to set building colors
```

To show the extruded 3D buildings in external software such as Google Earth or NASA WorldWind, we can export the vector maps into KML format using v.out.ogr. However, this requires that the building map is first reprojected into a latitude-longitude location to match the projection used in these software packages. See Section 3.3.2 for details how to use v.proj accordingly. Alternatively, you can use ogr2ogr to reproject the map.

In general, GRASS is able to handle even more complex 3D geometries than just blocks, but these data are usually generated in CAD software. Such CAD drawings can then be imported and geocoded if needed.

Adding DEM-derived elevation to 2D vector maps You can modify 2D vector points, lines or areas to 3D with v.drape by extracting the elevation values from a digital elevation model or other raster surface. In this example, we convert a 2D vector line to a 3D line:

```
g.region rast=elevation -p
v.drape busroute_a rast=elevation method=bilinear \
        out=busroute_a_3d
v.info busroute_a_3d
```

The new map busroute_a_3d now includes elevation data for the bus route. Slopes for each segment can be calculated and uploaded to the attribute table with v.to.db (slope option) after splitting into shorter segments with v.split. This type of 3D route analysis may be of interest for bicycle tour planning, too. You can also visualize the 3D vector map in nviz (see Chapter 7).

Converting point data elevation attributes to z-coordinate   Vector point maps with elevation stored as attributes can be converted to real 3D geometry (from (x,y) and z as attribute to (x,y,z) geometry) using v.in.db. In this example, we create a new 3D map from the 30 years precipitation normals map precip_30ynormals. The procedure requires to upload the x and y coordinates as attributes to the table, and then re-import a new map from the updated table with the z-coordinate set to the attribute elev:

```
g.copy vect=precip_30ynormals,precip_30ynormals_tmp
v.info -c precip_30ynormals_tmp
v.db.addcol precip_30ynormals_tmp \
        col="x double precision, y double precision"
v.to.db precip_30ynormals_tmp option=coor col=x,y

db.tables -p
# we assume that the table name corresponds to input map name
v.in.db precip_30ynormals_tmp out=myprecip_30ynormals \
      x=x y=y z=elev key=cat
v.db.select myprecip_30ynormals
```

We can now remove the superfluous 'x' and 'y' columns from the table:

```
v.db.dropcol myprecip_30ynormals col=x
v.db.dropcol myprecip_30ynormals col=y
v.info -c myprecip_30ynormals
v.db.select myprecip_30ynormals
v.info myprecip_30ynormals
g.remove vect=precip_30ynormals_tmp
```

The new map myprecip_30ynormals now contains 3D geometry.

6.5.7 Convex hull and triangulation from points

Triangulation and point-to-polygon conversions can be done with v.delaunay, v.hull, and v.voronoi. To generate an outer convex boundary for a set of points, the *convex hull*, we can use the v.hull module. As an example, we generate the convex hull map from schools map of Wake county schools_wake, to compare it with the extent of municipalities and the county boundaries:

```
v.hull schools_wake out=schools_wake_hull
g.region vect=schools_wake_hull -p
d.erase
d.vect boundary_municp fcol=yellow type=area col=yellow
d.vect schools_wake icon=basic/circle
d.vect schools_wake_hull col=blue type=boundary width=2
d.vect boundary_county type=boundary col=red

# find out why hull and county boundary have big gap NW and SW
d.vect lakes type=area fcol=cyan col=cyan
```

The resulting vector area `schools_wake_hull` shown in blue follows the SW-NE shape of the county boundary with the largest gaps in the north-west and south-west corners where large water reservoirs are located and development is limited.

We can generate Voronoi polygons for the Wake schools to show the distribution and shape of nearest neighbor areas for the given distribution of schools. For illustration, we also generate Delaunay triangulation to show the relationship between its triangles and Voronoi polygons:

```
v.voronoi schools_wake out=schools_vor
g.region swwake_10m
d.vect -c schools_vor type=area
d.vect schools_wake icon=basic/circle siz=10 fcol=yellow
d.vect streets_wake

# show relation between Voronoi and Delaunay triangulation
v.delaunay schools_wake out=schools_del
d.erase
d.vect -c schools_del type=area
d.vect schools_vor type=boundary
```

The Voronoi polygons can be combined with streets and census data to assign students to schools close to their home.

6.5.8 Find multiple points in same location

Sometimes maps contain multiple points in same location within a given horizontal resolution. This is common for data acquired by real time kinematic GPS surveys, lidar maps and others. Here we describe a procedure for counting the number of such points, taking advantage of the multi-layer concept (see Section 6.2.4). Note that for this kind of analysis, you have to use a SQL driver other than DBF. We generate a few replicated points for this example (note that the survey data usually come in cm or mm and to count the replicated points within 1m resolution, you would have to convert the coordinates to integer meters):

```
# generate 3 + 2 points in the same locations
echo "637885|225271|1
637885|225271|2
637885|225271|3
637810|224885|4
637810|224885|5" | v.in.ascii out=multi_pts

# we set region to map, enlarged by 100m for better view
g.region vect=multi_pts n=n+100 s=s-100 w=w-100 e=e+100
d.erase
d.vect multi_pts
```

```
# look at attributes
v.db.select multi_pts

# remove duplicates from map
v.clean multi_pts out=multi_simple tool=rmdupl
```

At this stage, a one-to-many relationship remains between geometry and attributes. We can now verify the category numbers which are defining the link between vector object and attribute. Since we use the map with one point per position, we effectively assign an individual ID (category number) to each of the points:

```
v.category multi_simple op=report
 LAYER/TABLE 1/multi_simple:
 type          count          min          max
 point            5             1            5
 [...]
 all              5             1            5
```

The idea is now to add a second layer which carries different attributes, in this case the point count per position:

```
# add second layer with 1 cat/point
v.category multi_simple out=multi_sl layer=2

# create and link new table to layer 2
v.db.addtable multi_sl layer=2 \
          columns="cat integer, count integer"
v.info -c multi_sl layer=2

# check layer 1 attributes
v.db.select multi_sl
 cat|int_1|int_2|int_3
 1|637885|225271|1
 2|637885|225271|2
 3|637885|225271|3
 4|637810|224885|4
 5|637810|224885|5

# check layer 2 attributes
v.db.select multi_sl layer=2
 cat|count
 1|
 2|
```

The table connected to layer 2 is not yet populated with values. We use the SQL count() function to count the number of features in the original map for each category (this is not supported by the DBF driver):

```
v.to.db multi_sl option=query layer=2 qlayer=1 column=count \
        qcolumn="count(*)"
v.db.select multi_sl layer=2
 cat|count
 1|3
 2|2
```

```
# print number of points in map
d.vect multi_sl layer=2 disp=shape,attr attrcol=count
```

This shows the points and the number of points falling into the same position.

6.5.9 Length of common polygon boundaries

Land use management and planning tasks may require information about the length of common boundaries shared by neighbouring polygons in various types of vector maps. As an example, we perform this task using the Census map of Wake county:

```
# display the map and create your own modifiable copy
g.region vect=census_wake2000 -p
d.erase
d.vect census_wake2000
g.copy vect=census_wake2000,cens2000wake
```

The underlying idea is that we are retrieving the IDs (category numbers) of the polygons left and right from each boundary and store them into the attribute table linked to a new layer 2 which we add to the map. For this, we first generate an ID (category) for each boundary in vector geometry that will be later used to attach a second attribute table:

```
v.category cens2000wake out=cens2000wake2 layer=2 \
           type=boundary option=add
```

The planned table structure will then be:

```
cat_boundary|cat_left_polyg|cat_right_polyg|length_boundary
```

A boundary may consist of several vectors. As we only want one category per boundary, we will use a sides check to keep the same category number also for multi-line boundaries. Next we create a new attribute table and link it to the new layer 2 of the vector map:

```
v.db.addtable cens2000wake2 layer=2 \
    col="left integer,right integer,length double precision"
```

Now we perform the query of the polygon/boundary relationships and store it into the attribute table linked to layer 2. This will provide us with unique categories for the boundaries, allowing us to calculate the lengths and update the related attribute table column:

```
v.to.db cens2000wake2 option=sides col=left,right layer=2
v.to.db cens2000wake2 option=length col=length layer=2
```

We can now check the new attribute table containing the boundary lengths, and graphically verify the result, for example by measuring the boundary with category number 125:

```
v.db.select cens2000wake2 layer=2
d.vect cens2000wake2 cat=125 layer=2 col=red type=boundary
d.zoom
d.measure
 LEN:        12968.22 meters
 ...
v.db.select cens2000wake2 layer=2 where="cat=125"
 125|47|48|12969.214886
```

The result is reasonably close to our manual screen measure and acceptable since interactive screen measureements are usually not very precise.

Retrieving neighbors of vector polygons The same procedure can be used to retrieve the neighbors of a given polygon. The resulting table linked to the second layer contains all relevant information. For example, to find the zip code neighbors of Cary 27511, we perform the following steps:

```
v.category zipcodes_wake out=zipwake_cat layer=2 type=boundary\
           option=add
v.db.addtable zipwake_cat layer=2 \
    col="left integer,right integer"
v.to.db zipwake_cat option=sides col=left,right layer=2

# find category number for Cary 27511
v.db.select zipwake_cat where="ZIPNUM=27511"
 cat|WAKE_ZIPCO|PERIMETER|ZIPCODE_|ZIPCODE_ID|ZIPNAME|ZIPNUM|...
 34|595719341.254|135382.61154|40|10|CARY|27511|...
 [...]

# find related neighbor polygon category numbers
v.db.select zipwake_cat where="left=34 OR right=34" layer=2 \
           col=left,right
 left|right
 -1|34
 17|34
 -1|34
 [...]
```

The returned values are the neighbor polygon category numbers which can be matched to the names in layer 1. A negative number indicates that no neighbor was found.

6.6 Vector network analysis

GRASS provides support for vector network analysis, an important tool for routing, search for a closest facility along the route, travel directions and time, and calculation of service areas. The analysis is based on graph theory and topology. A graph or vector network consists of a set of nodes (also called centers) which are connected by edges. The latter are represented by vector lines in GRASS.

6.6.1 Network analysis

Network analysis covers a set of methods which are implemented in GRASS as follows:

- Vector network maintenance: `v.net`;
- Shortest path: `d.path` and `v.net.path`;
- Traveling salesman (round trip): `v.net.salesman`;
- Allocation of sources (create subnetworks, e.g. police station zones): `v.net.alloc`;
- Minimum Steiner trees (star-like connections, e.g. broadband cable connections at mimimum costs): `v.net.steiner`;
- Iso-distances (from centers): `v.net.iso`.

It is important to note that network analysis is based on heuristic algorithms which means that probably only a suboptimal solution is found (tradeoff between finding an optimal solution and limiting the computational time to search for it). Vector lines directions are defined by the digitizing direction (see Section 6.5.1). Both backward and forward directions are supported as all network modules provide parameters that allow us to assign attribute columns defining the costs to move along the forward and backward direction. Blocked direction is represented by negative cost (e.g., one way roads, street closure due to maintenance). You can find practical examples for network analysis modules in the related manual pages.

To reduce computational time for our example, we create a subset of the Wake county street map for the Raleigh area and its close neighborhood, using the `census_wake2000` map:

```
g.region vect=census_wake2000 -p
d.erase
d.vect census_wake2000
d.vect census_wake2000 disp=cat

v.extract census_wake2000 out=census_raleigh list=1-33
v.select ain=streets_wake bin=census_raleigh \
        out=streets_rlgh op=overlap
g.region vect=streets_rlgh
d.erase
d.vect streets_rlgh
```

We will use this "lean" vector map in our following examples.

Iso-distances and iso-chrones maps While staying in a downtown Raleigh hotel near -78.638259E, 35.77465N (found with GPS), we seek a nice park within walking distance. We first convert our GPS position to current LOCATION coordinates (here: NC State Plane metric):

```
# east north
echo "-78.638259 35.77465" | m.proj -i
 [...]
 642306.78 224657.80 0.00
```

We now digitize the starting point as node and merge it into the streets map:

```
echo "642306.78|224657.80|1" | v.in.ascii out=myhotel
```

In order to be able to navigate from this vector point on the network, we have to connect it with an additional vector line. We can selectively connect within a threshold, so we first determine the distance of the hotel GPS position to the network:

```
v.distance -p from=myhotel to=streets_rlgh upload=dist \
         column=dist
 from_cat|dist
 1|13.249617
 [...]
```

The hotel point is already very close to the network. To connect it to the closest street, we run the network maintenance tool **v.net** and create a new map **streets_net** which then contains the network and the point as node on layer 2:

```
v.net streets_rlgh points=myhotel out=streets_net \
     op=connect thresh=30

g.region vect=myhotel n=n+500 s=s-500 w=w-500 e=e+500 -p
d.erase
d.vect streets_net col=green
d.vect streets_rlgh
d.vect myhotel col=red icon=basic/pushpin
```

The hotel point is now connected to the network.

As final preparation step, we assign costs to navigate on the network. As pedestrians, we dislike to walk along dangerous streets and we want to avoid highways altogether. The existing maximum speed information helps us to easily assign costs in a new cost column (you can also distinguish between backward and forward moving by simply generating two different columns). We penalize the big streets by assigning them high cost:

```
v.db.addcol streets_net col="navcost double precision"
v.db.update streets_net col=navcost val=100 \
         where="SPEED <= 25"
```

```
v.db.update streets_net col=navcost val=100000 \
          where="SPEED  > 25"

g.region vect=myhotel n=n+2000 s=s-2000 w=w-2000 e=e+2000 -p
d.erase
d.vect streets_net col=green where="navcost=100"
d.vect streets_net col=red   where="navcost=100000"
d.vect myhotel col=red icon=basic/pushpin size=15
```

In case of separate backward and forward cost columns, use the display of native vector directions to assign the directions properly (see Section 6.5.1).

Now the vector network is ready to navigate. We define a set of walking distances to make it easier to choose the nearest park. The parameter ccats allows you to calculate the iso-distance map only for selected nodes (centers). To simplify the request, since we have only one center, we can just specify a large range to use it:

```
# the hotel point is on nlayer 2
# specify range of center cats (easier to catch all centers)
v.net.iso in=streets_net out=streets_net_iso ccats=1-1000000 \
        nlayer=2 costs=200,400,600,800 afcolumn=navcost
```

The column **afcolumn** is not only used for forward movements but also for bi-directional movements. This calculation may take a while depending on how complex the network graph is. Once the result is generated, we can create a report and show the iso-chrones map:

```
v.category streets_net_iso option=report
# ... reports 5 categories.
```

The network is categorized as:
cat   distance from point
1   0 - 200
2   200 - 400
3   400 - 600
4   600 - 800
5   > 800

To see the result, you can run (we select a FreeType font, see Section 7.1.1):

```
d.erase
d.font FreeSansBold
# show high traffic streets to be avoided
d.vect streets_net col=grey where="navcost=100000" width=4

# show colorized iso-distance network
d.vect streets_net_iso col=green  cats=1 width=4
d.vect streets_net_iso col=yellow cats=2 width=4
d.vect streets_net_iso col=orange cats=3 width=3
```

Fig. 6.8. Iso-distances map from the hotel starting point to nearest parks

```
d.vect streets_net_iso col=red     cats=4 width=3
d.vect streets_net_iso col=grey    cats=5 width=2
d.vect myhotel col=red fcol=green icon=basic/pushpin size=15
```

To find a park, we look at the points of interest map and select all parks therein:

```
d.vect geonames_wake where="FEATURECOD='PRK'" fcol=green \
   icon=basic/diamond bgcol=white bcol=black lsize=10 \
   yref=bottom xref=left
d.vect geonames_wake disp=attr attrcol=NAME fcol=green \
   where="FEATURECOD='PRK'" bgcol=white bcol=black lsize=10 \
   yref=bottom xref=left
```

The resulting map (see Figure 6.8 shows that the parks "Nash Square" and "Moore Square" are within the walking distance from the hotel under the defined conditions.

Shortest path maps A common vector network analysis task is shortest path routing with the goal to find an optimal routing on the network according to given constraints such as traffic density (attributes stored in a PostgreSQL server could be automatically updated in real-time), road closure, landscape attractiveness or other. While using **d.path** is rather straightforward, we explain the **v.net.path** command which also saves the result.

As an example, we want to calculate the shortest path between two hospitals to transport a patient. First we display the streets map and add the vector points map `hospitals` to the view:

```
g.region vect=streets_net -p
d.erase
d.vect streets_net
d.vect hospitals fcol=red icon=basic/diamond
d.vect hospitals disp=attr attrcol=NAME
```

In the next step, we connect the hospital nodes to the network on layer 2 in a given threshold:

```
v.net streets_rlgh points=hospitals out=streets_hnet \
    op=connect thresh=200
d.erase
d.vect streets_hnet col=red
d.vect streets_rlgh
```

The next step is to assign navigational costs to the streets. We use the inverse of the speed limit multiplied by a factor to obtain the costs:

```
v.db.addcol streets_hnet col="navcost double precision"
v.db.update streets_hnet col=navcost val="1.0/SPEED * 100."
v.db.select streets_hnet col=navcost,speed
```

Then we can calculate the shortest path between two given nodes. The patient has to be transported from the Dorothea Dix Hospital to the Rex Hospital Hospital. Before calculating the shortest path, we need to find the node category numbers of these two hospitals:

```
d.erase
d.vect streets_hnet layer=2
d.vect streets_hnet layer=2 disp=cat
# Rex: 8; Dorothea Dix: 5

# ID as first number, then cat1 and cat2
echo "1 8 5" | v.net.path streets_hnet out=spath \
    afcolumn=navcost

d.vect streets_hnet
d.vect spath col=red width=2
```

The map shows the shortest path on the network according to the costs and distance. Care must be taken to use vector maps which do not contain nodes at vector intersections where bridges are (as in this example). Using v.**edit**, superfluous nodes can be removed from 2D maps to obtain a perfect shortest path map.

6.6.2 Linear reference system (LRS)

A Linear Reference System (sometimes also called dynamic segmentation) uses linear features and distance measured along those features to position objects or events. It is useful for the management of roads (e.g., for geocoding of accidents), pipelines, and other linear systems. LRS is a double referenced system which gives the possibility to preserve most of the existing referenced points (mileposts) or linear events if modifications are needed due to real world changes. For example, if a street is modified, most of the milestones along the street are preserved and new milestones may be inserted only in the modified section of the street. The same can be done in LRS since it supports the addition/removal of mileposts in relative distance to the existing mileposts.

GRASS provides a set of LRS commands: `v.lrs.create` to create the linear reference system, `v.lrs.label` to create stationing on LRS, `v.lrs.segment` to create points/segments from LRS, and `v.lrs.where` to find the line ID and real km+offset for given points in vector map using linear reference system.

We illustrate LRS with a public transportation exercise using the "Wolfline" bus data for the NC State University bus service. To process the data, we need to use a real SQL driver for attributes management to take advantange of more sophisticated SQL queries which are not supported by the DBF driver. For this task, we create a new MAPSET `wolfline_lrs` in the `nc_spm` LOCATION and select SQLite as driver:

```
# in new MAPSET wolfline_lrs
db.connect driver=sqlite \
        database='$GISDBASE/$LOCATION_NAME/$MAPSET/sqlite.db'
db.connect -p
db.tables -p
# ... should not show any tables.
```

To start, we display the available bus route data:

```
# several bus lines available
g.region vect=busroutesall -p n=n+100 s=s-100
d.erase
d.vect -c busroutesall

# bus stops for all lines
d.vect -c busstopsall icon=basic/triangle

# look at the attributes
v.db.select busstopsall
```

In our example, we will work with the bus route 1 (Avent Ferry Rt.). Latest map and schedule, including real-time location of the buses, are available online.[5] In the NC data set, the bus route 1 is already available as a sep-

---

[5] Route 1 "Avent Ferry Rt." map and schedule,
  http://www.gotriangle.org/Bus/ncsuMapsAndSchedules.html

arate `busroute1` map, but we have to extract the related bus stops from
`busstopsall`. To use the SQL capabilities of SQLite, we copy the vector maps
into the current mapset. During this procedure the attribute tables are converted from the original DBF format (as provided in the LOCATION `nc_spm`)
to SQLite format in the current mapset:

```
g.copy vect=busroute1,route1
g.copy vect=busstopsall,stopsall

# verify map-database connections for SQLite connection
v.db.connect -p route1
v.db.connect -p stopsall
```

We use `v.extract` with a SQL query to extract the bus stops of route 1, by
applying pattern matching to the `ROUTES` column of the attribute table of the
map `stopsall` (see Section 6.2.2). As mentioned, this requires the SQLite or
other DBMS driver (not DBF):

```
# check the selection and extract stops to a new map
v.db.select stopsall \
  where="ROUTES LIKE '%1%' AND ROUTES NOT LIKE '%11%'"
v.extract stopsall out=stops1 \
  where="ROUTES LIKE '%1%' AND ROUTES NOT LIKE '%11%'"
```

To simplify the milepost definition (here: the bus stops order) for LRS, we
build a polyline from the lines in map `route1` using `v.build.polylines`. This
removes nodes along the route and allows us to create a single vector line:

```
v.build.polylines route1 out=route1tmp

# substitute old 'route1' with new map (but we lose the table)
v.category route1tmp out=route1 op=add --o
g.remove vect=route1tmp
```

As the next step, we prepare input maps for LRS, by adding an integer column
to indicate to which route the bus stop belongs to and defining the order
(optionally offsets) of the bus stops:

```
# add column to link the route with the bus stop
v.db.addtable route1 col="lid integer"
v.db.update route1 col=lid val=1
v.db.select route1
 cat|lid
 1|1

v.db.addcol stops1 col="lid integer"
v.db.update stops1 col=lid val=1
v.db.select stops1
 cat|ROUTES|UPDATED|STREET_1|STREET_2|CAMPUS|...
 7|1,5,7a,8,9,A,B|2006/11/14|Dan Allen Dr.||North|...
 [...]
```

```
# define the order, offsets of the bus stops, preparing table
v.db.addcol stops1 col="start_mp double precision, \
        start_off double precision, end_mp double precision,\
        end_off double precision"

# check direction of the vector line for the bus stops order
g.region vect=route1 n=n+100 s=s-100 -p
d.erase
d.vect route1 disp=shape,dir,topo col=grey lcol=blue
d.vect stops1 disp=attr attr=cat size=10 bgcolor=white
d.vect stops1 icon=basic/circle fcol=green
```

We see that the bus stops are not perfectly located on the bus route (for cartographic reasons, as they are also associated with other routes). This can be later addressed since the LRS tools permit to define a threshold to assign a point to a line.

Note that the intersection in the `route1` map influences the vector direction. Some effort is needed to get the order right according to the vector topology. We update segment per segment. The map contains three segments; and there is no node in the line intersection. To query points for their category numbers, use `d.what.vect`.

The bus route 1 is oriented from vector map nodes $n1$ to $n2$. Next we update the attribute column `start_mp` to indicate order of the bus stops along the bus tour:

```
# from node n1 to n2 - line cat 1
v.db.update stops1 col=start_mp where="cat=7" val=1
v.db.update stops1 col=start_mp where="cat=13" val=2
v.db.update stops1 col=start_mp where="cat=14" val=3
v.db.update stops1 col=start_mp where="cat=22" val=4
v.db.update stops1 col=start_mp where="cat=83" val=5
v.db.update stops1 col=start_mp where="cat=30" val=6
v.db.update stops1 col=start_mp where="cat=55" val=7
v.db.update stops1 col=start_mp where="cat=56" val=8
v.db.update stops1 col=start_mp where="cat=82" val=9
v.db.update stops1 col=start_mp where="cat=58" val=10
v.db.update stops1 col=start_mp where="cat=38" val=11
v.db.update stops1 col=start_mp where="cat=37" val=12
v.db.update stops1 col=start_mp where="cat=36" val=13
v.db.update stops1 col=start_mp where="cat=35" val=14
v.db.update stops1 col=start_mp where="cat=34" val=15
v.db.update stops1 col=start_mp where="cat=103" val=16
v.db.update stops1 col=start_mp where="cat=31" val=17
v.db.update stops1 col=start_mp where="cat=29" val=18
v.db.update stops1 col=start_mp where="cat=24" val=19
v.db.update stops1 col=start_mp where="cat=28" val=20
v.db.update stops1 col=start_mp where="cat=96" val=21
v.db.update stops1 col=start_mp where="cat=27" val=22
```

```
v.db.update stops1 col=start_mp where="cat=60" val=23
v.db.update stops1 col=start_mp where="cat=10" val=24
v.db.update stops1 col=start_mp where="cat=9" val=25

# verify route
v.db.select route1
 cat|lid
 1|1

# verify stops
v.db.select stops1 \
         col=cat,ROUTES,start_mp,start_off,end_mp,end_off,lid
 cat|ROUTES|start_mp|start_off|end_mp|end_off|lid
 7|1,5,7a,8,9,A,B|1||||1
 9|1,4,5,7a,9,A,B|25|||||1
 ...
 96|1,5,7,7a,8,8a,9,A,B|21|||||1
 103|1,A|16|||||1
```

The data are now almost prepared to build the LRS. If you would run the
`v.lrs.create` command now, it would report that the bus stop with category
number 30 is not in the right order. The reason is that it is closer to the
line segment between stops 29 and 103 than to the segment between stops
55 and 63 to which it really belongs to. GRASS offers a special tool for this:
`v.edit` which supports direct vector editing on a map via command line. With
`d.measure` we can roughly estimate the shift in East and South direction to
shift this bus stop onto the related vector segment. This does not have to be
perfect since `v.lrs.create` will operate with a threshold. With `d.measure` we
find out that we have to roughly shift 18m to East and 12m to South. For
`v.edit`, we select the "move" tool and the category number of the vector we
want to modify:

```
v.edit stops1 tool=move cats=30 move=18,-12

# redraw to verify updated map
d.redraw
```

Since the bus stops are not perfectly located on the route, we have to define a
reasonable threshold to capture all bus stops. To find the maximum distance
that we will use later as a threshold, we run `v.distance`:

```
# find maximum distance between bus stops and route1
v.distance from=stops1 to=route1 upload=dist column=dummy -p
# the highest reported value is 44.819408m

# the "start_mp" column is used to indicate the bus stops order
v.lrs.create in_lines=route1 points=stops1 out=route1_lrs \
 err=lrs_error lidcol=lid pidcol=lid rstable=route1_lrs thre=45
# the error map should be empty
```

```
# verify new LRS table
db.select route1_lrs
```

Now we have a complete linear reference system that we can display as follows:

```
d.erase
# show route and nodes
d.vect route1 disp=shape,topo col=grey lcol=blue
d.vect stops1 icon=basic/circle fcol=green

# show bus stop numbers (bottom right labels)
d.vect stops1 disp=attr attr=cat size=10 bgcolor=white      \
      lcol=green yref=top
# show milepost numbers (top right labels)
d.vect stops1 disp=attr attr=start_mp size=10 bgcolor=white \
      lcol=red yref=bottom
```

The green numbers show the bus stop numbers (equal to node category numbers), the blue numbers show the increasing milepost numbers which indicate the bus route direction. Depending on the GRASS monitor size, you will have to zoom to better identify the number labels.

Querying the LRS We can now query the linear reference system. For example, we are interested in how far it is to the next bus stop. We can get our current position from our GPS or GPS-enabled cellphone (or simply **d.where**):

```
# these coordinates can be retrieved via GPS
echo "638632|224857" | v.in.ascii out=position

g.region vect=route1 n=n+100 s=s-100 -p
d.erase
# show route and nodes
d.vect route1 disp=shape,topo col=grey lcol=blue
# show bus stop numbers (bottom right labels)
d.vect stops1 disp=attr attr=cat size=10 bgcolor=white \
      lcol=green yref=top
# show milepost numbers (top right labels)
d.vect stops1 disp=attr attr=start_mp size=10 bgcolor=white \
      lcol=blue yref=bottom
# show markers
d.vect stops1 icon=basic/circle fcol=green
d.vect position col=red icon=basic/marker size=20

v.lrs.where line=route1_lrs point=position rstab=route1_lrs
 pcat|lid|mpost|offset
 1|1|6.000000+134.532728
 [1] points read from input
 [1] positions found
```

```
# check to which bus stop the milepost 6 belongs to
# a) get corresponding bus stop number "upstream"
v.db.select stops1 col=cat,start_mp where="start_mp=6"
 cat|start_mp
 30|6
```

```
# b) get next bus stop number along the tour
v.db.select stops1 col=cat,start_mp where="start_mp=7"
 cat|start_mp
 55|7
```

The corresponding bus stop numbers are 30 from which the bus is coming and 55 the next stop in direction of the tour. Our position is 134.5m after bus stop 30.

The LRS table contains even more information:

```
db.select sql="SELECT * FROM route1_lrs WHERE start_mp=6"
 rsid|lcat|lid|start_map|end_map|start_mp|start_off|\
                         end_mp|end_off|end_type
 6|1|1|1946.147074|2411.275995|6|0|7|0|2
```

Relevant entries are: **start_map** (here: 1946.1m) which is the distance in map units from the beginning of the vector line to the milepost found before the queried position in LRS direction; **start_mp** (here: 6) which is the milepost before the queried position in LRS direction; **end_mp** (here: 7) which is the next milepost to be reached; and **end_map** which is the distance in map units from the beginning of the vector line to the milepost found before the queried position in LRS direction.

We finally want to figure out if we are closer to bus stop 30 or 55:

```
# what is the distance between bus stop 30 and 55
db.select sql="SELECT (end_map - start_map) as dist_30_55 \
         FROM route1_lrs WHERE start_mp=6"
 dist_30_55
 465.128921
```

```
%HM3 overflow - can we skip the spaces? not sure about sql rules
# distance from our position to bus stop 55 (MP 7)
db.select sql="SELECT (end_map - start_map - 134.5) as \
         dist_to_55 FROM route1_lrs WHERE start_mp=6"
 dist_to_55
 330.628921
```

In the first **db.select** call, we subtract the real world distances from the beginning of the vector line to the two mileposts from each other to obtain their distance. In the second command, we subtract our actual offset from milepost 6 (bus stop 30). The result shows, that we are much closer to bus stop 30 than bus stop 55. All queries could be put into a script for routine operations.

Visualization of the LRS The bus route LRS can be visualized in a dedicated way as follows:

```
v.lrs.label route1_lrs rstable=route1_lrs labels=labels \
     col=red size=50 xoffset=100 output=route1_lrs_labels

g.region vect=route1 n=n+100 s=s-100 -p
d.erase
d.vect route1_lrs
d.vect route1_lrs_labels col=grey type=line
d.vect stops1 disp=attr attr=cat size=10 bg=white lcol=green \
     yref=bottom
d.vect stops1 icon=basic/circle fcol=green
d.labels labels

# simple PNG output
d.out.file route1_lrs format=png res=2
display route1_lrs.png
```

Once the LRS is established, various tasks can be accomplished, including adding a new milepost or relocating a bus stop.

6.7 Vector data transformations to raster

We have already explained how to convert existing raster data (points, lines, areas) to vector data. This can either be done manually by digitizing in v.digit or by an automated procedure using r.to.vect for discrete geometry features or by r.contour for continuous fields, as described in the Section 5.3.1. Here we focus on transformation from vector to raster maps. This transformation is useful, for example, for performing map algebra – a task that is easily done in the raster model. Also the speed of data processing (at the price of lower spatial precision) is much higher in the raster model.

Depending on the type of geographic phenomenon that the vector maps represent, we distinguish two types of transformation from vector to raster model:

- for geometric features (points, lines, areas), we use direct transformation from vectors to raster lines/areas;
- for continuous fields (points, isolines, contours), we need spatial interpolation to transform from vector lines to raster representation.

Figure 6.9 shows an overview of available conversion techniques.

Direct transformation of vector data to raster data The module v.to.rast generates a raster map from an input vector map by assigning the vector data values/attributes or their function to the corresponding grid

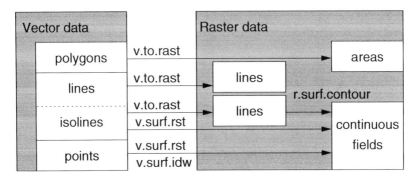

Fig. 6.9. Methods for transforming and interpolating vector data to raster data

cells, an operation sometimes called binning. Transformation of vector objects to raster cells depends on the current resolution. The resolution settings can be defined with g.region using the res (resolution) parameter. You may try different resolutions to see the effect. You have several choices to assign vector data values/attributes to the resulting raster map. The source of raster values can be (options for use parameter):

- attr: read values from attribute table (with column parameter);
- cat: use category values (i.e, vector IDs);
- val: use specified value (with value parameter);
- z: use z coordinate (points or contours only);
- dir: output as flow direction (lines only).

For discrete geometric features, the transformation of vector points to raster is also performed with v.to.rast. It creates an output raster map with the selected point attribute value in a cell where the point is located, while inserting NULLs elsewhere. If more than one point falls into one raster cell, the module will continue to read the points and the last imported point value will determine the cell value. Lines and areas are transformed likewise:

```
g.region swwake_10m -p

# points: use point elevation
v.info -c geodetic_swwake_pts
v.to.rast geodetic_swwake_pts  out=geodetic_swwake_pts \
         use=attr col=zval

# lines: use category number
v.to.rast roadsmajor out=roadsmajor use=val

# areas: use attribute column
v.info -c zipcodes_wake
v.to.rast zipcodes_wake out=zipcodes_dbl use=attr col=ZIPNUM
```

```
# areas with labels
g.region swwake_30m -p
v.to.rast geology out=geology_30m use=attr layer=1 \
  type=point,line,area col=GEOL250_ID labelcolumn=GEO_NAME
```

As a special case, we show how to extract boundaries from a vector map and convert them to raster lines. We must first run **v.category** to add category numbers to the boundary type because boundaries are usually without category numbers and the module **v.extract** works only with objects that have a category number. After extraction, we convert the boundary type to line type to ensure topological consistency:

```
g.region vect=soils_wake -p
v.category soils_wake output=soils_bnd_cats type=boundary
# extract boundaries only
v.extract soils_bnd_cats output=soils_bnds type=boundary
# vector type conversion
v.type soils_bnds out=soils_lines type=boundary,line

d.erase
d.vect soils_lines
```

Then we can transform the resulting vector lines map to raster (we use the -a align flag to the desired resolution, see Section 4.1.2):

```
g.region -pa res=30 vect=soils_lines
v.to.rast soils_lines out=soils_borders use=cat
d.erase
d.rast soils_borders
```

As before, we define the target raster resolution before converting the vector map to raster. The resulting raster map **soils_borders** contains only the rasterized outlines of the soils areas.

Generating raster density map from vector point data   You can compute a raster density map from vector point data using a moving 2D isotropic Gaussian kernel with **v.kernel**. As an example, we will compute the density map for schools in the Wake county:

```
g.region vect=schools_wake res=30 -p
d.erase
v.kernel schools_wake out=schoolskernel stddev=800 mult=10000
d.rast schoolskernel
d.vect schools_wake icon=basic/circle
d.vect roadsmajor
```

The highest density of schools is in central and east Raleigh with some clusters of schools in subsurbs with elementary and middle schools on neighboring campuses. The `sstdev` parameter effectively controls the smoothness of the map, for example, at 100m the resulting map has narrow local maxima practically at each school site.

6.8 Spatial interpolation and approximation

Spatial interpolation transforms vector points or isolines representing continuous phenomena, such as elevation or temperature to the raster representation using a function which passes through or close to the given points. The function that passes exactly through the given points is used for interpolation; when this condition is relaxed and the function passes close to the given points (for example, if we are working with noisy data), we describe it as approximation. Because there exists an infinite number of functions which fulfill one of these requirements, additional conditions have to be imposed, leading to a large number of different interpolation and approximation techniques. GRASS offers interpolation and approximation functions which are based on conditions of locality (Voronoi polygons, inverse distance weighted method) and smoothness (splines). The methods based on geostatistical concepts can be applied by taking advantage of the link with the Open Source geostatistical tools (see Chapter 10). It is important to keep in mind that different methods, and often even the same method with different parameters, can produce quite different surfaces (see for example, Figures 6.10, 6.11). A good knowledge of the modeled phenomenon is needed to evaluate which one is closest to reality. Statistical measures of accuracy do not always ensure that the properties of the interpolated surface are adequate representation of the behavior of the modeled phenomenon.

6.8.1 Selecting an interpolation method

GRASS provides several different methods for interpolation of raster surfaces from vector point or isoline data (Figure 6.10). Depending on the method, the surface can be interpolated directly from vector data, or the vector data must be first transformed to raster points or lines and interpolated with the related raster module. The following modules can be used:

- `v.voronoi` converts vector point data to vector polygons and it can be used for vector to raster conversion in special cases (e.g., for qualitative data) when the resulting raster map should have only those discrete values assigned to the points. The resulting "surface" is a discontinuous, rasterized version of Voronoi polygons (see Figure 6.10 a).
- `v.surf.idw` and `r.surf.idw` is based on *inverse distance weighted* interpolation (IDW, Figure 6.10 b). Isolines need to be transformed to vector points or rasterized beforehand (see Appendix A.1 for equations);

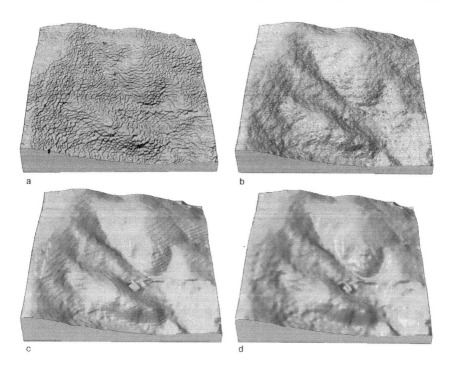

Fig. 6.10. Interpolation methods available in GRASS and the resulting surfaces: a) v.voronoi generated discontinuous surface from randomly distributed point data; b) v.surf.idw applied to point data, note the small peaks and pits around the data points; c) r.surf.contour, specially designed for rasterized contours; d) v.surf.rst applied to thinned contour data

- r.surf.contour requires conversion of contours or isolines to raster lines and then it linearly interpolates between contour lines (Figure 6.10 c);
- r.surf.nnbathy is a surface interpolation program, which uses "nn" – a natural neighbor interpolation library.[6] It is available from the GRASS AddOns Wiki;
- v.surf.rst interpolates the raster surface directly from the vector point or isoline data using the *regularized spline with tension* method (RST, see Section 6.8.3 and Appendix A.1, Figure 6.10 d).

Before interpolating in GRASS, it is necessary to set the resolution of the resulting raster map with g.region (see Section 4.1.2). To highlight the properties and behavior of various interpolation and approximation functions in the following sections, we have selected a resolution of the resulting raster map that is much higher than the distance between the given points or contours (2m compared to 10m-20m). In practical applications, it is recommended to

---

[6] "nnbathy" Web site, http://www.marine.csiro.au/~sakov/

use a resolution that is close to the data density to minimize the possible artifacts in the resulting surface.

Voronoi polygons   This method is suitable for transformation of qualitative vector point data to vector polygons or to a raster map when the condition of continuity is not required. Each point attribute is simply assigned to all cells within its natural neighborhood defined by a Voronoi polygon (Fortune, 1987). The module v.voronoi generates these polygons as a vector map with each polygon carrying the point attribute. You can then transform this vector map to a raster using v.to.rast, resulting in a surface composed of discontinuous, horizontal patches (see Figure 6.10 a). The procedure, using the random samples of NC elevation data in the rural subregion given by the vector map elev_lid792_randpts, is performed as follows:

```
# compute and display the polygons with 2 pixel centroids
g.region rural_1m res=2 -p
d.erase
v.voronoi elev_lid792_randpts out=elev_vor
d.vect -c elev_vor siz=2

# convert to raster and display with standard elevation color
v.info -c elev_vor
v.to.rast elev_vor out=elev_vor_2m col=value
r.colors elev_vor_2m col=elevation
d.rast elev_vor_2m
d.vect elev_lid792_randpts siz=2

# check the result using aspect
r.slope.aspect elev_vor_2m aspect=asp_vor_2m
d.rast asp_vor_2m
```

The resulting raster map depends on the spatial distribution of input data. For example, the raster surface generated from contour points would be quite different from the one generated from randomly distributed samples of the same surface used in our example (Figure 6.10a). The figure clearly demonstrates that Voronoi polygons are not a good choice for continuous fields, but they may be appropriate for numerous applications in ecosystem studies or geomarketing.

Inverse distance weighted average (IDW)   This approach calculates the value for each grid point as a weighted average of values at the n closest points (Burrough and McDonnell, 1998), see the equation in Appendix A.13. In the GRASS module v.surf.idw, weights are inversely proportional to a power $p = 2$ of distance and the default $n = 12$. It is a simple approach, however, the results are less accurate compared to other methods such as splines, kriging or multiquadrics (Mitas and Mitasova, 1999). Often the method does not reproduce the local shape implied by data and produces local extrema at the

data points (Figure 6.10 b), also noticeable as small circular contours around
the given points. The module is useful for rough interpolation of smaller data
sets, especially at lower resolutions, when the density of points is higher than
the density of the resulting grid points. We illustrate the method using random
points generated from the lidar-based DEM `elev_lid792_randpts`:

```
# set resolution and find column name where z is stored
g.region rural_1m res=2 -p
v.info -c elev_lid792_randpts
v.surf.idw elev_lid792_randpts out=elev_idw_2m col=value
r.colors elev_idw_2m col=elevation
d.erase
d.rast elev_idw_2m

# check the interpolated elevation surface using aspect map
r.slope.aspect elev_idw_2m aspect=asp_idw_2m
d.rast asp_idw_2m
d.vect elev_lid792_randpts siz=2 col=red
```

You can see that the resulting surface is continuous over most of the area
(compare with the Voronoi polygon representation), although it is quite noisy
with small peaks and pits around the given points and discontinuous patches in
areas with larger gaps in the data. Increasing number of points for computing
the average from n=12 to n=24 or reducing the resolution leads to a smoother
surface but you may start losing some detail.

As an alternative, we can convert the input vector points to raster
point representation and use the raster version of the IDW implementation
(`r.surf.idw`) that also offers computation of predictive error using cross-
validation when run with the flag -e. See the next section for details about
cross-validation. We need to convert the data to centimeters using `r.mapcalc`
to avoid steps in the resulting surface that is computed as integer raster map:

```
g.region rural_1m res=2 -p
v.info -c elev_lid792_randpts
v.to.rast elev_lid792_randpts out=el_lid792_randpts col=value
r.mapcalc "el_lid792_randpts100=100.*el_lid792_randpts"
r.surf.idw el_lid792_randpts100 out=el_ridw_2m100
d.erase
d.rast el_ridw_2m100

# check the interpolated elevation surface using aspect map
r.mapcalc "el_ridw_2m=el_ridw_2m100/100."
r.slope.aspect el_ridw_2m aspect=asp_ridw_2m
d.rast asp_ridw_2m

# perform cross-validation
r.surf.idw -e el_lid792_randpts100 out=el_ridwer_2m100
r.mapcalc el_ridwer_2m=el_ridwer_2m100/100.
```

```
r.colors el_ridwer_2m col=differences
d.rast el_ridwer_2m
r.univar el_ridwer_2m
```

Without the multiplication, the result would be an integer raster with a noisy contour-like aspect map. A map computed in centimeters does not have this artifact but shows small peaks or pits in the data points. The module uses an efficient search for the neighboring grid points and also supports interpolation of data in geographic coordinates (latitude-longitude). The cross-validation results show that the maximum predictive error can be as much as 1m in some points but overall the interpolation is statistically very accurate with mean absolute error less than 2cm and RMSE less than 4cm.

Linear interpolation between contours   The module r.surf.contour is specially designed for interpolation from isolines or contours and requires the vector lines map to be converted to a raster lines map using v.to.rast. To interpolate the raster surface, we run r.surf.contour with the name of the raster lines map and a name for the resulting raster surface. The module has not yet been upgraded to support floating point data output, but we can avoid potential steps in the resulting surface by converting the elevation input from meters to centimeters. In our example (Figure 6.10c), we first generate the raster contour map using the 1m interval contours elev_lid792_cont1m and then run the interpolation:

```
g.region rural_1m res=2 -p

# convert vector lines to raster and multiply by 100
v.to.rast elev_lid792_cont1m out=el_lid792_cont1m_2m col=level
r.mapcalc "el_lid792_cont1m_2m100=100.*el_lid792_cont1m_2m"

# interpolate the DEM and convert back to meters
r.surf.contour el_lid792_cont1m_2m100 out=el_rcont_2m100
r.mapcalc "el_rcont_2m=el_rcont_2m100/100.0"
d.erase
r.colors el_rcont_2m col=elevation
d.rast el_rcont_2m

# check the result using aspect map
r.slope.aspect el_rcont_2m aspect=asp_rcont_2m
d.rast asp_rcont_2m
d.vect elev_lid792_cont1m col=white
```

The module linearly interpolates the elevation at a given cell from the uphill and downhill contour values and you may be able to see slight discontinuities perpendicular to the contours in our resulting aspect map. To obtain good results, make sure that the contour lines extend to the edge of the current region and there are no disjointed contour lines. Since a flood fill algorithm is used, the running time grows exponentially with the distance between contour lines. Without the conversion to centimeters, the result would be an integer raster map with step-like pattern, but the contours reflected in the aspect map would be smoother than the result from r.surf.idw.

Natural neighbor   Natural neighbor interpolation is implemented in add-on module r.surf.nnbathy. It uses "nn" – a natural neighbor interpolation library and allows the user to define breaklines such as stream lines or ridges as 2D lines without the need to define their elevation.[7] It is available from the GRASS AddOns Wiki.

Regularized Spline with Tension (RST)   The RST method computes the values at grid points using a function which simulates a thin flexible plate passing through or close to the data points (Figure 6.10 d). It is the most general and accurate method currently available in GRASS but it may require tuning of parameters to achieve optimal accuracy. Optionally, it also computes topographic parameters and partial derivatives of the modeled surface. The bivariate (2D) version is called v.surf.rst and the trivariate (3D) version is v.vol.rst. There is also a quad-variate experimental version available (e.g., for 3 spatial dimensions and time) called v.volt.rst for those who are interested in development of multivariate interpolation capabilities. The method, its properties and examples are described in more detail in the following sections.

6.8.2 Interpolation and approximation with RST

The v.surf.rst module runs with both vector point and vector isoline data; internally, all lines are converted to vector points and interpolated using the same algorithms. To interpolate DEM from the elevation random points elev_lid792_randpts and from the contours elev_lid792_cont1m, you can simply run the module with its default parameters (Figures 6.11 a, 6.10 d). We will also output the aspect map so that we can visually check the resulting elevation surface geometry. The computation may take a few minutes, but you can run it in background by using & at the end of the command (in the Section 6.8.5 we explain how to speed-up the computation when data are homogeneously distributed, in the example below we set segmax=30 npmin=140). The module optionally computes topographic parameters, so we can compute the aspect directly and use it to check the surface geometry:

```
# interpolate from points
g.region rural_1m res=2 -p
v.info -c elev_lid792_randpts
v.surf.rst elev_lid792_randpts elev=elev_rstdef_2m zcol=value\
          aspect=asp_rstdef_2m segmax=30 npmin=140
d.erase
```

---

[7] nnbathy Web site, http://www.marine.csiro.au/~sakov/

```
d.rast elev_rstdef_2m
d.rast asp_rstdef_2m

# change the color table to grey aspect
r.colors asp_rstdef_2m col=aspect
d.rast asp_rstdef_2m
d.vect elev_lid792_randpts siz=2 col=red
```

While the results may be satisfactory for many applications, it is worth exploring additional capabilities provided by this module, including tuning the character of the resulting surface, computation of topographic parameters and speeding up the computation using segmentation parameters.

When interpolating from contour data (vector lines) that were obtained by scanning or computed from a dense TIN (Triangulated Irregular Network) or a high resolution raster, it is important to recognize that the density of points along the lines is often very high (in our case 1 point per 1-2m) while the contours are usually much farther apart (in our case 10m or more), and there may be large areas between contours (especially in flat terrain) without any data (we have a 90m section without contours on the top of the hill). This type of strongly heterogeneous spatial distribution of data points presents substantial challenge for most interpolation methods, which tend to create waves or steps along the isolines. Reducing the number of points on the lines (for example, by increasing the dmin parameter in v.surf.rst), adding points between the contours, and tuning the interpolation parameters (e.g., lowering tension for v.surf.rst or increasing the number of points for IDW) helps to minimize the problem. We reduce the density of points on contour lines in the following example by setting the dmin parameter to 3m, so that all points closer to each other than 3m are eliminated (see more in Section 6.8.5):

```
# interpolate from contours
g.region rural_1m res=2 -p
v.surf.rst elev_lid792_cont1m elev=elev_rstcontd_2m zcol=level\
        aspect=asp_rstcontd_2m dmin=3
d.erase
d.rast elev_rstcontd_2m
r.colors asp_rstcontd_2m col=aspect
d.rast asp_rstcontd_2m
d.vect elev_lid792_cont1m col=yellow
```

The use of 3m minimum distance between the given points reduces the number of points used for interpolation from over 40000 to little less than 10000, closer to the 6000 points used for the interpolation from points. In spite of this reduction, you can see visible square pattern on the top of the hill where we have the largest area without data points. We will discuss its elimination in Section 6.8.5.

6.8.3 Tuning the RST parameters: tension and smoothing

To take the full advantage of the v.surf.rst module, understanding the principles behind the method is important. The mathematical description is given in the Appendix A.1; here we provide only a verbal description with illustrations. The RST function minimizes a specific measure of surface smoothness (also called smoothness seminorm or roughness penalty) and simulates a flexible sheet forced to pass through the data points while minimizing its energy (Mitas and Mitasova, 1999; Wahba, 1990; Talmi and Gilat, 1977). Properties of this function can be controlled by the *tension* and *smoothing* parameters.

Tension parameter   Tension tunes the surface from a stiff plate to an elastic membrane (Figure 6.11, Mitasova and Mitas, 1993). For very high tension, the surface resembles a rubber sheet with cusps at the data points (Figure 6.11 c). For low tension, the surface behaves like a stiff (hard to bend) plate, creating a very smooth surface (Figure 6.11 b). Due to its stiffness, it can overshoot in the areas of sharp gradient change (especially if zero smoothing is used); if the interpolated values exceed the range of given values over 15%, the program gives a warning and increase in tension or smoothing is suggested.

The role of the tension parameter can be also interpreted as a control of the range over which the given point influences the resulting surface. For high tension, each point influences only its close neighborhood and the surface goes rapidly to trend between the points. This may create cusps around the data points (Figure 6.11 c) or steps along the contours (Figure 6.14). With very low tension, each point has a long range of influence, so it is suitable for interpolation of areas with relatively flat terrain with data points spaced far apart. On the other hand, it may cause visible segments in large data sets (see Section 6.8.5 for a solution). The default input parameters try to adjust the tension to a suitable value based on the analysis of the data point density; however, to fully optimize the tension a more complex procedure based on cross-validation may be used (see Section 6.8.4).

To explore the impact of tension (Figure 6.11), you can interpolate elevation surfaces from the random point elevation data with different tension parameters. We also compute an aspect map so that we can visually check the surface geometry:

```
g.region rural_1m res=2 -p
v.surf.rst elev_lid792_randpts elev=elev_rstt10_2m \
    asp=asp_rstt10_2m zcol=value ten=10 seg=30 npmin=140
v.surf.rst elev_lid792_randpts elev=elev_rstt160_2m \
    asp=asp_rstt160_2m zcol=value ten=160 seg=30 npmin=140

# check the interpolated elevation surface using aspect maps
r.colors asp_rstt10_2m col=aspect
r.colors asp_rstt160_2m col=aspect
```

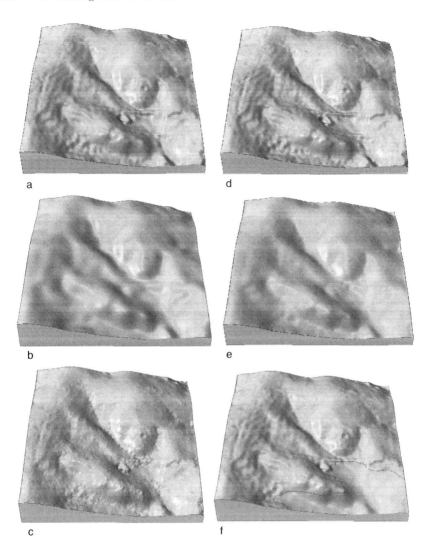

a   d

b   e

c   f

Fig. 6.11. Tuning the geometry of interpolated surface by tension and smoothing parameters: a) default values of tension=40, smoothing=0.1; b) tension=10, smoothing=0.1: smooth surface with low level of detail useful for modeling major trends; c) tension=160, smoothing=0.1: surface with cusps in data points but smooth in between; d) tension=40 and smoothing=0.0: surface passes exactly through the data points (result close to the one with default parameters, but overshoots are possible); e) tension=40 and smoothing=10.0: surface is very smooth and does not pass exactly through each data point; f) tension=40 and smoothing is 0.1 for $z > 113$ and 10.0 for $z < 113$

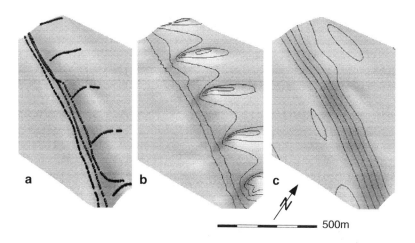

Fig. 6.12. RST interpolation of a beach surface surveyed by Real Time Kinematic GPS: a) given data points; b) default parameters in v.surf.rst; c) anisotropic tension with angle **theta**=160 degrees and scaling **scalex**=0.25

```
d.erase
d.rast asp_rstt10_2m
d.rast asp_rstt160_2m
d.vect elev_lid792_randpts siz=1 col=red
```

You can compare these results to the surface computed with the default tension=40 and smoothing=0.1 in our first example in the Figure 6.11 a.

Because the tension parameter is scale dependent, it can have different values in different directions, supporting modeling of anisotropic surfaces and volumes (Figure 6.12, Hofierka et al., 2002). Two additional parameters, angle and scale, were added to v.surf.rst in GRASS 6 for computation of surfaces with uniform anisotropy, where the new parameters **theta** and **scalex** represent the direction and ratio (scaling) of the anisotropic features.

Smoothing parameter The functionality of smoothing can be illustrated using springs attached to the "pins" representing data points. The higher the smoothing the "softer" the springs and the more is the surface allowed to deviate from the data point in its effort to minimize its energy, that means we perform approximation rather than interpolation. GRASS implementation supports spatially variable smoothing parameter – each point can have different softness of its spring. The interpolation function will pass exactly through the data points that have smoothing set to zero. Uniform smoothing is given as a constant parameter **smooth**, while variable smoothing is given as a floating point attribute in the vector point map. Smoothing is important when

using low tension to prevent overshoots, as well as for removing noise which
may be present in data.

To explore the impact of smoothing, we can interpolate three surfaces from
our random vector points using the default tension ten=40 and the smoothing
set to a) smo=0, b) smo=10, and c) to spatially variable values (Figure 6.11.
We then compare the root mean square deviation rmsdevi for each surface,
computed from the differences between the given point elevations and the
elevations in the resulting surface at the same point. The value of rmsdevi
can be retrieved from the history file using r.info (see also Subsection 6.8.4):

```
v.surf.rst elev_lid792_randpts elev=elev_rstsm0_2m zcol=value \
                              asp=asp_rstsm0_2m smo=0
v.surf.rst elev_lid792_randpts elev=elev_rstsm10_2m zcol=value \
                              asp=asp_rstsm10_2m smo=10

# compare the standard deviation printed as rmsdevi
# smoothing is zero (interpolation)
r.info elev_rstsm0_2m
 [...] rmsdevi=0.000000

# default smoothing 0.1 (close to interpolation)
r.info elev_rstdef_2m
 [...] rmsdevi=0.047771

# smoothing 10 (approximation)
r.info elev_rstsm10_2m
 [...] rmsdevi=0.956931

r.colors asp_rstsm0_2m col=aspect
d.rast asp_rstsm0_2m
r.colors asp_rstsm10_2m col=aspect
d.rast asp_rstsm10_2m
```

The standard deviation increases with the increasing smoothing parameter
and the effect of smoothing is clearly reflected also on the aspect maps. We can
use the SQL support to add variable smoothing to our point data, for example,
we can use low smoothing 0.1 for elevations > 113 and higher smoothing 10.0
for elevations < 113:

```
# add variable smoothing parameter to input data
g.copy vect=elev_lid792_randpts,myelev_randpts
v.db.addcol myelev_randpts col="smooth double precision"
v.info -c myelev_randpts
v.db.update myelev_randpts col=smooth val=10. wher="value<113."
v.db.update myelev_randpts col=smooth val=0.1 wher="value>113."
# verify
v.db.select myelev_randpts
# compute the elevation surface and aspect
v.surf.rst myelev_randpts elev=ele_rstsmvar_2m zcol=value \
          scol=smooth asp=asp_rstsmvar_2m
```

```
# check the result
r.colors asp_rstsmvar_2m col=aspect
d.erase
d.rast asp_rstsmvar_2m
d.vect elev_lid792_cont1m where="level = 113" col=red
```

The resulting DEM has a rougher surface in areas with elevation greater than 113m, while the surface with lower elevation is much smoother (Figure 6.11 f).

The current implementation of the RST method links the smoothing and tension parameters in such a way that the smoothing automatically increases when tension is lowered (see more details in Mitasova et al., 2005a). This approach reduces the potential for overshoots when low tension is used. Some examples of cases when we need to use low tension are in the Section 6.8.6 about topographic analysis with RST. We will later discuss how we can use the tension and smoothing parameters to control the level of detail in the resulting surface and minimize artificial features such as waves along contours or overshoots. The tension and smoothing parameters can be selected empirically, based on the knowledge of the modeled phenomenon, or automatically by minimization of the predictive error estimated by a cross-validation procedure (Mitasova et al., 1995; Hofierka et al., 2002) described in the following section.

6.8.4 Estimating RST accuracy

Several measures can be used to estimate accuracy of spatial interpolation. The module v.surf.rst computes deviations of the resulting surface from the given vector points that can be output to a vector point map devi for further analysis. For example, you can compare the deviations of the surfaces generated by the RST default settings (smoothing 0.1) and with a smoothing of 10 by adding the output of the deviation map to interpolation and then computing the summary statistics as follows:

```
g.region rural_1m res=2 -p

# compute elevation raster and deviations vector point map
v.surf.rst elev_lid792_randpts el=elev_rstdef_2mb zcol=value \
        devi=elev_rstdef_devi
v.surf.rst elev_lid792_randpts el=elev_rstsm10_2mb zcol=value \
        smo=10 devi=elev_rstsm10_devi
```

We can now compute the statistics for deviations for the two values of smoothing:

```
# smoothing 0.1
v.info -c elev_rstdef_devi
```

```
v.univar elev_rstdef_devi col=flt1 type=point
 [...]
 Number of values: 5874
 Minimum: -0.079855
 Maximum: 0.078443
 Range: 0.158298
 Mean: 2.95603e-05
 Arithmetic mean of absolute values: 0.0072427
 Variance: 0.000115238
 Standard deviation: 0.0107349
 Coefficient of variation: 363.153

r.info elev_rstdef_2mb
 [...]
 rmsdevi=0.047771

# smoothing 10 (approximation)
v.info -c elev_rstsm10_devi
v.univar elev_rstsm10_devi col=flt1 type=point
 [...]
 Number of values: 5874
 Minimum: -0.770713
 Maximum: 1.646673
 Range: 2.41739
 Mean: 0.00276214
 Arithmetic mean of absolute values: 0.128166
 Variance: 0.032859
 Standard deviation: 0.18127
 Coefficient of variation: 65.6269

r.info elev_rstsm10_2m
 [...]
 rmsdevi=0.956931

# compute and display deviations maps using same color table
v.surf.rst elev_rstdef_devi elev=elev_rstdef_devi zcol=flt1
v.surf.rst elev_rstsm10_devi elev=elev_rstsm10_devi zcol=flt1
r.colors -n elev_rstsm10_devi col=differences
r.colors elev_rstdef_devi rast=elev_rstsm10_devi
d.erase
d.rast.leg elev_rstdef_devi
d.rast.leg elev_rstsm10_devi
```

Note that the rmsdevi computed by v.surf.rst and the standard deviation output from v.univar are computed differently. The first is computed from differences between the given elevation data and the resulting surface, while the second is computed using the mean value of differences and their values in the given points. They are equal if the mean value of differences is zero (it should be close to zero if smoothing is unbiased). When you compare the range and the mean of absolute values of differences you can clearly see that the surface with the lower smoothing is closer to the data points than the surface with the high value of smoothing. The deviations maps show areas where smoothing has "eroded" elevation as red (mostly ridges and the elevated section of the road), and where it was filled-in as blue (mostly valleys and the bottom of the pond). We have used the flag -n to revert the default color scheme in the color table differences to get the negative values of deviations show in red. Your values of deviations may be slightly different from those published here because your input map, generated by a random procedure, will be slightly different, too. In addition to the root mean square deviation written into the *history file* of the computed raster map, you can also check the minimum and maximum of the given and interpolated values to measure the level of smoothing. Note that the interpolated minimum and maximum values can be higher or lower than those at the given points, especially if the interpolation is performed outside of the area covered by the input data set. The contents of the history file can be retrieved by r.info.

The predictive error of the RST interpolation for the given set of parameters can be estimated by a cross-validation procedure (Mitasova et al., 1995; Hofierka et al., 2002). The method is based on a procedure that removes one data point at a time, performs interpolation for the location of the removed point using the remaining samples, and calculates the residual between the actual value of the removed data point and the approximated value for this point obtained from the remaining samples. This procedure is repeated until every sample has been, in turn, removed. The overall performance of the interpolator is then evaluated as the root-mean of squared residuals. Low root-mean-squared error (RMSE) indicates an interpolator that is likely to give more reliable estimates in the areas between the data points. The cross-validation can also be used to find optimal interpolation parameters by minimizing the RMSE (Mitasova et al., 1995; Hofierka et al., 2002) or by identifying areas that need more samples. To perform cross-validation, run the module v.surf.rst with the flag -c (we have set npmin and segmax to lower values, to speed-up the computation, see Section 6.8.5):

```
# perform cross-validation (no raster output)
v.surf.rst -c elev_lid792_randpts zcol=value \
           cvdev=elev_rstdef_cv npmin=120 segmax=25
v.univar elev_rstdef_cv col=flt1 type=point
 [...]
 Number of values: 6000
 Minimum: -0.682364
 Maximum: 1.561201
 Range: 2.24357
 Mean: 0.000827084
 Arithmetic mean of absolute values: 0.0368782
 Variance: 0.00506539
 Standard deviation: 0.0711715
```

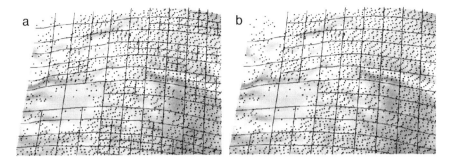

Fig. 6.13. Segmented processing using quadtrees: segments adjust to the number of points given by parameter segmax: a) segmax=25, b) segmax=40

```
# compute raster map of predictive errors
v.surf.rst elev_rstdef_cv elev=elev_rstdef_cv zcol=flt1
r.colors -n elev_rstdef_cv col=differences
d.rast.leg elev_rstdef_cv
d.vect elev_rstdef_cv siz=2
```

The mean of absolute values of the cross-validation error as well as the standard deviation are much smaller than the published accuracy of the lidar data (0.15m) and the mean value is close to zero, so there is no bias. However, there are few locations where the cross-validation error is high, indicating that at these locations additional points would be useful. The highest values of predictive error are in location with a single point surrounded by unsampled area and with rapidly changing surface – the pond is a good example. On the book Web site you can find a script to easily compute the cross-validation error for a range of parameters that can be used to find the optimal values.[8]

6.8.5 Segmented processing

Digitized contours or lidar elevation data sets can have over million data points with the resulting DEMs with thousands of rows and columns. To support processing of such large data sets, v.surf.rst and v.vol.rst were implemented with a segmented processing procedure (Mitasova et al., 2005a). The segmented processing is based on the fact that splines have local behavior, i.e., impact of data points which are far from a given location diminishes rapidly with increasing distance (Powell, 1992). The segmentation uses a decomposition of the studied region into rectangular segments with variable size dependent on the density of data points (Figure 6.13), using *quadtrees* for 2D

---

[8] Script to calculate cross-validation error, http://www.grassbook.org/, Code examples of 3rd Edition

and *octtrees* for 3D interpolation (Mitasova et al., 1995). For a given segment, the interpolation is carried out using the data points within this segment and from its neighborhood, selected automatically depending on their spatial distribution (Mitasova et al., 2005a). Because the tension inversely controls the range of influence of data points, this approach requires large neighborhoods to achieve smooth connection of segments for very low tension. The number of points in the segment is controlled by segmax, the number of points used for interpolation (within the segment and its neighborhood) is controlled by npmin. Default values for these parameters usually work, in case that the segments are visible, npmin should be increased. Note that while v.surf.idw uses 12 given points for computation of a single grid value (that means each grid point is computed using an independent function), the approach in v.surf.rst uses at least npmin points (the default is npmin=300) for computation of tens or hundreds of grid points within a given segment (that means, all points within a segment are computed using the same function). If the points are dense and homogeneously distributed (for example as is often the case with lidar data), both segmax and npmin can be set to lower values, leading to substantially faster computation. You can compare the computational time using the following examples (remember that you can run your slow computation in background by putting & at the end of the command) :

```
g.region rural_1m res=2 -p

# set layer=0 because elevation is stored as z-coord.
v.surf.rst elev_lid792_bepts elev=elev_rst_fast layer=0 \
      asp=asp_rst_fast tree=segments_n120 npmin=120 segm=25
v.surf.rst elev_lid792_bepts elev=elev_rst_slow layer=0 \
      asp=asp_rst_slow tree=segments_n300
d.erase
d.rast asp_rst_fast
d.vect segments_n120
d.vect elev_lid792_bepts siz=1
d.rast asp_rst_slow
d.vect segments_n300

# check the differences between the resulting raster maps
r.mapcalc "diffelevseg=elev_rst_slow-elev_rst_fast"
r.univar diffelevseg
 [...]
 minimum: -0.49794
 maximum: 0.27153
 range: 0.76947
 mean: -0.000387165
 mean of absolute values: 0.0102965
 [...]
```

You can see that the computation is much faster when smaller segments are used, while the mean absolute value of differences between the two DEMs

is only 1cm, over one magnitude less than the vertical accuracy of this set
of lidar data (15cm in open areas, 60cm in areas with vegetation); however,
there are locations where the difference is relatively high (see the min and max
difference). To get practically the same result, you need to adjust the tension
according to the ratio of the normalization factors that are derived from the
average size of segments Mitasova et al. (2005a) and that are used to rescale
the coordinates to ensure numerical stability. Use r.info to find the value of
the normalization factor dnorm and decrease the tension for the interpolation
with smaller segment by multiplying it by the ratio $dnorm_2/dnorm_1$, in our
case, $t_{adjusted} = 40. * (52.1498/82.4561) = 25.298$:

```
r.info elev_rst_slow
 [...] dnorm=82.456082 [...]
r.info elev_rst_fast
 [...] dnorm=52.149805 [...]
v.surf.rst elev_lid792_bepts elev=elev_rstt25_fast \
     asp=asp_rstt25_fast layer=0 treefile=segments_n120 \
     npmin=120 segmax=25 ten=25.298

# check the differences between the resulting raster maps
r.mapcalc "diffelevsega=elev_rst_slow-elev_rstt25_fast"
r.univar diffelevsega
 [...]
 minimum: -0.0412674
 maximum: 0.0512772
 range: 0.0925446
 mean: 2.62179e-05
 mean of absolute values: 0.000205665
 [...]
```

To learn more about segmentation, see also Mitasova et al. (2005a).

The high density of points typical for the modern mapping technologies,
leads to substantial oversampling, therefore, we can further speed up the computation by reducing the number of data points to the minimum necessary
for a given level of detail (size of the features we want to preserve), accuracy
and resolution. The density of points used for interpolation in v.surf.rst
is controlled by the parameter dmin representing the minimum distance allowed between the data points – points that are closer to each other than
this distance are considered identical and not included into interpolation. We
have already used this parameter to reduce the number of points on contours,
you can explore its effect by interpolating, for example a 5m resolution DEM
from the vector map elev_lid792_bepts (set the region to this map to get a
big enough DEM and use dmin=5 to reduce the number of points, along with
segmax=25 and npmin=120).

Interpolating from contours   We have already explained that it is useful
to reduce the number of points on the contour lines to improve the distribution

of points by increasing the dmin parameter. You have also seen that in a larger area with sparse contours (such as the flat top of the hill in our test area), rectangular segments may become visible, especially in the aspect map. While the error in the elevations due to the segments is usually negligible, it is not acceptable for shaded maps (see Section 7.1.2). The problem can be eliminated using two step interpolation. First interpolate the surface using v.surf.rst. If segments are visible, generate additional points sparsely but homogeneously distributed over the elevation surface using r.random. Transform the contours to vector points using v.to.points and merge with the vector points map generated by r.random using the module v.patch. Finally, interpolate this merged vector points map using v.surf.rst. The surface should be without segments. We have already interpolated the first surface elev_rstcontd_2m from contours in the Section 6.8.3. Here we add the random points and re-interpolate the surface to remove the visible segments:

```
g.region rural_1m res=2 -p
# generate random points and patch them with contours
# converted to points
r.random elev_rstcontd_2m n=300 vector_out=elev_add300_pts
v.to.points elev_lid792_cont1m out=elev_lid792_contpts
v.info -c elev_add300_pts
v.info -c elev_lid792_contpts

# we need the same column name for patching
v.db.renamecol elev_add300_pts col=value,level
v.patch -e elev_lid792_contpts,elev_add300_pts out=el_contpts
d.erase
d.rast asp_rstcontd_2m
d.vect el_contpts
v.db.select el_contpts

# reinterpolate new DEM
v.surf.rst el_contpts elev=elev_rstcontptsd_2m zcol=level \
           asp=asp_rstcontptsd_2m dmin=3
d.rast asp_rstcontptsd_2m
```

The resulting map does not have any visible segments.

6.8.6 Topographic analysis with RST

We have already explained computation of topographic parameters from raster DEMs using r.slope.aspect (see Section 5.4) which computes the topo-graphic parameters at a grid point using the elevation at this point and its 3×3 neighborhood. This approach works well for smooth surfaces where local polynomial approximation is adequate. However, for high resolution data, the small neighborhood may not be sufficient to adequately capture the geometry of topographic features and the resulting map may represent the geometry

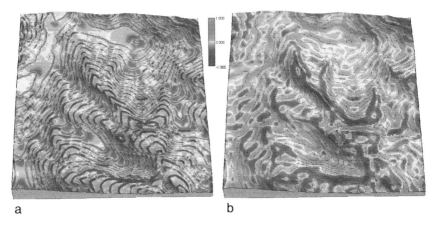

a   b

Fig. 6.14. Profile curvature draped over 1m resolution DEM computed from contours: a) with tension too high pattern follows contours, b) lower tension leads to pattern reflecting main terrain features

of noise or sampling pattern instead. The module v.surf.rst allows us to compute the topographic parameters simultaneously with the computation of elevation surface directly from data points using the partial derivatives of the RST function and principles of differential geometry (Mitasova and Hofierka, 1993; Mitasova et al., 2005a). The explicit form of the RST derivatives can be found in Appendix A.1. Entire segment that includes tens or hundreds of grid cells is used to compute the RST function and its derivates, allowing us to capture geometry of much larger features than the standard, raster-based methods. We can also use the smoothing parameter to reduce noise that often makes extraction of surface geometry difficult and the tension parameter to tune the level of detail of the topographic features. The module computes slope, aspect, profile and tangential curvatures (see Appendix A.1 for equations), or first and second order partial derivatives that can be used to compute additional parameters, such as slope or curvature in a given direction.

Topographic parameters are used for wide range of analyses and as inputs for modeling and planning. Here we show how to use these parameters for visual inspection of surface geometry and identification of artifacts, that can negatively influence results of modeling or lead to misleading results of analysis. We have already used simultaneous computation of aspect to evaluate the surfaces; in the following example, we use the contour data as input and compute the slope and profile curvature along with the elevation surface:

```
g.region rural_1m res=2 -p
v.surf.rst elev_lid792_cont1m zcol=level dmin=3 npmin=250 \
  el=elev_rstcontd_2mt slo=slp_rstcontd_2m pc=pc_rstcontd_2m
d.erase
d.rast.leg slp_rstcontd_2m
```

```
d.rast.leg pc_rstcontd_2m
d.vect elev_lid792_cont1m
```

When we run v.surf.rst with the default parameters the resulting surface
has waves along contours that are reflected by changing convex and concave
shape along each contour except for the most steep areas (Figure 6.14). This
means that the tension is too high and the curvatures follow the pattern of
contours that can lead, for example, to a false pattern of erosion and deposition, see Mitasova et al. (1996). The geometry of the resulting surface thus
reflects the geometry implied by the distribution of data points rather than
real terrain (we will illustrate a similar effect for lidar data in the next section). To minimize this bias we can lower the tension from the default value
of 40 to 10:

```
g.region rural_1m res=2 -p
v.surf.rst elev_lid792_cont1m zcol=level el=elev_rstct10_2m \
    slo=slp_rstct10_2m asp=asp_rstct10_2m pcur=pc_rstct10_2m \
    tcu=tc_rstct10_2m ten=10 dmin=3 npmin=250
d.rast.leg pc_rstct10_2m
d.vect elev_lid792_cont1m
d.rast.leg tc_rstct10_2m
```

We will lose some detail, but we will also reduce the artificial patterns, as you
can see by overlaying the profile curvature with contours (Figure 6.14). You
can further explore the results by displaying the DEM using the nviz module
and draping the computed topographic parameters and contours over it (see
Section 7.3).

Landscape processes have multiscale character and different processes are
dominant at different scales, so it is sometimes useful to extract topographic
features and patterns at different levels of detail. In v.surf.rst, this can
be achieved by changing tension and smoothing (lower tension and higher
smoothing produces smoother topography with curvatures representing main
features, higher tension captures more detail and smaller features).

6.9 Working with lidar point cloud data

Point cloud data, as a new type of representation of 3D surfaces, are usually
produced by airborne or on-ground laser scanning, also known as Light Detection and Ranging (lidar, http://www.lidarmap.org/lidar/). The data
are often provided as sets of very dense (x, y, z) points or in a more complex,
public file binary format called LAS (see http://www.lasformat.org/) that
may include multiple returns as well as intensities. In this section, we will
explain how to perform basic lidar data processing and analysis, first creating
elevation surfaces with given properties from pre-processed bare ground data,
followed by more complex feature extraction.

Lidar point data create almost continuous coverage of the mapped surface (Figure 6.15) and lower resolution surfaces can be rapidly created by on-fly import and conversion from points to raster using the r.in.xyz module (a method often called binning) as we have shown in the Section 4.1.3. This tool also provides useful capabilities for analysis of point cloud data properties by computing the number of points per grid cell (useful for selecting an appropriate resolution), or range of values, standard deviation, coefficient of variation (helpful when assessing the need for smoothing) and other statistics. You can eliminate some of the most extreme outliers by using the filter method **range** for the z-values and preprocess the data for interpolation by computing the mean elevation for high resolution grid cells that are then used as input points for interpolation (see the example in the module's manual page). We will illustrate the application of this module using the preprocessed bare ground data available for our area and provided in the **ncexternal/** directory in a file BE3720079200WC20020829m.txt:

```
g.region rural_1m res=2 -p

# compute a raster map representing number of points per cell
r.in.xyz BE3720079200WC20020829m.txt out=lid_792_binn2m meth=n
23162 points found in region
d.erase
d.rast.leg lid_792_binn2m
r.report lid_792_binn2m unit=p
 0 .... 82.36
 1 .... 17.64
 2 ....  0.01
```

These are older (2001) lidar data acquired from relatively high altitude, so their density is a little bit lower than what is currently common. At 2m resolution, 82% of cells have no points and only 17% have a single point, indicating that our data points are farther apart than 2m. We need to reduce the resolution to 6m to get at least one point per cell for most of our study area:

```
g.region rural_1m res=6 -ap
r.in.xyz BE3720079200WC20020829m.txt out=lid_792_binn6m meth=n
d.rast.leg lid_792_binn6m
r.report lid_792_binn6m unit=p
 0 ....  8.05
 1 .... 35.79
 2 .... 46.59
 3 ....  8.89
 4 ....  0.68
```

```
# compute raster maps representing mean elevation for each cell
r.in.xyz BE3720079200WC20020829m.txt out=lid_792_binmean6m  \
         meth=mean
d.rast.leg lid_792_binmean6m
```

```
# compute range and variation
r.in.xyz BE3720079200WC20020829m.txt out=lid_792_binrange6m \
        meth=range
d.rast.leg lid_792_binrange6m
r.in.xyz BE3720079200WC20020829m.txt out=lid_792_binvar6m    \
        meth=coeff_var
d.rast.leg lid_792_binvar6m
d.vect streets_wake
d.vect lakes type=boundary
d.vect streams
```

At 6m resolution most cells have at least one or two points but there are
still many gaps, so interpolation is needed to create DEM. Range map
lid_792_binrange6m shows that at 6m resolution we may lose important
smaller features as several 6m cells have points with elevation differences more
than 1m. Coefficient of variation map overlayed by the road, stream and lake
maps shows that the highest variation is in locations with sharp change in
elevation such as along the edges of the pond and elevated road with culvert
under it. To compute the full coverage, high resolution (1m) DEM you can either import the points through r.in.xyz, convert from raster to vector points
using r.to.vect and interpolate by v.surf.rst as suggested in the manual
page for the r.in.xyz module or you can import the data as vector points using v.in.ascii and interpolate them directly. The data can be imported with
elevation stored as attribute or as a third z-coordinate (see the manual for
more details). If the data set including its topology does not fit into memory
(roughly if you have half million points and 1GB RAM memory available) use
the flags -tbzr to skip building topology and creating a table, store elevation
as z-coordinate and import only data in the current region, if desired.

We will use the bare ground lidar data from the above example to illustrate
the import and interpolation of high resolution DEM. At the same time, we
will also work with the multiple return data set provided in the vector point
map elev_lidrural_mrpts. First we import and display the data and check
their elevation using conversion to raster by v.to.rast:

```
# find how many points (number of lines) with system tool
wc -l BE3720079200WC20020829m.txt
 319214 BE3720079200WC20020829m.txt

# we import only points in the rural area  without building
# topology and using z-coordinate for elevation
g.region rural_1m -p
v.in.ascii -ztbr BE3720079200WC20020829m.txt \
          out=elev_lidrural_bepts z=3
 [...]
 Skipping 296052 of 319214 rows falling outside of current reg.

# display bare ground and multiple return points over orthophoto
d.rast ortho_2001_t792_1m
```

```
d.vect elev_lidrural_bepts siz=2 col=red
d.vect elev_lidrural_mrpts siz=1 col=green

# build topology so that we can use all vector tools
# we can do it because our number of points is small
v.build elev_lidrural_bepts
v.info elev_lidrural_pts

# check the imported data by creating low resolution DEM
g.region res=6 -ap
v.to.rast elev_lidrural_bepts out=elev_lidrural_6m use=z
v.to.rast elev_lidrural_mrpts out=elev_lidrural_mr6m use=z
d.rast elev_lidrural_6m
d.rast elev_lidrural_mr6m
```

You can see that the bare ground data are missing points in the areas with buildings, forest and the pond; multiple return data have more complete coverage and include top of buildings and trees, but they too do not capture water as there is no return on water surfaces (Figure 6.15.

Computing a high resolution DEM and topographic parameters   To compute a high resolution DEM, we set the resolution to 1m and use approximation with RST (Figure 6.11) – don't forget to set lower **npmin** and `segmax` for faster computation:

```
g.region rural_1m  -p
# send command to background
v.surf.rst elev_lidrural_bepts elev=elev_lidrural_1m \
  asp=asp_lidrural_1m pcu=pc_lidrural_1m tcu=tc_lidrural_1m \
  npmin=120 segmax=25 layer=0
v.surf.rst elev_lidrural_mrpts elev=elev_lidruralmr_1m \
  asp=asp_lidruralmr_1m pcu=pc_lidruralmr_1m \
  tcu=tc_lidruralmr_1m npmin=120 segmax=25 layer=0
 Percent complete: WARNING:
 Overshoot -- increase in tension suggested.
 Overshoot occures at (8,631) cell
 The z-value is 158.649880,zmin is 103.512798,zmax is 153.0655

# when the jobs have finished
d.erase
d.rast elev_lidrural_1m
d.rast pc_lidrural_1m
d.rast elev_lidruralmr_1m
d.rast pc_lidruralmr_1m
d.vect elev_lidrural_mrpts siz=1
```

The resulting surface is very rough and there is a strong influence of the scanning pattern visible in the tangential curvature map. The multiple return data

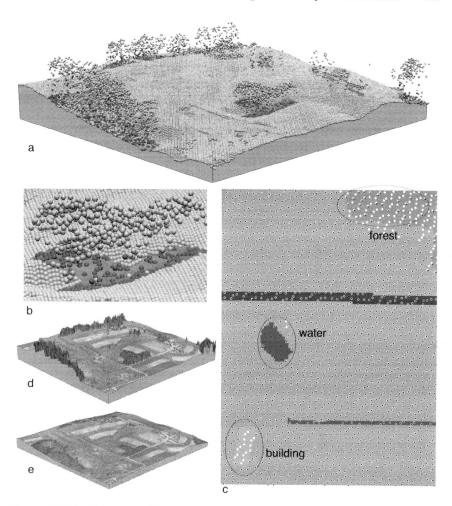

Fig. 6.15. Multiple return lidar point cloud: a) displayed in 3D by `nviz` (first return lighter color, second return darker color); b) 3D points: zoomed into the forested area; c) lidar returns and land use: no return from water, first return above ground from buildings, multiple returns in forest, and first return on ground in open areas; d) DSM from first return points; e) DEM from bare ground points; orthophoto is draped over the surface for reference

interpolation gives warning about overshoots that are due to sharp change in the surface gradient along buildings or trees (Figure 6.16). The surface is best viewed using the 3D visualization tool `nviz` described in the next chapter and used in the Figure 6.15, here we just use elevation and curvature to evaluate the surface geometry. The next example shows how to smooth the surfaces

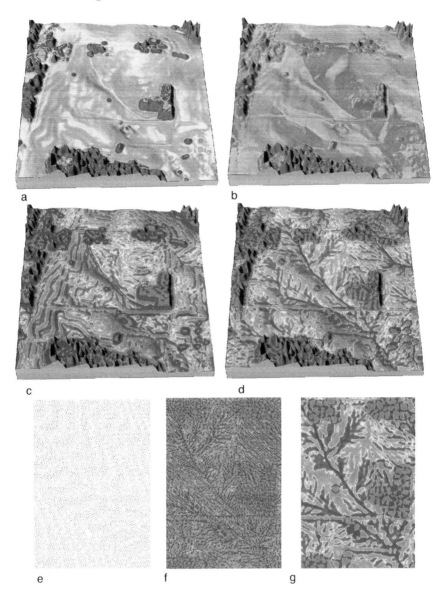

Fig. 6.16. Topographic analysis from first return lidar data using RST: a) slope, b) aspect, c) profile curvature, and d) tangential curvature

and extract topographic parameters at a desired level of detail, by changing the tension and smoothing parameters:

```
g.region rural_1m  -p
v.surf.rst elev_lidrural_bepts elev=elev_lidruralt15_1m \
  asp=asp_lidruralt15_1m pcu=pc_lidruralt15_1m \
  tcu=tc_lidruralt15_1m npmi=120 seg=25 lay=0 ten=15 smo=1.
v.surf.rst elev_lidrural_mrpts elev=elev_lidruralmrt15_1m \
  asp=asp_lidruralmrt15_1m pcu=pc_lidruralmrt15_1m \
  tcu=tc_lidruralmrt15_1m npmi=120 seg=25 lay=0 ten=15 smo=1.

# when the jobs have finished
d.erase
d.rast elev_lidruralt15_1m
d.rast pc_lidruralt15_1m
d.rast tc_lidruralt15_1m
d.rast elev_lidruralmrt15_1m
d.rast pc_lidruralmrt15_1m
d.vect elev_lidrural_mrpts siz=1
```

When you compare the profile and tangential curvatures with our previous example, you can see that only the main features of terrain geometry are captured with lower tension and the pattern is very different from the pattern of data sampling (Figure 6.16). You can use the curvatures to approximately outline the forested areas and structures, because these areas have substantially higher curvatures than open terrain surface:

```
d.rast ortho_2001_t792_1m
d.rast -o pc_lidruralmrt15_1m val=-2.--0.10
d.rast -o pc_lidruralmrt15_1m val=0.020-1.0

# for comparison display result with higher tension
d.erase
d.rast -o pc_lidruralmr_1m val=-2.--0.08
d.rast -o pc_lidruralmr_1m val=0.1-1.0
```

The curvatures obtained with lower tension lead to more complete boundary definitions but would require further thinning. The higher tension result provides sharper boundaries but with more gaps. This lidar data set does not provide sufficiently dense data for accurate extraction of buildings and further processing would be needed to improve the building geometry.

Using multiple return lidar data   In the previous examples, we have computed bare ground DEM (no buildings or vegetation) using preprocessed point data and a digital surface model (DSM) from multiple return data that includes also buildings and vegetation and is useful, for example, for visibility analysis (Figure 6.15. We can also use the multiple return lidar data to separate buildings and vegetation from the DEM using a new module set (Brovelli et al., 2004). These lidar tools follow the following concept for multiple return data:

- if more than one return is found at a given location, then there are first and last return points;
- if only one return is found at a given location, then this point is considered as last point.

Due to these assumptions, last points are more likely to be ground points. Our data set includes each return (pulse) as a separate 3D point and ground points where only single return is found are assigned first return attribute – some preprocessing is therefore needed to fulfill the assumptions used in the lidar processing modules. We use the results of per cell statistical analysis performed by r.in.xyz to extract the first return points that have only single return and merge them with the last return points to create a preliminary bare ground point data set that can be used with the following procedure.

The general workflow is as follows: Outlier detection is done with v.outlier on both the first and last return data. Then, with v.lidar.edgedetection, edges are detected from last return data. The buildings are generated by v.lidar.growing from detected edges. The resulting data are post-processed with v.lidar.correction. Finally, the DEM and DSM are generated with v.surf.bspline. We apply this procedure to our data:

```
# find out where we have multiple returns
g.region rural_1m -p
d.erase
d.rast ortho_2001_t792_1m
d.vect elev_lidrural_mrpts where="return=1" col=red siz=2
d.vect elev_lidrural_mrpts where="return=2" col=green siz=3
d.vect elev_lidrural_mrpts where="return=3" col=blue
d.vect elev_lidrural_mrpts where="return=4" col=yellow
```

For each emitted laser pulse, lidar that was used to acquire our data set returned up to four range values for location and elevation data. You can see that the first return data have practically continuous coverage while the additional returns are available mostly in forested areas where the laser beam penetrated the canopy (Figure 6.15. For the calculation, we use only the 1st and 2nd return:

```
v.extract elev_lidrural_mrpts out=elev_lidfirst_pts \
          where="return=1"
v.extract elev_lidrural_mrpts out=elev_lidlast_pts  \
          where="return=2"

# outlier detection and separation into two maps
# 1st return
v.outlier elev_lidfirst_pts output=elev_lidfirst_clean \
          outlier=elev_lidfirst_outl
d.erase
d.vect elev_lidfirst_clean siz=2
```

```
d.vect elev_lidfirst_outl col=red

# 2nd return
v.outlier elev_lidlast_pts output=elev_lidlast_clean \
         outlier=elev_lidlast_outl
d.erase
d.vect elev_lidlast_clean siz=2
d.vect elev_lidlast_outl col=red
```

Then we run an edge detection on cleaned last return:

```
v.lidar.edgedetection elev_lidlast_clean \
                    out=elev_lidlast_edges

# buildings/vegetation are generated from detected edges
v.lidar.growing elev_lidlast_edges out=elev_lidlast_grow \
               first=elev_lidfirst_clean
d.vect elev_lidlast_grow col=green
```

The resulting data are post-processed:

```
v.lidar.correction elev_lidlast_grow out=elev_lidlast_corr1 \
                  terrain=elev_lidlast_terr1
v.lidar.correction elev_lidlast_corr1 out=elev_lid_dsm \
                  terrain=elev_lid_dtm

# DEM and DSM are generated
# Estimation of lambda_i parameter with cross validation
v.surf.bspline -c elev_lid_dsm sie=100 sin=100
v.surf.bspline -c elev_lid_dtm sie=100 sin=100
```

From the cross-validation, we select **lambda** with minimal RMS error:

```
# generate raster surfaces at 1m resolution
v.surf.bspline elev_lid_dsm raster=lidar_dsm lambda=0.1
v.surf.bspline elev_lid_dtm raster=lidar_dtm lambda=0.01

d.rast lidar_dsm
d.rast lidar_dtm
nviz elev_lid_dsm,elev_lid_dtm \
    col=ortho_2001_t792_1m,ortho_2001_t792_1m
# with the position slider you can visually separate DSM and DEM
```

6.10 Volume based interpolation

Physical and chemical properties of air, water or earth mass can be measured at 3D points capturing the distribution of the monitored variable in

3D space. To transform such data into a continuous volume represented by voxels, trivariate interpolation or approximation is needed. In this section, we explain multivariate and volume interpolation methods using raster voxels.

6.10.1 Adding third variable: precipitation with elevation

Multivariate interpolation is a valuable tool for incorporating the influence of an additional variable. For example, to interpolate precipitation with the influence of topography, the trivariate version of RST v.vol.rst can be used. The approach is similar to the one proposed by Hutchinson and Bischof (1983), and it is described in more detail by Mitasova et al. (1995), and Hofierka et al. (2002). The approach requires 3D precipitation vector points (x, y, z, p) and a raster DEM. The result is a precipitation raster map computed as an intersection of the precipitation volume model with the elevation surface.

As an example, we compute a 30 years mean annual precipitation raster map for North Carolina using data from over 137 meteorological stations and a statewide 500m resolution DEM. The input vector points precip_30ynormals_3d include 3D coordinates for each meteorological station and 30 year monthly and annual mean precipitation as floating point attributes. The output is a 2D raster map precip_anntopo90_500m representing spatial distribution of annual precipitation (Figure 6.17b). We set both the 2D and 3D region horizontal resolutions to the resolution of the input DEM (500m) and define a single depth layer between 0m and 2000m elevation (bottom and top) by setting the vertical resolution tbres to 2000m using g.region. To ensure that the interpolation is performed only for the NC state area we set a MASK using the provided raster file ncmask_500m. For comparison, we compute 2D precipitation raster maps without and with the influence of topography as follows (Figure 6.17 a):

```
# set the region and MASK
g.region rast=elev_state_500m -p
g.region t=2000 b=0 tbres=2000 res3=500 -p3
r.mask ncmask_500m


# compute precipitation raster map without elevation
v.info -c precip_30ynormals
v.surf.rst precip_30ynormals elev=precip_annual_500m \
        zcol=annual segmax=600
r.colors precip_annual_500m col=rules <<EOF
950 red
1000 orange
1200 yellow
1400 cyan
1600 aqua
1800 blue
2500 violet
EOF
```

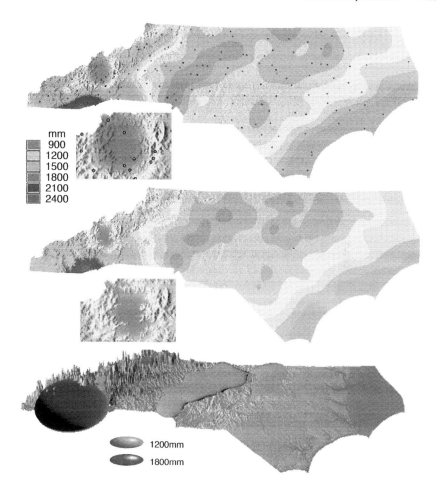

Fig. 6.17. Interpolation of precipitation with influence of topography: a) precipitation distribution using bivariate interpolation by v.surf.rst, points represent the climate stations; b) precipitation with influence of topography interpolated by v.vol.rst. c) elevation surface intersected by 1200mm and 1800mm precipitation isosurfaces

```
d.erase
d.rast precip_annual_500m
# compute precipitation raster map with elevation
v.info -c precip_30ynormals_3d
v.vol.rst precip_30ynormals_3d cellinp=elev_state_500m \
          cellout=precip_anntopo_500m \
          wcolumn=annual zmult=90 segmax=700
```

```
r.colors precip_anntopo_500m rast=precip_annual_500m
d.rast precip_anntopo_500m
r.shaded.relief elev_state_500m shaded=elev_state_shade
d.his h=precip_anntopo_500m i=elev_state_shade bright=60
```

We have skipped segmentation by setting the parameters segmax=600 for 2D and segmax=700 for 3D interpolation, because the number of input data points is only 137 and no segmentation is needed for such small data set. When you compare the results, you can see that the most significant difference in the spatial pattern of precipitation in the raster maps precip_annual_500m and precip_anntopo_500m is in the mountains (western NC) where the trivariate interpolation that takes into account topography produces much richer pattern. The impact of topography is controlled by the vertical scaling parameter zmult, as well as by the resolution and smoothing of the DEM as demonstrated by Hofierka et al. (2002). The module v.vol.rst internally computes a 3D (volume) precipitation function. When using the cellinp option, the precipitation values are extracted from the precipitation volume at the elevations given by the elev_state_500m raster map (see Figure 6.17c). This approach captures a more complex, spatially variable relation between precipitation and elevation than the traditional methods that are based on statistical correlation.

To illustrate the internal working of the method described above, we can compute the associated volume map by setting the vertical resolution to 200m (leading to 10 vertical levels). We will reduce the horizontal resolution to 1000m for speed-up the computation and run the module v.vol.rst with the parameter (with a slightly misleading name) elev set to the name of the resulting volume map precip_annvol_1000m:

```
# compute precipitation volume for illustration
# set region, compute the volume, convert to 2d
g.region nc_500m
g.region t=2000 b=0 tbres=200 res=1000 res3=1000 -p3
v.vol.rst precip_30ynormals_3d elev=precip_annvol_1000m \
        wcolumn=annual zmult=200 segmax=700
r3.to.rast precip_annvol_1000m out=precip_annvol_slice
# slice horizontaly through the volume
xganim view1="precip_annvol_slice*"

# see section about visualization how to work with volumes
nviz elev_state_500m volume=precip_annvol_1000m
```

You can view the result in nviz (see the next chapter on visualization for more details) or you can convert the volume to horizontal 2D raster slices and display them as animation using xganim.

6.10.2 Volume and volume-temporal interpolation

GRASS provides some limited experimental tools for working with volume data. We describe the functioning tools here to encourage further development. Some of the prototype applications are demonstrated at the Spatial interpolation Web site.[9] To illustrate the volume data processing tools, we use the Chesapeake Bay nitrogen concentration data that can be downloaded from the GRASS book Web site as LOCATION chesapeake (see also an example at the NCSU Multidimensional Spatial interpolation Web site). Start GRASS with this LOCATION and check the preset 3D region using g.region. You can then convert the provided 3D point data (x, y, z, w) to discrete voxel representation with v.to.rast3:

```
# set the vertical region
g.region b=-33 t=1 res=500 tbres=2 -p3
v.to.rast3 nitro3d out=nitro3d.vol
g.list rast3d
```

The flags -p3 also print the third dimension of the boundary coordinates. The last command allows you to list the volume raster files that are available. The continuous volume from 3D scattered point data can be created by interpolation to 3D raster using the IDW method implemented as v.vol.idw, or the RST method using v.vol.rst, with RST usually providing more accurate results. The mathematical description of the method, including the equations for computation of associated gradients and curvatures, is in the Appendix A.1. Trivariate RST has similar properties and parameters as the bivariate version, so the principles described in the previous sections apply here as well (see for example impact of tension in 2D and 3D in Mitas and Mitasova, 1999).

To illustrate the volume interpolation, you can create a volume model of spatial distribution of the Chesapeake Bay nitrogen concentrations. To limit the interpolation to the water body, we first create a mask for the Bay using the shoreline provided with the data set and v.to.rast. Then you can interpolate the masked volume using v.vol.rst:

```
v.in.ascii chesapeakeshore.asc out=shore
v.to.rast shore out=shore_mask use=val val=1
v.vol.rst nitro3d clev=nitro3d segmax=400 zmult=1000 \
                        maskmap=shore_mask
```

---

[9] Spatial interpolation:
  - Chesapeake Bay Nitrogen,
  http://skagit.meas.ncsu.edu/~helena/gmslab/viz/ches.html
  - Concentrations of chemicals,
  http://skagit.meas.ncsu.edu/~helena/gmslab/viz/vol1.html
  - Soil properties,
  http://skagit.meas.ncsu.edu/~helena/gmslab/gsoils/ccsoil2.html

Note that the vertical resolution of the resulting grid is much higher than the horizontal resolution. The parameter `zmult` is set so that the vertical distances between the data points are of the same magnitude as the horizontal distances to ensure the stability of interpolation. See the Section 7.3.4 for various possibilities to visualize the results.

Similarly to the bivariate version, trivariate RST can compute a number of parameters related to the gradients and curvatures of the volume model. An experimental version of RST interpolation for 4D data (volumes changing in time) `v.volt.rst` is also available for update and further development.

6.10.3 Geostatistics and splines

As we have described in the previous sections, GRASS provides fully integrated spline interpolation and a wide range of geostatistical tools, including kriging through the link with Open Source geostatistical software (see Chapter 10). Because the relation between splines and kriging is a frequently asked question, we provide here a brief explanation.

Several authors (e.g., Wahba, 1990; Cressie, 1993) have demonstrated that splines are formally equivalent to universal kriging with the choice of the covariance function determined by the smoothness seminorm (also called roughness penalty). Therefore, many of the geostatistical concepts can be exploited within the spline framework.

Kriging assumes that the spatial distribution of a geographic phenomenon can be modeled by a realization of a random function and uses statistical techniques to analyze the data (drift, covariance) and statistical criteria (unbiasedness and minimum variance) for predictions. However, subjective decisions are necessary (Journel, 1996) such as judgement about stationarity, and choice of a function for theoretical variogram (variogram model). Kriging is therefore successful for phenomena with a strong random component and/or for the problems where estimation of statistical characteristics (uncertainty) is the key.

Splines rely on a physical model with flexibility provided by change of elastic properties of the interpolation function. Often, physical phenomena result from processes which minimize energy, with a typical example of terrain with its balance between gravitation force, soil cohesion, and impact of climate. For these cases, splines proved to be rather successful. Moreover, splines provide enough flexibility for local geometry analysis, which is often used as input to various process-based models.

However, most of the surfaces or volumes are neither stochastic nor elastic media, but they are results of a host of natural (e.g., fluxes, diffusion) and/or socioeconomic processes. Therefore, each of the mentioned methods has a limited realm of applicability and, depending on the knowledge and experience of the user, proper choice of the method and its parameters can significantly affect the final results as illustrated by numerous examples throughout this book.

7

Graphical output and visualization

Visual analysis and communication based on graphical output is a core component of GIS. Graphical representation of georeferenced data and creation of cartographic models provides important means for understanding and communicating complex spatial relationships. GRASS includes a wide range of graphical tools from simple two-dimensional display to sophisticated visualization and animation.

7.1 Two-dimensional display and animation

The most common approach to viewing and visually exploring geospatial data is an interactive display of two-dimensional images using color and different area, line and point symbols. This approach is mostly based on traditional cartography, with current computer graphics tools offering greater flexibility in color, symbols, dynamics and interactivity that was not possible with the traditional maps.

7.1.1 Advanced map display in the GRASS monitor

We have already described and used the basic tools for viewing maps, such as d.rast and d.vect. You have also learned how to add a legend, scale, and text to your displayed map using d.legend, d.barscale, and d.text in Section 3.1.5. To give you more flexibility in creating graphical output from GRASS, we explain some additional tools and options.

Monitor size and frames   As with any other window, you can adjust the GRASS monitor size using the mouse. To change its default size, use the UNIX environment variables GRASS_HEIGHT and GRASS_WIDTH either in /etc/profile or locally in $HOME/.grass.bashrc or $HOME/.grass.cshrc.

To create more complex displays, each monitor can be split into several frames using the d.frame command. You can subdivide the GRASS monitor

NC shaded elevation   NC LANDSAT RGB and major roads

Fig. 7.1. Map display with d.frame: two frames with shaded DEM and LANDSAT RGB color composite map (North Carolina data set). The titles are written with d.text

into rectangular areas (frames) by mouse using the flag -c. Subsequent display commands will be applied only to the latest defined or selected frame. In case that you have several frames, select another one with d.frame -s and click with left mouse button into the desired frame, accept it with right mouse button. To remove all frames, run d.frame -e. As an example, we create an image shown in Figure 7.1:

```
g.region rast=elevation -p
d.mon x0
d.font FreeSans
d.frame -e
d.frame -c at="0,100,0,50"
d.his i=elevation_shade h=elevation
echo "NC shaded elevation" | d.text col=black

d.frame -c at="0,100,50.1,100"
d.rgb b=lsat7_2002_10 g=lsat7_2002_20 r=lsat7_2002_30
d.vect roadsmajor col=yellow
d.vect roadsmajor disp=attr attrcol=ROAD_NAME
echo "NC LANDSAT RGB and major roads" | d.text col=black
```

The frame coordinates are defined in percent of the monitor size (bottom, top, left, right). Using this approach you can define frame regions independently from the true monitor pixel size. If you want to include a displayed map in a presentation or a report, you can use d.out.file to export the image into a PNG, PostScript, TIFF or JPEG file. Later, in Section 7.1.4, we show how to generate high resolution output as an image file with the PNG driver.

Legends and labels   You can display a legend in a separate monitor, frame, or you can just stretch your current monitor to make space for the legend and place it with a mouse using the -m option. The module d.legend will use continuous colors for the floating point (DCELL, FCELL) raster maps or when there are too many categories for current window height for integer (CELL) maps. Otherwise, discrete colors are used (see examples in Section 5.1.1), although you can force continuous colors for any raster map with the flag -s. You can list only a subset of categories or range of values for which you want to display the legend and control number of text labels. Use the d.font command to change the font for legend labels and other textual output in the GRASS monitor:

```
# list available fonts
d.font -l
# select font from list
d.font FreeSans
```

There is no legend tool yet for vector data – you need to use external graphical software to add it to the snapshot image although the new GUIs provide some enhanced capabilities.

Labels can be added to the vector features displayed in the GRASS monitor with v.label/d.labels. We illustrate vector labeling with rotated labels along lines:

```
g.region rast=lsat7_2002_10 -p
d.erase -f
d.rast lsat7_2002_10
d.vect roadsmajor col=yellow
v.label -a roadsmajor label=roads_labels column=ROAD_NAME \
        size=200 color=red ref=center,upper back=white
d.labels roads_labels
```

An improved label tool with collision mitigation is under development.

Color tables   For raster data, working with color provides a powerful tool for extracting important spatial information and communicating it effectively. As we have explained in Section 5.1.1, a raster color table can be defined by r.colors. Besides selection from a substantially extended set of predefined color tables you can define the colors by their names using the option color=rules, or you can copy a color table from another raster map using the option raster=mymap (see examples in Section 5.1.1). For a refined definition of colors you can use the red, green, blue (RGB) color description (see Section 8.5.1). To find suitable RGB values for a desired color, use any graphics tool provided by your system. For example in gimp, find a palette under File ⤳ Dialogs ⤳ Palettes. Select a palette suitable for your map and then click on the individual colors to get the RGB values which you can then use in r.colors. Another nice online tool is the ColorBrewer[1], designed to help

---

[1] ColorBrewer Web site, http://colorbrewer.org

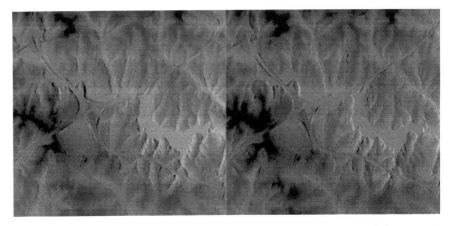

Fig. 7.2. Shaded elevation maps: shade map with sun azimuth=270° from north (left) and shade map with sun azimuth=90° (right), sun altitude=30° above horizon (North Carolina data set)

people select good color schemes for maps and other graphics. You can find additional examples of rules for creating the color tables in the Sections 5.4.3 and 5.5.2 and, of course, in the manual page for r.colors.

7.1.2 Creating a 2D shaded elevation map

To enhance the perception of topography represented by a DEM, a shaded elevation map can be generated quite easily as you have already learned in Section 5.4.4. A special color transformation is used to prepare a translucent view of the DEM (or any other raster map) and the shade map. It is based on the IHS color transformation which is explained in greater detail in Section 8.5.1. First, we generate a shade map based on the sun position using the script r.shaded.relief. By default, the name of the resulting shade map is created by adding .shade name extension to the name of the elevation file. This map is then used to display the elevation map with shaded topography by d.his:

```
g.region rast=elevation -p
r.shaded.relief elevation shadedmap=myelev_shade \
                altitude=30 azim=270 zmult=3
d.erase
d.rast myelev_shade
d.his h=elevation i=myelev_shade

# display a brighter map
d.shadedmap rel=myelev_shade drape=elevation bright=60
```

You may experiment with different values for altitude, azimuth, and zmult when creating the shade map to highlight various topographic features. Figure 7.2 shows the effects for different sun azimuth angles. You can also apply shading to other types of surfaces when studying their structure.

If you want to save the shaded map composite into a new raster map, use r.his instead of d.his. It creates three maps representing the red, green and blue channels (because the original map is 24bit, it writes three 8bit maps). In our next example, we call them el.b, el.g, el.r. You can then use the module r.composite to combine the three color maps within GRASS into a single shaded elevation map dem.shaded:

```
r.his h=elevation i=myelev_shade b=el_b g=el_g r=el_r
r.composite b=el_b g=el_g r=el_r out=myelev_shaded_10m
d.rast myelev_shaded_10m
```

The module r.composite provides optional parameters to control the color levels to be used for each color component (default color levels per channel: 32). This default number of levels results into a total of 32768 possible colors (equivalent to 15bit per pixel). Due to limitations in the GRASS display color model both r.composite and d.rast will significantly slow down if more colors are used. However, for human eye, this number of grey shades for each channel is quite sufficient. You can also export the three maps and compose them into 24bit shaded elevation image using external graphics tools. Alternatively you can export the map using r.out.ppm3 which writes a 24bit PPM file. Note that the composite shaded elevation map is only usable for visualization purposes as the elevation cell values are modified due to the shading.

7.1.3 Using display tools for analysis

We have already used GRASS monitor for query of raster and vector maps using d.what.rast and d.what.vect. Additional display tools can be used to view results of statistical analysis or to add graphical coordinate information to your map. Correlation between a pair of raster maps, in our case bare ground elevation models (NED) and SRTM DSM, and histograms can be displayed as follows:

```
g.region swwake_30m
d.erase
d.correlate layer1=elev_ned_30m layer2=elevation
d.correlate layer1=elev_ned_30m layer2=elev_srtm_30m

g.region swwake_10m
d.histogram elevation
d.polar aspect undef=0
```

The results of correlation show that the 30m and 10m NED have excellent correlation as they come from the same data source (bare earth lidar DEM).

The second example shows correlation between the NED and SRTM data, with the SRTM apparently shifted due to vegetation and buildings (as we have already explained, it is DSM). The polar diagram for the aspect map shows balanced distribution of values in all directions. We have set the parameter undef=0 to skip cells with zero aspect values (mostly lakes) that would create a spike in the diagram in the eastern direction.

You can add coordinate information to your map as a projected or geographic coordinate grid or a ruler along the map borders using the command d.grid:

```
d.rast elevation
d.grid size=5000 col=brown
d.grid -n size=1000
d.grid -g size=0:02 col=black
```

You can also examine your raster data at pixel level by displaying the pixel values as numbers or, in case of aspect, as arrows:

```
g.region rural_1m res=30
d.erase
d.rast elev_ned_30m
d.rast.num elev_ned_30m dp=0
# use -f to erase the numbered raster
d.erase -f

# display aspect with arrows and flowlines
d.rast.arrow aspect
r.mapcalc "elev_nedrural_30m=elev_ned_30m*1."
r.flow elev_nedrural_30m fl=flowline skip=3
d.vect flowline
```

Vector features can be interactively extracted using graphical extractor in the GRASS monitor d.extract. In the following example, we will extract a forested area from the wake county planimetry map and store it as a separate vector map:

```
g.region rural_1m
d.erase
d.rast ortho_2001_t792_1m
d.vect P079215 col=yellow

# extract the forest by clicking on its boundary
# note that it has two segments
d.extract P079215 out=forest_rural
Select vector(s) with mouse
- L: draw box with left mouse button to select
- M: draw box with middle mouse button to remove from display
- R: quit and save selected vectors to new map
L: add M: remove R: quit and save
```

```
d.vect forest_rural col=green
```

You may find it easier just to click twice on the segment that you want to extract rather than draw a box, which may include the features that you don't want to include.

Animations in 2D space   If you have a series of data (temporal, spatial, 3D cross-sections) you can animate them using xganim. We have already used xganim to display time series from raster-based modeling in Section 5.5 and to view volume data in Section 5.6. Up to four different map series can be animated simultaneously – a task often needed when analyzing outputs from simulations of landscape processes. You can either list all the maps by name or use wildcards, however, you need to be careful about the numbering system that you use to ensure proper order of maps, and keep the possibility to insert additional maps. The program provides a simple interface with controls for speed, looping, direction of play, running and stopping the animation, and stepping through the frames. If the animation is "jumpy", it is usually because your pattern does not change smoothly – you can add additional frames by interpolating between the maps. Often, a simple average between two raster maps is sufficient (use r.mapcalc). Animation is also useful for browsing through a larger set of maps and as a preview tool when preparing data for animation in 3D using nviz. To save your animation slides as an MPEG file, use r.out.mpeg (see the manual how to use wildcards). This command requires an additional encoding program from the netpbm tools (see Section 4.1.4).

7.1.4 Monitor output to PNG or PostScript files

Besides the GRASS monitor, it is possible to output the map display to other types of graphics drivers such as PNG (read more about the drivers in the GRASS User manual[2], or by running g.manual displaydrivers).

PNG file driver   While the regular GRASS monitor displays the raster map at the resolution given by your display system, the PNG driver was implemented to create a user defined, high resolution output in PNG image format. It uses the PNG library[3]. True color output is supported. The PNG format (Portable Network Graphics) is a lossless, highly compressing image format designed as a replacement for GIF and TIFF which may contain patented algorithms (note that the JPEG algorithm compression is lossy and is often inadequate).

The use of the driver is similar to the use of the GRASS monitor with the output stored in a PNG file when the PNG driver is stopped. The resolution

---

[2] GRASS 6 online user manual, http://grass.itc.it/gdp/manuals.php

[3] PNG library, http://www.libpng.org/pub/png/

of the resulting image, its background color, true color support, and the name of the output file are controlled by GRASS variables and UNIX environment variables (see the manual page for the PNG driver). For example, you can create a PNG image called myimage.png with the size of 1500 × 1350 pixels with elevation and major roads maps as illustrated by the following sequence of commands (here we use syntax for bash and similar shells):

```
g.region swwake_10m -p
GRASS_TRUECOLOR="TRUE"
export GRASS_TRUECOLOR
GRASS_WIDTH=1500
GRASS_HEIGHT=1350
GRASS_PNGFILE=myimage.png
export GRASS_WIDTH GRASS_HEIGHT GRASS_PNGFILE
d.mon PNG
```

Then, display a raster map and a vector map from our NC sample data set and save the image by stopping the driver:

```
d.rast elevation
d.vect -c roadsmajor
d.mon stop=PNG
```

The PNG file called myimage.png will be automatically written into your current directory. If you do not define the related environment variables, your output will use the default settings; that means 8bit colors, the 640 × 480 image size, and the file name map.png.

PostScript file driver   The use of the driver is similar to the use of the PNG driver with additional variables for paper size and orientation. For example, you can create a postscript file for printing a rich content, high quality image called myimage.png with the size of 3000 × 2700 pixels that includes shaded elevation, streams and street maps as illustrated by the following sequence of commands (here we use syntax for bash and similar shells):

```
g.region swwake_10m -p
GRASS_TRUECOLOR="TRUE"
GRASS_PSFILE=myimagelarge.ps
export GRASS_TRUECOLOR GRASS_PSFILE
GRASS_WIDTH=3000
GRASS_HEIGHT=2700
GRASS_LANDSCAPE="TRUE"
export GRASS_WIDTH GRASS_HEIGHT GRASS_LANDSCAPE
d.mon PS
```

Then, display the selected raster and vector maps from our NC sample data set and save the image by stopping the driver:

```
d.shadedmap rel=elevation_shade drap=elevation bright=60
d.vect streets_wake
```

```
d.vect streams col=blue
d.mon stop=PS
```

The PNG file called `myimagelarge.ps` will be automatically written into your current directory.

7.2 Creating hardcopy maps with ps.map

The graphical tools for creating hardcopy maps in GRASS are relatively limited because of its focus on modeling and spatial analysis rather than computer cartography. The hardcopy maps can be created by a text-oriented but powerful Postscript graphics tool or in combination with other Open Source graphics programs. Hardcopy maps can be created by printing Postscript graphics generated by `ps.map`. The Postscript graphics is produced from a control file describing the layout and map style. All raster and vector data are supported, as well as coordinate grids, user defined icons, and a bar scale.

You may try `ps.map` with the North Carolina data set and draw a raster and a vector map at a given map scale. The map definitions are saved in a text file. A sample text file `psmap.def` for the geology map of North Carolina may look like this (find an extended version at the GRASS Book Web site):

```
paper a3
    end
raster geology_30m
outline
    color black
    width 1
    end
colortable y
    where 1 6.5
    cols 4
    width 4
    font Helvetica
    end
setcolor 6,8,9 white
setcolor 10 green
vlines roadsmajor
    width 0.1
    style 0111
    color grey
    masked n
    end
labels roads_labels
    font Helvetica
    end
text 30% 100% NC Wake Geology map (State Plane metric, 1:100000)
```

```
    color red
    width 1
    hcolor black
    hwidth 1
    background white
    size 500
    ref lower left
    end
vlegend
  where 4 0
  font Courier
  fontsize 8
  end
point 40% 60%
  color purple
  size 0.5
  symbol basic/diamond
  masked n
  end
scale 1:100000
grid 2500
  color grey
  numbers 2 grey
  end
end
```

The definition uses special labels for the roads generated with v.label (see Section 7.1.1). The module is run on command line with these map definitions to produce a Postscript map file:

```
ps.map input=psmap.def output=geology_wake_nc.ps
```

The module generates the Postscript map which may require some time depending on the input map size, the map scale and the selected paper size. Please refer to the manual page to learn more about this module. After the map is generated, you may preview it with a Postscript interpreter such as **ghostscript** (a convenient graphical user interface for ghostscript is **gv**). If the map is as desired, you can send it to the printer:

```
lpr -s -P printer geology_wake_nc.ps
```

The optional flag -s avoids generating another temporary copy of the file in the printer queue. Enter the correct printer name for **printer**. You can also convert the map into PDF format using **ps2pdf** or a similar tool available on most systems.

7.3 Visualization in 3D space with NVIZ

The advanced, interactive visualization tool nviz can be used to view the data in 3D space and to perform visual analysis of multiple raster surfaces, vector maps and volumes. The module is fully integrated with the GRASS data structure and runs directly from the GRASS prompt. It also supports scripting for producing dynamic visualizations via animation. The module provides the following capabilities:

- visualization of 2D raster maps as multiple surfaces in 3D space, with the capability to use different raster maps for surface topography, surface color and transparency;
- interactive positioning, zooming, and z-scaling;
- interactive lighting with adjustable light position, color, intensity and surface reflectivity;
- display of multiple vector maps draped over selected surfaces or flat at a selected height;
- multiple vector point data maps with attributes displayed in their 3D position or draped over selected surfaces;
- animation capabilities with two options:
  - key-frame animation for creating fly-by's,
  - scripting for automatically generating complex animations from series of maps;
- interactive query of raster data displayed as a surface and color;
- interactive slicing through multiple surfaces using cutting planes;
- display of volume data using isosurfaces and cross-sections.

The nviz module is enhanced quite frequently, so some differences between the latest version and our description are possible. The most important updates will be posted on the GRASS Web site. You can also learn how to use nviz from *Nviz tutorial* that can be accessed by clicking on the help button of its interface.

7.3.1 Viewing surfaces, raster and vector maps

You can start the program without defining any maps by nviz -q and use the interface to load your data. Alternatively, you can define the maps that you want to visualize on the command line. For example, to view the elevation surface with major roads, and schools from our North Carolina data set, run (use elev_ned_30m for both region and elevation if you have less than 1GB memory):

```
g.region rast=elevation -p
nviz elevation vect=roadsmajor points=firestations,hospitals
```

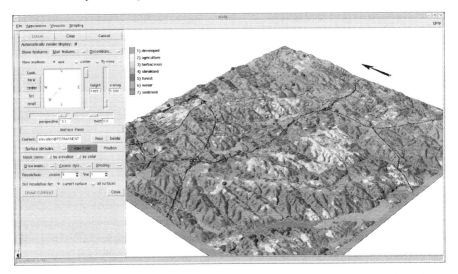

Fig. 7.3. Land use map draped over DEM with overlayed streams and roads as vector line data, and hospital and firestation locations as point symbols (pyramids and spheres, respectively)

The program opens a graphics window with a coarse model of the elevation surface and a control panel window. Depending on the size of your raster map, the surface may not be rendered at its full resolution and the view may not be optimal. In the following paragraphs, we explain how to adjust it to create a desired 3D view of studied area.

Controlling the view   The position of the viewing point, viewing direction, perspective (zoom-in, zoom-out) and tilt of the surface can be adjusted using the Controls menu (Figure 7.3). Use the left mouse button to move the puck around the viewing direction square to change the position of viewer and the direction of view – the coarse model of your DEM will move simultaneously making it easier to find the desired viewing position. The **perspective** slider allows you to zoom-in and zoom-out while the **height** slider controls the viewing height. The **z-exag** slider is used to interactively modify the z-exaggeration (it effectively multiplies the elevation data); note that it also changes the height of the view so you may need to re-adjust it. If you cannot achieve the desired view with the sliders or if you want to define an exact value for any of the viewing variables, you can type them into the related field and type <ENTER> to continue.

To focus on an area off the center of your map, use the **Look here** button to select a new center of view with a mouse click (pin your surface in your focus area). All movement of the surface will be centered around this point. To put the focus back on the center click the **center** button. To get an ortho

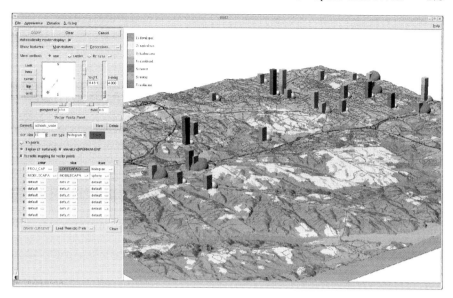

Fig. 7.4. Displaying schools as 3D point symbols sized according to the school capacity

view, click on the top button and use the perspective slider to zoom in and out. With the reset button you will get back to the default behavior. The twist slider allows you to tilt the displayed surface to simulate the view from a turning airplane. The surface with all other maps will be rendered after each change. If you are using multiple steps to adjust the view of your surface, you can switch off this automatic rendering using the buttons from the Show features menu in the upper part of the main panel.

Modifying properties of surfaces   The viewed surfaces are managed using the options provided by the Visualize ⤳ Raster Surfaces menu (Figure 7.3). In the lower part of this panel you can adjust the drawing style as well as the level of detail for the rendered surface. By default, the surface is rendered as colored polygons with Gouraud (smoothed) shading, using coarser resolution while the surface is interactively manipulated. To change the surface display to a mesh (wire) or colored surface with a mesh, select the desired option from the panel under Coarse style. For fast interactive manipulation, you can select wire. To render the surface at the current region resolution (as given by g.region), set the Resolution: fine to "1". Note that if the current resolution is higher than the resolution of the raster file used for topography, the raster file is automatically resampled leading to the discontinuous surface shown by Figure 5.4 (see Chapter 5). To speed up rendering while exploring the viewing parameters, you can lower the rendering resolution (increase the cell size) by choosing a higher value of polygon resolution. If you are using any style that involves wire, you can adjust its grid spacing with Resolution: coarse.

To drape a new color map over the surface, select a new raster map using the color option from the Surface Attribute menu, for example landclass96 from the PERMANENT MAPSET in our North Carolina example. After loading the raster map, use DRAW to render your elevation surface with the new color map (Figure 7.3). To overlay an additional raster map, you can use transparency. However, for its meaningful application, the raster map for transparency should be fairly simple. A possible use may be to suppress (lighten) the areas outside a studied watershed, you can try it out by setting the transparency to the raster map basin_50K. You can remove the transparency by setting it to a new constant 0. To render only a subset of the surface, you can define a raster map to be applied as a mask using the mask option from the Surface Attribute menu.

Displaying vector lines and polygons   If you would like to add a vector line or polygon map or select a different color for your vector map, open the vector panel by choosing Visualize ⤳ Vector Lines/3D Polygons. You can load a vector map using the New button, e.g. streams. After loading, the switch Display on surface(s) is automatically activated for all surfaces (when working with multiple surfaces, switch off those on which you do not want to drape your new vector data). You can adjust the line width and color by using the appropriate buttons. For example, select Color and pick a lighter blue for the currently loaded streams and set the width to 1, then select roadsmajor as Current vector map and assign it black color, keep the width to 2. If the lines are not fully rendered, move them slightly above the surface using the vector height above surface slider, if it is too high, you can type in the number (0.5 works well both for the streams and roads in our example). You can use Draw current in the vector panel to drape only the selected vector file. To render the surface with all vector maps draped over it, use the Draw button on the top of the movement panel. We have already shown how to generate 3D vector representation of buildings and view them in nviz in the previous chapter, Section 6.5.6.

Displaying vector points   To modify the symbols used for the vector point data, go to Visualize ⤳ Vector Points (Figure 7.4). It is similar to the vector lines and polygons panel – you can use it to add or delete a vector points map, select the surface on which the current vector point data should be draped, and select the symbol, including its color and size. For example, we can select hospitals, change the icon to diamond, color to purple, and size to 300 to clearly distinguish them from firestations. If you have 3D vector point data, you can also display them in 3D position rather than draped over the surface. This is particularly useful when evaluating interpolation and smoothing of surfaces as you can see how much the surface deviates from the given data and where the highest deviations are, or for visualization of multiple return lidar data (Figure 6.15).

You can create more sophisticated displays for quantitative point data using the option thematic mapping for vector points. When you click it, you will get a set of buttons that allow you to adjust the color and size of the selected icon according to values of the selected attribute. For example, start nviz as follows:

```
g.region rast=elevation -p
nviz elevation col=landclass96 vect=roadsmajor \
     points=schools_wake
```

Open Visualize ↝ Vector points, click the thematic mapping button, select histogram for icon type, set icon size to 80 (this will adjust the size of the symbol footprint). Then click on size in the first row and select an attribute that will be used to adjust the height of the histogram point symbol – we choose the school capacity CORECAPACI. In the size selection panel, we select Auto option, but you can highlight or suppress certain values by specifying the symbol height for selected attribute values or just change the maximum height from 10 to 6 to get a better image. You can further enhance the symbols by adjusting the color (we select light blue to to dark blue for PROJ_CAP). You can add aditional symbols to the histogram based on other variables, for example, select the size for second variable using the attribute representing the number of mobile classrooms MOBILECAPA. Adjust the maximum to 3, set the colors for this variable from yellow to red and change the symbol in the third column from default to sphere to clearly distinguish it from the core capacity variable. You can see that the schools with the larger number of mobile classrooms are smaller schools, many of them in suburbs with many young families with children. You can learn more about the multiatribute point data visualization from the related workshop material presented at the FOSS4G2006 conference http://www.foss4g2006.org/contributionDisplay.py?contribId=45&sessionId=59&confId=1.

Controlling light Interactive light manipulation is useful for detecting noise or small errors in surfaces as well as for enhancing the 3D perception of surface topography and creating special effects. The panel Appearance ↝ Lighting is used to change the lighting parameters such as the light color, brightness, ambient (dispersed) light and a position of light source (height and direction) by using the appropriate sliders and a square with puck. Interactive adjustment of lighting is made easier by a sphere which appears in the center of the graphics window and shows the current lighting effects as the changes are made (Figure 7.5). The sphere disappears when the surface is rendered, for example with the Draw button.

Adding legends and labels To add a title or other text to the images displayed by nviz, open the Appearance ↝ Labels panel or that allows you to

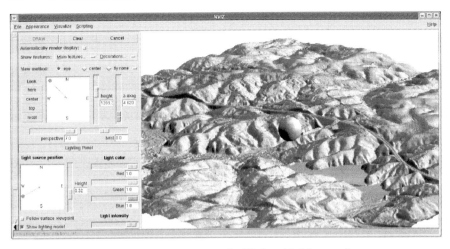

Fig. 7.5. Interactive control of light aided by a sphere

type in a short text and place it in the graphical window with a mouse. The legend can be created through the Legends panel with options similar to the command d.legend. You can again place it with a mouse – click left button to identify the upper left corner and right button for the lower right corner. Our displayed land cover map, loaded in the previous examples (Figure 7.3) has 7 classes, click show labels to add the land cover type for each class. The legend is automatically redrawn with the image, to remove the legend, click Erase legend. You can also add north arrow and a simple scale - place it in a location close to you and then farther from you to explore the difference in scale in the 3D view. Adding the fringe further enhances the 3D model, you can set the bottom elevation to 50m and click the buttons on all sides, the click on Draw Fringe will add it to the displayed surface.

Saving images, state and view settings To save the created image, go to File ↝ Save image as and select one of the formats that you want to use (TIFF, PPM). You can transform the saved image to JPEG, Postscript or other formats using a graphics program such as gimp. To render and save an image at high resolution (for example, at 5000 × 5000 pixels), use the Maximum Resolution PPM option. The image will be split, rendered piece by piece and then patched together.

Using File ↝ Save state, you can save the settings of your current 3D visualization in your current directory. You can restore the settings by File ↝ Load state. Using this capability, you can save and quit your work and then start it again without losing just the right light and view that you have found. It is highly recommended to save the state regularly so that you can go back and restore your settings at later time. You can also save your 3D view settings in a GRASS 3d.view file within the current MAPSET using the Save

3d settings option from the File menu or load the settings from a previously created file to get specific viewing parameters.

7.3.2 Querying data and analyzing multiple surfaces

To use nviz for both qualitative and quantitative analysis, you can perform 3D queries. Choose Visualize ↝ Raster Query or Vector Query and activate the query on/off switch. You can select which attributes you want to be included in the query using the button Attributes. By pointing the mouse at a location of interest, you will get its coordinates including the elevation, draped raster map category, label, and color expressed as RGB triplet, and a distance from previous queried location. You can also use nviz to perform small digitizing tasks in 3D by querying and piping the result into a file defined file using the Send results to button and then editing the saved text file to desired format. When you choose Vector Query, again click query on/off switch and then select vector maps by clicking on Choose map(s). Select from the listed maps using the >> button to add a map and << button to remove a map from query. We select roadsmajor and firestations, click OK and then click on a displayed line or a point symbol – the results are shown in a special window for each vector map. The principles used in the algorithm for 3D spatial query are described by Brown et al. (1995).

Working with multiple surfaces   Multiple surfaces are useful for visual analysis of terrain change (Mitasova et al., 2005b), cut-and-fill during construction or display of multiple soil horizons or geological layers. We illustrate the tools using the lidar-based DSM elev_lidruralmr_1m and the bare ground DEM elev_lidrural_1m that we have created in Section 6.9 (Figure 7.6):

```
g.region rural_1m
nviz elev_lidrural_1m,elev_lidruralmr_1m
```

Both surfaces are by default displayed based on their 3D coordinates – you can see that the DSM surface is above the bare ground DEM. The relative position of the displayed surfaces can be changed using the position menu, which opens after clicking on the Visualize ↝ Raster Surfaces ↝ Position button. You can change the relative vertical position of the current surface using the z slider or you can move the surface around using the cross in a horizontal position square. Using this interface, you can arrange your surfaces within your graphical window in a way useful for visual analysis; for example, next to each other, as shown in Figure 7.6. To get the surfaces in their original position click Reset for each of them.

When comparing multiple surfaces, cutting planes can be very useful. Choose Cutting plane from Visualize menu, select a cutting plane from Active cutting plane (you can have up to 8) and set its appropriate orientation using the Rotate slider (Figure 7.6). You can also slide the rotate slider to interactively cut through your surfaces, or you can slide the cross to move the cutting plane through the surfaces in fixed direction. The color of the cutting plane can be selected from the set shading menu to the color of the top or the bottom surface, blend of those two colors, transparent grey or clear. Make sure that all your surfaces have the same resolution, otherwise the cutting plane won't be filled with color.

When working with multiple surfaces and cutting planes, keep in mind that in geoscience applications, the vertical spatial variability requires resolutions much higher than are the resolutions typically used in a horizontal plane. Therefore, a global vertical exaggeration (z-exag) factor is applied for better visual perception of terrain. However, to visualize vertical relationships with sufficient detail, relative exaggeration of depths has to be used. For example, for multiple surfaces representing soil horizons, the depths often need to be exaggerated relative to terrain surface (Brown et al., 1995), you can find a related tool under Visualize ⤳ Scaled difference.

7.3.3 Creating animations in 3D space

Animation is a powerful tool for exploring large data sets and for analyzing time series observations or modeling results. Fly-by's over digital elevation models or multiple surfaces can be created by using two types of key frame animations. Scripting with file sequence tool provides capabilities to build more complex animations with multiple maps including combination of surfaces, vector data and volumes.

Creating a fly-by You can directly control the position and movement of the displayed surface using the View method ⤳ fly by selecting the basic, simple and orbit options. The relevant panel offers fly help button that explains how to use mouse buttons to move the surface in different modes and directions.

Animations simulating flying over a surface can be created using animation panels which allow you to define *key frames*, representing key positions defining your path and then render and save the surface views along this path. Basic control of the fly-by is provided by the menu Visualize ⤳ Simple Animation (see Figure 7.7). First, enter the number of images to be rendered as set max. frames. The default is 25, a larger number is suggested to get an interesting flight, for example, type in 100 followed by <ENTER>. Now select your initial viewing position using the viewing direction puck and perspective, height or twist sliders in the upper movement control panel. Click on Add to save this position. Then select the next position on the key frame (time) axis by dragging the thick vertical blue line to the right and define the second key (viewing) position by adjusting the direction, perspective, height or twist. Save your second key frame by clicking on Add. You can continue moving your viewing position and adding key frames until you reach the end of the time line. Additional frames will be automatically interpolated between the key frames as indicated by the black bars on the time axis.

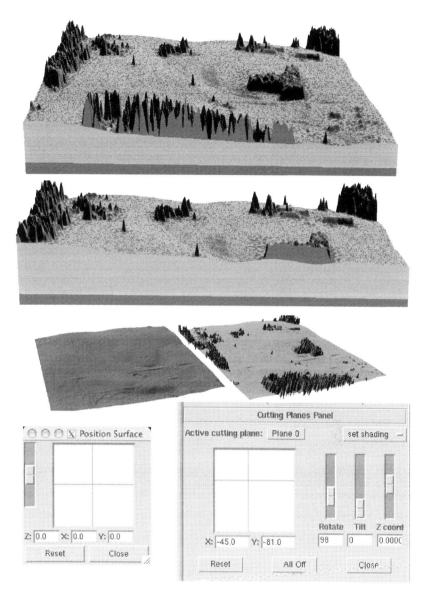

Fig. 7.6. Viewing multiple surfaces next to each other and in their relative position with a cutting plane (elevation surfaces computed from the first return and bare ground lidar elevation data, two bottom surfaces represent the hypothetical ground water maps generated in the raster modeling section)

To avoid a jumpy flight, it is useful to start with a small number of key positions, for example, by locating four key frames after each 25% segment on

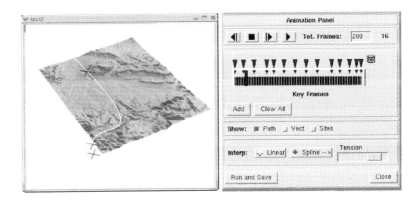

Fig. 7.7. Fly-by animation menu in `nviz`

the time axis. After defining the key positions and a number of key frames, run the coarse resolution movie by clicking on the play button (black right arrow) to get some feeling for how the controls work and see how fast and smoothly you are flying. If the flight is too fast, add more frames. You can also control the number of frames between the key positions by dragging the green triangle on the **Key Frames** axis. To make the fly path smoother, activate the **Spline** button and control the sharpness of the curves on your path using the **Tension** slider. You can add vector maps to your surface by switching on the **Show lines** and **Show points** buttons.

After previewing the animation by running the coarse resolution fly-by, you can render the full resolution images for your movie by selecting the **Run and save** and providing a filename (e.g. `film`). You can select MPEG-1 encoding if your GRASS installation was configured with ffmpeg support, otherwise the result will be a series of image files, which are automatically numbered (e.g., `film00000.ppm - film00099.ppm`). The images are then used to create a movie file using external tools. Depending on the resolution, image size, and speed of your computer, the procedure may take some time, especially if fine rendering mode is used.

If you just want to try out animation, the simple animation tools described above are enough; however, for a serious animation work it is worth learning and using the **Keyframe animation** panel which provides control over the frame rate and key frame time as well as refined control of the camera (viewing position and direction). The use of **Keyframe animation** is described in detail in the *Nviz tutorial*.

Converting series of images to movies When not using MPEG-1 encoding, the rendered images are saved in the PPM or TIFF format. To create an animation, you can merge them into an animation file using external tools such as ppmtompeg (netpbm tools, see Section 4.1.4). To create an animated GIF (suitable for animations from a smaller number of frames – 300 and less) you can use the command convert available on most systems (see the manual for all options):

```
convert -delay 10 -loop 3 water*.ppm animation.gif
```

You can find additional advice and links to various tools for encoding series of images into animations at GRASS Wiki help page for Movies[4].

Creating animations using scripting   Complex animations involving multiple surfaces, vector and volume data, cutting planes, changing views and light parameters can be created using the scripting capabilities of nviz. Animations using dynamic maps can be used, for example, to view and analyze the following models and data (Mitas et al., 1997):

- results of dynamic simulations such as water, sediment, pollutant transport, fire spread, migration of animals, traffic, and urban growth;
- time series of observed data from monitoring and remote sensing, such as movement of pollutants, change in rainfall or temperature, past urban growth or vegetation change;
- behavior of a method or algorithm, for example, by animating the results of parameter scans (impact of tension parameter on an interpolated surface, or impact of land cover factor on erosion and deposition pattern, etc.).

You can create scripts for creating animations by using the basic scripting tools in nviz. While the scripting is turned on, the performed visualization tasks are saved in a script file, so that they can be repeated as desired.

Dynamic surfaces can be created using the file sequence tool available under Scripting ↝ Script Tools. The use of these tools is described by a step-by-step example in the *Nviz tutorial* and in a more up-to-date workshop material on Multiple Surface Visualization presented at the FOSS4G2006 conference[5]. Many examples of animations created by scripting can be found on the Spatial modeling and visualization Web site.[6]

7.3.4 Visualizing volumes

We have already browsed through volume raster data using xganim in Section 5.6. Volume data can also be visualized using isosurfaces and cross-sections in nviz, see Figure 7.8. In the next example, we display the volume

---

[4] GRASS Wiki help page for Movies,
  http://grass.gdf-hannover.de/wiki/Movies
[5] "GRASS 3D and visualization" workshop material, http://www.foss4g2006.org/
  contributionDisplay.py?contribId=45&sessionId=59&confId=1
[6] Spatial modeling and visualization,
  http://skagit.meas.ncsu.edu/~helena/gmslab/

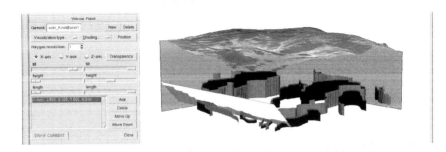

Fig. 7.8. Volume (3D raster) visualization integrated in `nviz`: isosurfaces (K=0.20 and K=0.24 – violet and red, respectively) representing the soil erodibility volume created in the Section 5.6. The volume is shifted lower for better visibility. Tilted cross-section plane illustrates the relation between isosurface and voxel space

data generated in Section 5.6 (`soils_Kvol`). First make sure that you have the 3d region set, check the range of values in your volume map (see 5.6), and then start nviz:

```
g.region rural_1m res3=3 t=155 b=102 tbres=2 -ap3
r3.info soils_Kvol
nviz elev_lid792_1m
```

Then load your volume data Visualize ⤳ Volumes, click New and select `soils_Kvol`. Set Polygon resolution to 1 and click Add. The Visualization type is isosurface, so with New constant you set its value, we select 0.27, click Accept and your level shows up in a list (Figure 7.8). We can add level 0.20 and then draw the image. The isosurfaces for this data set are ribbons with different heights, because the volume was created from a 2D raster. You can change the relative position of your volume and terrain surface using the Position panel for Volumes or for Raster surfaces and adjust your viewing position, z-exaggeration and light the same way as for 2D raster surfaces. You can explore display of closed isosurfaces by applying the procedure to `elev_state_500m` in combination with annual precipitation volume that we have computed in Section 6.10. For more sophisticated visualization, we describe the export to external tools.

7.4 Coupling with an external OpenGL viewer Paraview

After exporting with `r3.out.vtk`, GRASS 3D data can be displayed in `paraview`[7] visualization software. This tool offers various methods to render

---

[7] Paraview software, http://www.paraview.org

Fig. 7.9. Raster and vector visualization with Paraview

semi-transparent volumes, isosurfaces, movable cutting planes and isolines. Also 2D raster maps can be exported (r.out.vtk) as well as 2D/3D vector data (v.out.vtk). We export the 2D/3D data from Section 6.5.6:

```
g.region rast=elev_lid792_1m -p
# drape orthophoto over DEM for export
r.out.vtk ortho_2001_t792_1m out=ortho_dem_1m.vtk \
          elevation=elev_lid792_1m
v.out.vtk bldg_resid_3d out=bldg_resid_3d.vtk
v.out.vtk bldg_cmcl_3d out=bldg_cmcl_3d.vtk

# make P079215 a 3D map, using larger 6m DEM
v.drape P079215 rast=elevlid_D792_6m out=P079215_dem
v.out.vtk P079215_dem out=P079215_dem.vtk

paraview --data=ortho_dem_1m.vtk
# - load the maps bldg_resid_3d.vtk, bldg_cmcl_3d.vtk,
#   and P079215_dem.vtk via menu
# - set the color table of ortho_dem_1m to grey shade or color
```

Figure 7.9 illustrates the use of Paraview. You can learn more about using Paraview with GRASS from the related workshop material presented at the FOSS4G2006 conference[8].

_____
[8] "GRASS 3D and visualization" workshop material,
http://www.foss4g2006.org/contributionDisplay.py?contribId=
45&sessionId=59&confId=1

8

Image processing

Remote sensing, as a rapidly advancing technology for gathering environmental data using a wide range of satellite and airborne platforms, plays a major role in spatio-temporal earth surface monitoring. Throughout this chapter, we introduce the basic remote sensing methods and explain their use in GRASS. The tools for image processing and remote sensing applications will be illustrated using LANDSAT-TM5/7 scenes available in the North Carolina data set (MAPSET `landsat`). Furthermore, an annual time series of daily MODIS Land Surface Temperatures (LST) from Aqua and Terra satellites is available in MAPSET `modis2002lst`. These image scenes were projected and a subset was created for the Wake county.

GRASS provides several modules devoted to image processing. Image data are processed using the raster data model; therefore, all raster modules can be applied. To distinguish the specialized image processing tools from the others, they are prefixed with "i." (example: `i.class`).

8.1 Remote sensing basics

Before describing numerous methods implemented in GRASS in detail, we will explain basic concepts of satellite remote sensing. As this relatively short section cannot replace related textbooks, references will be given where appropriate.

8.1.1 Spectrum and remote sensing

In principle, there are different remote sensing approaches: *optical (passive)*, *thermal (passive)*, and *microwave (active)* systems. Optical remote sensing is based on the measurements of radiation reflected from surfaces. It usually covers the visible (VIS) and infrared (IR) range of the spectrum. The reflected radiation in near (NIR) and middle infrared (MIR) spectrum behaves similarly to visible light, while thermal radiation (TIR) is surface emitted radiation. Longer wavelengths are in far infrared (FIR) range and in the important microwave range. See Figure 8.1 for a portion of the spectrum. The optical region spans at wavelengths from 0.3-15μm where energy can be collected through lenses. A subdivision of this optical region is the reflective region, 0.4-3.0μm. The adjacent subdivision of the optical spectral region is the thermal spectral range which is between 3-15μm, where energy is primarily emitted from surfaces rather than reflected. Far infrared ranges from 15μm-1mm, microwaves from 1mm-1m. Optical scanners always operate in a limited spectral range per channel.

As opposed to optical systems, *radar systems* "actively" emit microwaves and measure the backscattered energy. The major advantage of radar is the relative independence from weather and solar illumination effects. In case of an overcast sky, the earth surface is hidden by clouds for optical satellites. However, radar satellites can continue to deliver usable images since microwaves pass through the cloud cover (this is of special interest in the tropics). Radar analysis is not covered in this book because it is fairly complex. For details please refer to the microwave remote sensing literature, e.g., the book by Oliver and Quegan (1998). Additional tutorials are available on the World Wide Web.[1] Besides tools for optical data, GRASS also provides basic capabilities to process radar and thermal data.

Another active remote sensing technique is lidar (Light Detection and Ranging) which is one of the most recent technologies in 3D surveying and mapping. A laser onboard a plane sends out laser pulses to the ground in order to determine the distance to an object or surface. This distance is determined by measuring the time delay between transmission of a pulse and detection of the reflected signal. The horizontal and vertical accuracies are in the centimeter range. Multiple returns or entire waveform can be acquired for each pulse. The lidar toolset in GRASS provides methods to compute the digital elevation surface (DEM or DSM) based on radial basis functions and spline functions with Thykhonov regularizer (Brovelli et al., 2004). We already discussed lidar data analysis in Section 6.9.

In this chapter, we focus on images acquired by optical systems because they are widely used and their interpretation as well as data processing is easier than for radar data.

Reflected radiation and atmospheric effects Optical remote sensing systems are measuring sun energy reflected from earth's surface. While the sun is emitting a full range spectrum with a special energy distribution, only part of the energy reaches the earth's surface. The reason are various absorption and scattering processes within the atmosphere. Figure 8.1 outlines the

---

[1] SAR User Guide from Alaska SAR Facility,
`http://www.asf.alaska.edu/reference/general/SciSARuserGuide.pdf`
Remote Sensing Core Curriculum (RSCC), `http://www.r-s-c-c.org/`
ISPRS tutorial collection, `http://www.isprs.org/links/tutorial.html`

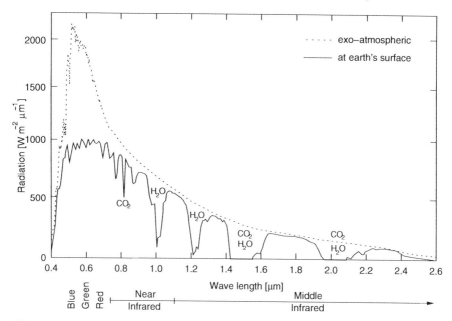

Fig. 8.1. Distribution of solar radiation (reflective portion of the spectrum) on upper boundary of atmosphere and at earth's surface with gaseous absorption (solar zenith angle 45°, curves as defined in 6S source code, Vermote et al., 1997)

solar radiation on top of atmosphere and at earth's surface. Before reflected solar energy reaches a sensor, it has passed the atmosphere twice at different angles. Correction of such atmospheric effects often requires image preprocessing. It is important to know that for some ranges of the spectrum, radiation cannot pass the atmosphere at all. Within these absorbed wavelengths, optical remote sensing platforms are unable to receive reflected radiation from earth. Therefore, the spectral filters of satellite sensors are defined accordingly. Schowengerdt (1997), describes these issues at a greater detail.

Remote sensing of the environment considers the sun energy's reflection in the visible and infrared range of the spectrum. Depending on the material of the observed object, the amount of reflected radiation varies. The reflectance curves of three basic materials are shown in Figure 8.2. While water absorbs most radiation in the visible spectrum and all in near-infrared, the radiation reflection of unvegetated sandy soil increases from the visible spectrum to the infrared range. The curve for green vegetation is highly dependent on the contents of chlorophyll. With a small peak for green wavelengths, the overall amount of reflection in the visible spectrum is much lower than for the infrared, especially in the near-infrared. The curve depressions at higher wavelength depend on the water contents of the plant. In case of a plant disease or dormant stage the reflection in infrared is dramatically decreased. Figure 8.2 shows the

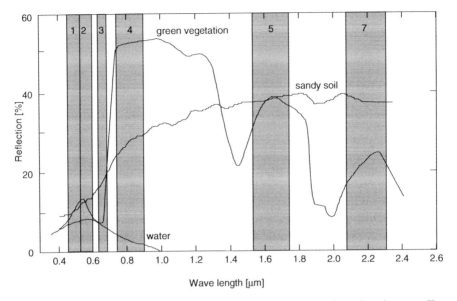

Fig. 8.2. Idealized reflection curves of green vegetation, sandy soil and water. For illustration the LANDSAT-TM5 channel filter functions except the thermal channel 6 are shown in background (curves as defined in 6S source code, Vermote et al., 1997)

spectral coverage of the LANDSAT-TM5 channels. The different reflectance curves allow us to distinguish the observed objects by multispectral remote sensing. For more theoretical details, please refer to Richards and Xiuping (1999).

Resolution   An important aspect of satellite data is the image resolution. In particular, we distinguish between:

- spatial (geometric) resolution;
- spectral resolution;
- radiometric resolution.

Spatial or geometric resolution is the spatial extent of each pixel as known from the raster data. For environmental satellite data, this resolution typically ranges from 1m to 30m. Spectral resolution refers to the bandwidth of each channel. The bandwidth is the range of the spectrum measured by one channel. Figure 8.2 shows the bandwidths of LANDSAT-TM5 channels with respect to the spectrum and object spectra. Both the higher number of channels and the bandwidth of each channel improve the potential to distinguish objects in an image. For example, when studying crop conditions, phenology-driven signal variations are found in a narrow spectral range between the red and the near-infrared spectrum. Only, when this range is covered by two or more channels with narrow bandwidth, the effects can be studied. Later on, we will show how to merge channels of high spatial resolution with those of high radiometrical resolution.

The radiometric resolution describes the signal dynamics within one image (bit resolution). For data distribution, the 8 bit, 16 bit, and 32 bit formats are common. Note that level numbering starts with 0 (no signal, usually colored black). Accordingly, an 8 bit image contains 256 levels numbered from 0 (black) to 255 (white) with different grey levels in between.

8.1.2 Import of image channels

Aerial and satellite data are delivered in a variety of formats. Most of them are imported easily into GRASS since the underlying GDAL library supports the most important raster and imagery formats. Some issues can arise if the maps are not oriented to north. Also important metadata are often available only from the original image file, requiring the use of GDAL tools for displaying them.

We show an example for a LANDSAT-TM7 data set in EOSAT FAST Format (thermal channel 6, low and high gain):[2]

```
# thermal channel metadata
gdalinfo p016r035_7t20000331.hdr
Driver: FAST/EOSAT FAST Format
Files: p016r035_7t20000331.hdr
Size is 7676, 6541
Coordinate System is:
PROJCS["UTM Zone 17, Northern Hemisphere",
    GEOGCS["WGS 84",
 [...]
GeoTransform =
  559939.132539704, 28.14122143322479, -4.530800840978595
  4100246.646970852, -4.53138436482102, -28.13755733944946
Metadata:
  ACQUISITION_DATE=20000331
  SATELLITE=LANDSAT7
  SENSOR=ETM+
  BIAS1=-6.978739990024116
  GAIN1=0.778740180758979
  BIAS2=-7.198818993380689
  GAIN2=0.798818898013258
 [...]
Corner Coordinates:
 [...]
```

---

[2] LANDSAT-TM7 data set "p016r035_7x20000331",
ftp://ftp.glcf.umiacs.umd.edu/glcf/Landsat/WRS2/p016/r035/p016r035_
7x20000331.ETM-EarthSat/

In case that "GeoTransform" tag is present, the gdalwarp program has to be applied to rotate the image from orbit orientation to north up orientation before importing it into GRASS. We define the appropriate target resolution for the channel to avoid introduction of irregular values:

```
# rotate to north up, write GeoTIFF, enforce 28.5m x 28.5m res.
gdalwarp -tr 28.5 28.5 p016r035_7t20000331.hdr \
         p016r035_7t20000331.tif
gdalinfo p016r035_7t20000331.tif
```

The GeoTIFF file can now be imported with r.in.gdal as shown in Section 4.1.3.

8.1.3 Managing channels and colors

A multispectral data set consists of various channels which represent portions of the spectrum. In the case of LANDSAT-TM5 and TM7, the visible spectrum with base colors blue, green and red is mostly covered as well as part of the infrared and thermal spectrum. Other satellites such as SPOT and ASTER do not provide the blue channel. To visually explore the imagery, we often need to analyze and modify its colors.

In a RGB color composite, three channels (each grey colored) are assigned to the colors red, green and blue; the result is a pixel-wise combined new image with a color table based on the input values.

When generating color composites from multispectral data, we need to determine which channels contain most information. For example, when considering LANDSAT-TM5 data, the following problem arises. Figure 8.3 shows the solar spectrum and the filter functions of LANDSAT-TM5. The color filter functions of channels 1, 2, and 3 partly overlap which leads to slightly correlated channels. Generally, 20 color composites can be produced from the six reflective LANDSAT-TM5 channels (not using the emissive thermal channel). Due to the slight correlations, the information contents is reduced when the first channels are combined. A simple method to find out the combinations with the highest information content is the *Optimum Index Factor* method (Chavez et al., 1984), which is based on correlation analysis. It is implemented in the i.oif script. It calculates a rank of all reflective LANDSAT-TM5/7 band combinations and outputs a sorted combination table.

Simple color composites   To obtain a color image from grey-colored channels, several channels have to be combined by assigning each channel to a different base color. For example, to generate a near-natural colored image, the blue, green, and red channels (each only grey-colored) have to be assigned to blue, green, and red color of the GRASS color composition module. In our example, we apply grey scale color tables to the channels; then, we can generate a near-natural color composite:

```
g.region rast=lsat7_2002_10 -p
r.colors lsat7_2002_10 col=grey
r.colors lsat7_2002_20 col=grey
r.colors lsat7_2002_30 col=grey
d.mon x0
d.erase
d.rgb b=lsat7_2002_10 g=lsat7_2002_20 r=lsat7_2002_30
```

This composite is drawn by `d.rgb` into the GRASS monitor.

Color composites can be stored as new map with `r.composite`. For example, to generate and store a near-natural color image, the satellite channels covering the red, green, and blue spectrum have to be assigned to `<r>`, `<g>` and `<b>`, respectively. The number of color levels to be used for each base color is predefined to a limited number of levels (number of resulting colors = specified levels[3]). For example, 10 color levels will lead to 1000 colors in the composite image. Due to speed limitations in the current GRASS display color model (GRASS monitor) for large color tables, we recommend not to generate 24 bit image composites inside GRASS (use `r.out.ppm3` to export to a 24 bit PPM image). Finally, the output name for the composite has to be specified.

A false color composite for LANDSAT-TM7 can be done in a similar way, but preferably from a histogram-equalized grey scale color tables calculated in advance:

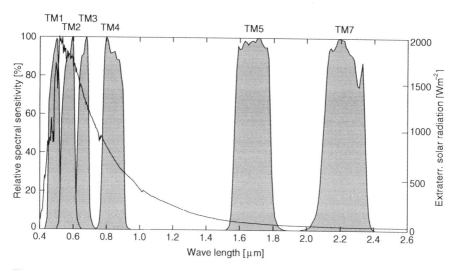

Fig. 8.3. Exo-atmospheric solar radiation (in W/m² on top of atmosphere, see right axis) and relative spectral sensitivity of LANDSAT-TM5 channel filter functions, thermal channel 6 is not shown (curves as defined in 6S source code, Vermote et al., 1997)

```
g.region rast=lsat7_2002_30 -p
r.colors lsat7_2002_30 col=grey.eq
r.colors lsat7_2002_40 col=grey.eq
r.colors lsat7_2002_70 col=grey.eq
d.erase
d.rgb b=lsat7_2002_30 g=lsat7_2002_40 r=lsat7_2002_70
```

This will display a false color image in the GRASS monitor, as we have effectively assigned the red, near-infrared and mid-infrared channels to blue, green, and red display colors. Green vegetation remains green while unvegetated areas are reddish. Likewise, other channel combinations can be visualized.

To access additional images, we add the `landsat` MAPSET to the MAPSET search path (see Section 3.1.6). Then we visually compare the RGB composites of LANDSAT coverages of Wake county, from 1987 and 2002:

```
# add additional MAPSET to search path
g.mapsets add=landsat -p

g.region rast=lsat7_2002_10 -p
d.erase
# LANDSAT-TM5 from 1987
d.rgb b=lsat5_1987_10 g=lsat5_1987_20 r=lsat5_1987_30
# LANDSAT-TM7 from 2002, restore good color tables
i.landsat.rgb b=lsat7_2002_10 g=lsat7_2002_20 r=lsat7_2002_30
d.rgb b=lsat7_2002_10 g=lsat7_2002_20 r=lsat7_2002_30

# show metadata
r.info -h  lsat5_1987_10
r.info -h  lsat7_2002_10
```

Note that the color tables were optimized beforehand with `i.landsat.rgb` to gain relatively natural colorized composites (reverts above `r.colors` usage). We observe that urban areas have significantly grown in few years. A more detailed explanation of color models and image composites, we provide in Section 8.5.

The image histogram   The first step in analysing image data is to look at the channel histograms. Each histogram shows the frequencies of grey levels in an image representing the given channel. For each grey level, the number of pixels in the image is counted and drawn into a diagram. As noted above, the number of grey levels (or brightness levels) depends on the sensor. The x-axis of the diagram represents the grey levels, while the y-axis shows the number of pixels found at that grey level.

To calculate and display the histogram of a channel, open a monitor and run `d.histogram` with the name of the channel you are interested in as parameter. The histogram will be displayed within the monitor using the color coding from the image. If the histogram is displayed in dark colors, consider modifying the image color table (see above paragraph). In the following exam-

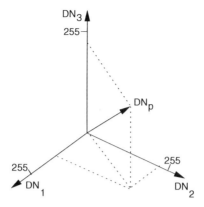

Fig. 8.4. Pixel in a three-dimensional feature space with digital number DN_p (adapted from Schowengerdt, 1997:119)

ple, we compare the histogram of the LANDSAT-TM7 panchromatic channel and that of an aerial ortho-photograph:

```
g.region rast=ortho_2001_t792_1m -p
r.info ortho_2001_t792_1m
d.erase
# show 1m orthophoto
d.rast ortho_2001_t792_1m
# show 15m panchromatic LANDSAT-TM7 channel
d.rast lsat7_2002_80

d.histogram ortho_2001_t792_1m
d.histogram lsat7_2002_80
```

The histograms are very different: While the distribution of the LANDSAT-TM7 panchromatic channel cell values is highly skewed, the histogram of the aerial ortho-photograph is more evenly distributed.

8.1.4 The feature space and image groups

The processing of data from a multispectral satellite sensor is based on the concept of feature space defined by the sensor channels. Together with the definition of an image group as a combination of multiple channels, this concept is a foundation for image classifications.

The feature space The channels of a multispectral satellite sensor are considered to span a multi-dimensional coordinate system called feature space. For example, the three channels covering visible light (blue, green, red) span a three-dimensional coordinate system. Within the coordinate system (or feature space), every pixel reaches a certain position depending on the bright-

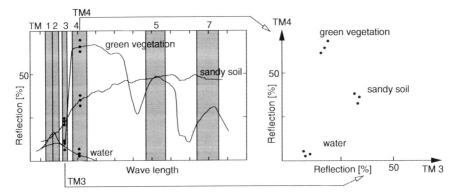

Fig. 8.5. Left: Spectrum showing typical spectral response of common objects with LANDSAT-TM5 channels 3 (red) and 4 (NIR); right: Two-dimensional feature space of channels 3 and 4 with pixel brightness levels. Three pixels for each observed object (water, sandy soil and green vegetation) are shown with their brightness levels in 3 and 4. The feature space scatterplot (right) represents reflection per channel as appear in channels 3 and 4 (adapted from Neteler, 2000:158)

ness levels in each channel. This position can be considered a vector in the multi-dimensional coordinate system. The brightness levels of the pixels in the different channels are called digital numbers (DN). Figure 8.4 shows the position of a pixel in a three-dimensional feature space, which may represent the blue (DN_1), green (DN_2), and red (DN_3) spectrum range.

The concept of feature space plays an important role in image classifications that are used to derive land use maps from satellite data. Classification methods are based on the idea that pixels containing the same land use are close to each other within the multi-dimensional feature space. The number of dimensions depends on the number of input channels. For example, a number of multispectral pixels which cover a forest will show similar spectral signatures and therefore should be assigned to land use class "forest". In the process of classification, many methods (e.g., cluster algorithm) "group" adjacent pixels in the multi-dimensional feature space and assign them to the same class. Figure 8.5 shows the relationship between the spectrum and a two-dimensional feature space spanned by the red and the near-infrared channel of LANDSAT-TM5. More details on image classification will be explained below.

For illustration, we display the feature space of two channels of the LANDSAT-TM5 satellite scene in the GRASS monitor (red and near-infrared channels):

```
g.region rast=lsat5_1987_30 -p
d.erase
d.correlate layer1=lsat5_1987_30 layer2=lsat5_1987_40
```

The pixel clouds from the red and near-infrared channels span the two-dimensional feature space, which would be partitioned in a classification to extract land use classes. However, in this scatterplot, it is difficult to visually distinguish clustered areas. If you zoom into a small subregion of one of the satellite channels and re-run the d.correlate script, the scatterplot will be different, with easier to distinguish pixel clusters. For higher level graphical data analysis external software like "R" (compare Section 10.2) or "GGobi"[3] are recommended.

Image groups in GRASS   Since we are dealing with multi-spectral data sets, we need a method to "bundle" the channels which belong together. This helps when operating on multiple channels with identical geolocation. GRASS offers a tool to "group" images by selecting the channels: i.group. Several multi-spectral image processing modules expect an image group, a few of them also need a subgroup (which may contain the same channels or a subset). Even when working only with a single image, this channel has to be assigned to a group. We are now ready to explore more complex remote sensing tasks.

8.2 Data preprocessing

In this section, we explain how to preprocess satellite data for further analysis. Since imagery data are usually already geocoded, we concentrate on radiometric preprocessing to statistically explore the data and to extract further information.

8.2.1 Radiometric preprocessing

Besides changes to the color lookup tables (LUTs) that are used to enhance the visual perception of an image, satellite data often have to be radiometrically preprocessed so that each pixel represents the apparent radiance measured at the satellite sensor. Up to three major effects have to be corrected, depending on the project goal, the image type, and the observed targets:

- the pixel values are usually a linear transformation of the original data performed by the data provider to fit into the range of e.g. 8bit (0 - 255). By applying "gain" and "bias" (also called "offset", see Section 8.1.2) values which are delivered in the image header files or available from the data provider, the DN values (DN: digital number) can be recalculated to apparent radiance at sensor values;
- optical data, depending on their spectral range, are influenced by atmospheric effects. To reconstruct the reflectance values at earth's surface (image includes only the values measured at the satellite), each satellite channel has to be atmospherically corrected;

---

[3] Xgobi/Ggobi software, http://www.ggobi.org

- when the observed target area contains hilly or mountainous regions, the
  slopes cause variations in the brightness reflectance (terrain effects), which
  can lead to wrong classification results. To overcome this problem, a terrain
  correction (illumination correction) based on the local slopes derived from
  an elevation model has to be applied. This issue is not covered here.

Image calibration from DN to apparent radiance at sensor The
gain/bias correction is applied channel-wise to the image data set. The values
for the gains and biases are available from the channel headers (leader file) or
the data providers. For the sample data sets, these values are included in the
raster map metadata and can be shown with r.info.

For LANDSAT-TM7, there are two gain states (low and high gain, see
GSFC/NASA, 2001). The rationale behind switching gain states is to maxi-
mize the instrument's 8bit radiometric resolution without saturating the de-
tectors. For thermal data, both low and high gain data are available by default.
For other bands (1 to 5, and 7) the satellite will acquire image data in one
of two possible gain settings. High gain measures a lesser radiance range with
increased sensitivity over areas of low reflectance. Low gain setting measures
a greater radiance with decreased sensor sensitivity for very bright regions to
avoid detector saturation.

The header/leader file of a satellite data set can be analyzed with gdalinfo
(delivered with GDAL library) for GDAL supported data formats. The program prints important metadata information, including the gain and bias
values, if they are present in the data set (see Section 8.1.2). From the
output of the original imagery file, we obtain low/high gain, bias (offset)
values and more. The units for gain/bias are usually $\frac{W}{m^2 sr \mu m}$. The general equation for calculation of the apparent pixel radiance at sensor is
(Schowengerdt, 1997:313, see also LANDSAT Handbook[4]):

$$L_j = gain_j * DN_j + bias_j \qquad (8.1)$$

with:
L_j: apparent pixel radiance of channel j [$\frac{W}{m^2 sr \mu m}$]
$bias_j$: offset of linear equation for channel j
$gain_j$: gain of linear equation for channel j

To apply a gain/bias correction, the module r.mapcalc can be used. For our
example, we use the LANDSAT-TM7 scene of 2002:

```
g.region rast=lsat7_2002_10 -p
# visual inspection
d.erase
d.rgb b=lsat7_2002_10 g=lsat7_2002_20 r=lsat7_2002_30
d.vect roadsmajor col=red
```

---

[4] LANDSAT Science Data Users Handbook,
   http://ltpwww.gsfc.nasa.gov/IAS/handbook/handbook_htmls/chapter11/
   chapter11.html

```
# show metadata
r.info lsat7_2002_10
```

Instead of gain and bias, the values are given here as QCALMIN, QCALMAX, LMIN, and LMAX. The related formula to calculate the spectral radiance at the sensor is (GSFC/NASA, 2001):

$$L_\lambda = ((LMAX_\lambda - LMIN_\lambda)/(QCALMAX - QCALMIN))$$
$$*(QCAL - QCALMIN) + LMIN_\lambda \qquad (8.2)$$

```
# convert pixel values to radiances: see p016r035_7x20020524.met
#   LMAX_BAND1=19
#   LMIN_BAND1=-6.200
#   QCALMAX_BAND1=255.0
#   QCALMIN_BAND1=1.0
#   QCAL is the quantized calibrated pixel value in DN
r.mapcalc "lsat7_2002_10.rad= ((19.0 - (-6.2))/(255.0 - 1.0)) \
          * (lsat7_2002_10 - 1.0) + (-6.2)"
r.info -r lsat7_2002_10.rad
 min=-2.132283
 max=19.000000
```

Note that the values depend on the data provider and the image acquisition date as gain/bias values regularly change for various reasons.

With additional calculations, it is also possible to convert apparent pixel radiance at sensor to planetary reflectance or albedo (see Mather, 1999:93 and Schowengerdt, 1997:317). These planetary reflectances can be computed to achieve a reduction in between-scene variability through a normalization for solar irradiance. Please refer to the remote sensing literature for details.

Correction of atmospheric effects   Satellite signal distortions are caused by several effects. Diffuse irradiance from sky may increase the radiance of an observed object. Path radiance (atmospheric intrinsic radiance) leads to haze effects. Local effects such as environmental radiance from neighborhood objects change the object's radiance, as well as a locally reduced upward transmittance. Finally, there is the adjacency effect, when a brighter adjacent object influences the surrounding object's radiation. All these problems are widely discussed in the remote sensing literature, see for example Schowengerdt (1997). Atmospheric effects are visible in color composites as a whitish-bluish haze.

The correction of such atmospheric effects is a complex issue. Using an atmosphere model like 6S (*Second Simulation of the Satellite Signal in the Solar Spectrum*[5], Vermote et al., 1997), the radiance at earth's surface can

---

[5] Atmosphere model 6S (Msix) software,
   http://www-loa.univ-lille1.fr/SOFTWARE/Msixs/msixs_gb.html

be reconstructed from the apparent radiance at sensor if the local weather conditions at image acquisition time are known. For a scripted method to use the 6S model within GRASS, see Neteler (1999). In GRASS 6.3, the 6S model has been implemented in a new module i.atcorr.

As detailed information about the local weather conditions and gaseous contents are often unknown, to some extent the atmospheric effects can be retrieved statistically from the image channels themselves. Known dark objects (e.g., water bodies or coniferous forest) can be used to do so. In an uncorrected image, these objects do not appear dark due to atmospheric effects. The amount of path radiance is approximately identified by calculating pixelwise the difference between the actual radiance for a dark object and zero (full absorption, given for water in infrared). This difference value can be removed from all pixels of the channel. Details are described in Moran et al. (1992) and Chavez Jr. (1996). The modules d.what.rast or r.what can be used to calculate the path radiance for dark objects, the subtraction can be done with r.mapcalc. A simpler method, not considered in detail here, is based on the Tasseled Cap transformation. It does not require the manual identification of dark objects and corrects the data set through a "haze" image (Tasseled Cap component TC4) and linear regression. The GRASS AddOns SVN repository contains a i.landsat.dehaze script for LANDSAT-TM.

8.2.2 Deriving a surface temperature map from thermal channel

Several satellites such as ASTER/TERRA, LANDSAT-TM5 and LANDSAT-TM7 provide thermal channels. The data delivered by a thermal channel (channel 6 for LANDSAT systems) can be calibrated to a surface temperature map. These surface temperatures must not be confused with air temperatures. Note that the methods are different for LANDSAT-TM5 and LANDSAT-TM7, as their gain/bias values are different. For an absolute calibration of satellite-derived temperatures, atmospheric correction has to be taken into account.

Surface temperature map from LANDSAT-TM5 channel 6   The following calculations derive the effective at-satellite temperatures (LANDSAT-TM5) of the viewed earth-atmosphere system under an assumption of unity emissivity and using pre-launch calibration constants. First, the gain/bias values are applied to the thermal channel to receive spectral radiances (Barsi et al., 2003), then the resulting pixel values are converted to absolute temperature in Kelvin. Optionally, the result can be recalculated to a degree Celsius temperature map. In MAPSET landsat, a LANDSAT-TM5 scene is available:[6]

---

[6] GLCF Maryland data set "p016r35_5t871014",
   ftp://ftp.glcf.umiacs.umd.edu/glcf/Landsat/WRS2/p016/r035/p016r35_
   5t871014.TM-EarthSat-Orthorectified/

```
# convert TM5/B6 digital numbers (DN) to spectral radiances
# (apparent radiance at sensor): radiance = gain * DN + offset
g.region rast=lsat5_1987_60 -p
r.info lsat5_1987_60


r.mapcalc "lsat5_1987_60.rad=0.0551584*lsat5_1987_60+1.2378"
r.info -r lsat5_1987_60.rad
 min=5.650472
 max=8.463550


# convert spectral radiances to absolute temperatures
# T = K2/ln(K1/L_l + 1))
r.mapcalc "tm5.temp_kelvin=1260.56 / \
         (log (607.76/lsat5_1987_60.rad + 1.))"
r.info -r tm5.temp_kelvin
 min=268.931214
 max=293.984778


# convert to degree Celsius
r.mapcalc "tm5.temp_celsius=tm5.temp_kelvin - 273.15"
r.info -r tm5.temp_celsius
 min=-4.218786
 max=20.834778
r.univar tm5.temp_celsius
 [...]
 range: 25.0536
 mean: 13.3068
 standard deviation: 1.65198
 [...]
```

The resulting Land Surface Temperature (LST) map for the 14 Oct. 1987 shows an average LST of 13.3° Celsius with a range of pixels with lower temperatures (such as cloud top temperatures) and higher temperatures (usually urban areas)[7] We apply a blue-green-yellow-red color ramp table before displaying the LST map:

```
# apply new color table, display
r.colors tm5.temp_celsius col=bgyr
d.rast.leg tm5.temp_celsius
d.vect roadsmajor
```

The map `tm5.temp_celsius` shows the distributed emitted thermal radiation in degree Celsius. The surface brightness temperature is the actual surface temperature only when the emissivity of the object in a particular waveband equals to 1.0. For most surfaces, where the emissivity is near, but not equal to 1.0, a calibration according to the Stefan-Boltzmann equation is needed when interpreting the results.

---

[7] NC Climatic Data, monthly normals,
   http://www.nc-climate.ncsu.edu/cronos/normals.php?station=317079

We can compare the LST map to the near-natural color composite (visible light channels) and extract only areas with high LST:

```
d.rgb b=lsat5_1987_10 g=lsat5_1987_20 r=lsat5_1987_30

# show areas with high LST
d.erase -f
d.rast -o tm5.temp_celsius val=15-50
d.vect streets_wake
```

Clearly, the urban areas show high surface temperatures. To get a more detailed comparison, we can compare to the landuse map landuse96_28m (generated a year before the LANDSAT overpass):

```
g.region rast=landuse96_28m -p
r.to.vect landuse96_28m out=landuse96_28m feat=area
d.erase
d.rast tm5.temp_celsius
d.vect landuse96_28m type=boundary

# interrogate selected polygons
d.what.vect

# generate report
r.report tm5.temp_celsius,landuse96_28m units=p,h nsteps=5
```

You may even re-generate a land use/land cover map through image classification as shown later in Section 8.6 where you apply the individual emissivity factors according to land use. With r.mapcalc, you can calibrate the landuse corrected temperature map from these maps.

Surface temperature map from LANDSAT-TM7 channel 6 As in the previous case, the LANDSAT-TM7 image data have to be converted from digital numbers to spectral radiances by applying the gain/bias values. Depending on the data format, these values may be retrieved from the image metadata with gdalinfo. In our example, we use the LANDSAT-TM7 scene as prepared in Section 3.3.3 for the nc_spm sample LOCATION. However, the original data are provided in GeoTIFF format. This requires to look up the gain and bias parameters from the accompanying metadata file which is separated from the image data.

We use the low gain thermal channel lsat7_2002_61 which is available in the landsat MAPSET. The conversion procedure is the same as outlined in Section 8.2.1:[8]

```
# show calibration metadata
r.info lsat7_2002_61
```

---

[8] Generating a surface temperature map from LANDSAT-TM7 channel 6, see *Landsat 7 Science Data Users Handbook*,
http://ltpwww.gsfc.nasa.gov/IAS/handbook/handbook_toc.html

```
# convert TM7/B61 digital numbers (DN) to spectral radiances
g.region rast=lsat7_2002_61 -p
r.mapcalc "lsat7_2002_61.rad=((17.0 - 0.) / \
        (255. - 1.))*(lsat7_2002_61 - 1.) + 0."
r.info -r lsat7_2002_61.rad

# convert spectral radiances to absolute temperatures
#   T = K2/ln(K1/L_l + 1))
#   K1: 666.09 W/ln(m^2 * sr * um)
#   K2: 1282.71 Kelvin
r.mapcalc "etm.temp_kelvin=1282.71 / \
        log(666.09 / lsat7_2002_61.rad + 1.)"

# calculate degree Celsius
r.mapcalc "etm.temp_celsius=etm.temp_kelvin - 273.15"
r.info -r etm.temp_celsius
 min=5.072581
 max=52.250418

# apply new color table, display
r.colors etm.temp_celsius col=bgyr
# display the map, overlay to ETM/PAN (B80) 14.25m channel
g.region rast=lsat7_2002_80 -p
d.his i=lsat7_2002_80 h=etm.temp_celsius bright=70
d.legend etm.temp_celsius
```

The resulting temperature map (in degree Celsius) represents the uncorrected surface temperatures at image acquisition time (around 9:30h local solar time, 24 May 2002), see notes above for emissivity correction. For deriving these maps from other satellites such as ASTER/TERRA, please refer to the related documents.[9]

8.3 Radiometric transformations and image enhancements

Various methods have been developed for the analysis of multi-channel satellite data using their multispectral nature for radiometric transformations and image enhancement. These techniques play a fundamental role in image interpretation. Most methods may either be applied to uncalibrated data sets or to preprocessed image data sets.

8.3.1 Image ratios

Image ratios are the basis of a simple algebraic method used for feature extraction, reduction of terrain illumination effects, image enhancement, computation of vegetation indices and more (this topic is widely discussed in various

---

[9] Asterweb (ASTER/TERRA), http://asterweb.jpl.nasa.gov

papers, for example, refer to Mather, 1999:117-124). To understand a particular channel ratio formula, the object reflectance curves have to be considered (sample curves for green vegetation, soil and water are shown in Figure 8.2). In general, the ratio result for pixels with very different values for the input channels is larger (brighter) than for pixels with similar values. The image ratio equations can be computed with **r.mapcalc**. It is important to include a multiplier of 1.0 at the *beginning* of the map algebra expression because we are dividing integer values. Otherwise, the result will become zero and not the expected floating point numbers. As an example, we calculate the ratio between the channels 7 and 4 of LANDSAT-TM7:

```
g.region rast=lsat7_2002_70 -p
r.mapcalc "ratio7_4=1.0 * lsat7_2002_70/lsat7_2002_40"
d.erase
d.rast ratio7_4
```

For more than 15 years, a variety of vegetation indices have been developed. To illustrate such a calculation, we can compute a NDVI map (normalized difference vegetation index) from LANDSAT-TM5/7:

```
r.mapcalc "ndvi=1.0 * (lsat7_2002_40 - lsat7_2002_30) / \
          (lsat7_2002_40 + lsat7_2002_30)"
r.colors ndvi col=rules << EOF
-1.0000 blue
-0.40 40 40 255
-0.310 220 220 250
 0.0000 150 150 150
 0.1000 120 100 51
 0.3000 120 200 100
 0.4000 28 144 3
 0.6000  6 55  0
 0.8000 10 30 25
 1.0000  6 27  7
EOF
d.rast.leg ndvi
d.vect roadsmajor col=red
# transparently over shaded DEM
d.his i=elevation_shade h=ndvi bright=70
```

The calculation of NDVI uses the pixel-wise differences between the red and the infrared channel to derive information about the land cover. When a pixel value in the near-infrared dominates over the red wavelength (as for green, healthy vegetation), NDVI is positive. NDVI for unvegetated soil is around zero; for water, below zero. To quickly classify these three (or more) landcovers, you may filter them with **r.mapcalc** (if-condition).

8.3.2 Principal Component Transformation

Multispectral image channels often contain correlations due to similarities of the spectral response of the observed objects or slightly overlapping filter functions of the spectral sensors. This leads to redundancies within the data set. The "Principal Component Transformation" (PCT) method has been developed to transform such a data set to a new data set without correlations between the channels. This will concentrate the image information in fewer image channels (reduction of image dimensionality), which is of particular interest for hyperspectral data. The PCT transforms the original multispectral data set to a new spectral coordinate system, the Principal Component axes, which are orthogonal to each other. Figure 8.6 shows the position of original multispectral pixels and the PCT coordinate system. In general, the first principal component (PC) image contains the maximum possible variance of the original images. The second principal component image contains the maximum possible variance not stored in the first PC image, as the second PC axis is orthogonal to the first PC axis (Schowengerdt, 1997:191). Accordingly, higher PC images explain remaining variances. The number of PC images is identical to the number of input channels. Since the amount of variance decreases from the first to the last PC, uncorrelated noise (and sometimes some remaining high frequencies) is found in the last PC image. As a result, the method is sometimes used for image compression, as it allows the image information to be concentrated in fewer channels. PCT is also sometimes used to generate additional channels to obtain more variables for later classification process. The scatterplot in Figure 8.7 shows the original spectral axes and, after transformation, the new rotated PC axes for a sample LANDSAT-TM5 pixel cloud (channel 3 and 4).

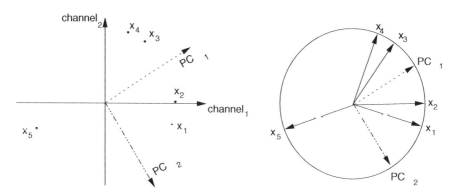

Fig. 8.6. Left: Multispectral pixel values shown as standardized data vectors x_1 to x_5 with related first and second orthogonal principal component vectors PCT_1 and PCT_2 in coordinates view. Right: Same data vectors in circle coordinates view

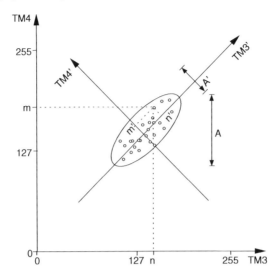

Fig. 8.7. Principal Component Transformation applied to channels tm3 and tm4 of a LANDSAT-TM5 data set. Both the original spectral axes (channels tm3, tm4) and the PC axes (PCT transformed channels tm3', tm4') are shown

In GRASS, the Principal Component Transformation is implemented in i.pca. The module requires the input channel names (at least two images) and a prefix for the transformed PC image files, which will be enumerated incrementally. Optionally, the data can be rescaled to a range different from the default range of 0-255:

```
g.region rast=lsat7_2002_10 -p
i.pca in=lsat5_1987_10,lsat5_1987_20,lsat5_1987_30,\
lsat7_2002_10,lsat7_2002_20,lsat7_2002_30 out=pca
d.erase
d.rast pca.1
d.rast pca.2
d.rast pca.3
d.rast pca.4
```

We can use the PCA to perform a simple change detection analysis:

```
d.rgb b=pca.1 g=pca.2 r=pca.3
d.rgb b=lsat5_1987_10 g=lsat5_1987_20 r=lsat5_1987_30
d.rgb b=lsat7_2002_10 g=lsat7_2002_20 r=lsat7_2002_30
r.univar -e pca.3
 [...]
 minimum: 0
 maximum: 255
 range: 255
 mean: 129.874
 [...]
```

```
3rd quartile: 134
90th percentile: 137

# simple change detection:
# consider high PCA values in high PCA component as change
r.mapcalc "changes=if(pca.3 > 134,1,null())"

# vectorize and remove small areas
# (3 pixel * 28.5m)^2 = 7310.2m^2
r.to.vect -s changes out=changes feat=area
v.clean changes out=major_changes tool=rmarea thresh=7300

# overlay to Oct/1987 map
d.rgb b=lsat5_1987_10 g=lsat5_1987_20 r=lsat5_1987_30
d.vect major_changes type=boundary col=red
d.vect lakes fcol=blue type=area

# overlay to May/2002 map
d.rgb b=lsat7_2002_10 g=lsat7_2002_20 r=lsat7_2002_30
d.vect major_changes type=boundary col=red
d.vect lakes fcol=blue type=area
```

We overlay the `lakes` map to omit those changes caused by water turbu-
lences etc. The resulting vector polygons capture significant land use changes,
especially conversion from vegetated area to developed area.

Another method not covered here is the Fourier transform, which is provided by `i.fft` and `i.ifft` (forward and backward transformation). It transforms image data from spatial to frequency domain. Among the important
applications of the Fourier Transformation are the identification and elimination of (periodic) noise or stripes in a satellite image.

8.4 Geometric feature analysis with matrix filters

The geometric (spatial) feature analysis applies local neighborhood operations
to raster data. Several methods are available for image smoothing: contrast
improvement, low- and high pass filtering, edge detection and more. Geometric
filters are user defined raster matrix templates ("moving window") that are
applied row- and column-wise over the image and are used to calculate the
new raster map. All raster cells which are covered by this moving window are
considered for the calculations.

Matrix filters can be used to locally modify the spatial frequency characteristics for an image. These modifications are based on calculations considering
the neighboring raster pixels in a 2D spatial convolution process (for theoretical details, refer to Richards and Xiuping, 1999:114-116). These spatial
convolution filters operate in spatial domain and are an alternative to fre-

quency domain filters (such as Fourier Transformation). Spatial convolution filtering is well suited for:

- high and low pass filtering (sharpening, blurring), averaging;
- edge detection by direction and gradient filters;
- morphological filters;
- preprocessing for image segmentation.

The high and low pass filtering can be performed using r.mapcalc or the module r.mfilter, as both are available to define matrix filters.

The use of r.mapcalc is less convenient, as every matrix element has to be addressed with relative coordinates to the central cell expressed as map[r,c], where r is the row offset and c is the column offset. For example, map[0,2] refers to the same row as the center cell and two columns to the right of the center cell, map[-2,-1] refers to the cell two rows up and one column to the left of the center cell. This syntax permits the development of neighborhood-type filters for one single map or across multiple maps. As a simple example, we define a 3×3 low pass filter. The filter equation for the each center cell x_c is:

$$x_c = \frac{1}{9} \sum_{i=1}^{9} x_i \qquad (8.3)$$

expressed in r.mapcalc as:

```
r.mapcalc "lowpass=(map[1,-1]+map[1,0]+map[1,1]+map[0,-1]+\
      map[0,0]+map[0,1]+map[-1,-1]+map[-1,0]+map[-1,1])/9."
```

You may try this example with the lsat7_2002_80 image. Further examples for r.mapcalc matrix operations are described in Shapiro and Westervelt (1992).

A convenient way to perform spatial convolution filtering is to use r.mfilter with a matrix template defined in an ASCII file. We extend our first example to a 7×7 median filter which filters existing sharp contrasts in a raster map. This is effectively a low pass filter. The filter definition has to be stored in an ASCII file, for example, lowpass.asc:

```
TITLE 7x7 Low pass
MATRIX 7
1 1 1 1 1 1 1
1 1 1 1 1 1 1
1 1 1 1 1 1 1
1 1 1 1 1 1 1
1 1 1 1 1 1 1
1 1 1 1 1 1 1
1 1 1 1 1 1 1
DIVISOR 49
TYPE P
```

The mean value is preserved if the sum of the filter values equals the line-number × column-number. Every cell within the "moving window" is multiplied by 1. The results are summed up and finally divided by DIVISOR 49 (product of 7 × 7 × 1). To see how this works, the filter may be applied to the LANDSAT-TM7 panchromatic channel:

```
g.region rast=lsat7_2002_80 -p
r.mfilter lsat7_2002_80 out=etm80.lowpass filt=lowpass.asc
r.colors etm80.lowpass col=grey.eq
d.erase
d.rast etm80.lowpass
```

The color table may be set to grey or grey.eq with r.colors as in the example above.

It is possible to define two types of filters: *sequential* and *parallel* filters. Sequential filters (TYPE S) use the modified neighboring raster cell values for calculation of the central cell, while the parallel filters (TYPE P) use the neighboring cell values of the original map. Directional filters should be set up as parallel filters. Further information related to these types can be found in the manual page for r.mfilter.

Another example is a high pass filter for sharpening an image. It can be defined as follows (we store it in an ASCII file highpass.asc):

```
TITLE 5x5 High pass
MATRIX 5
-1 -1 -1 -1 -1
-1 -1 -1 -1 -1
-1 -1 24 -1 -1
-1 -1 -1 -1 -1
-1 -1 -1 -1 -1
DIVISOR 25
TYPE P
```

In this example the central cell of the window is weighted by 24, while the other cells have weight -1. The entire matrix is finally divided by 25 and its values are stored in a new map. Again, we apply it to the map lsat7_2002_80:

```
g.region rast=lsat7_2002_80 -p
r.mfilter lsat7_2002_80 out=etm80.highpass filt=highpass.asc
r.colors etm80.highpass col=grey.eq
d.erase
d.rast etm80.highpass
```

The resulting map shows enhanced high frequencies (at the same spatial resolution). Note that a filter definition file may also contain multiple filters which will be applied to the image subsequently.

The only limitation of r.mfilter in comparison to r.mapcalc is that only integer numbers are accepted in a filter matrix. If you want to use floating point numbers or trigonometric functions, r.mapcalc must be used instead.

The latter is also well suited for a thresholded binarization used to extract selected features (if-condition).

An application for advanced edge detection with vectorization based on segmentation is explained for aerial photographs in Section 8.8.

8.5 Image fusion

Often, satellite data sets with high radiometric resolution (multispectral channels) lack high geometric resolution and vice versa. However, for an accurate image interpretation, both radiometric and geometric resolution should be high. Image fusion is a method to geometrically enhance images with high radiometric resolution by merging the multispectral channels with a panchromatic image. Different image fusion methods have been developed; two basic methods will be described in the following sections.

8.5.1 Introduction to RGB and IHS color model

To understand image fusion methods operating in color space, it is important to have basic knowledge about the RGB (red, green, blue) and IHS (also referred to as HIS or HSI: intensity, hue, saturation) color spaces. Similarly to geometrical data, the color spaces span their own coordinate systems. Due to their definitions, it is possible to convert images lossless from one color model to the other. The RGB model is an additive color model, where new colors are derived by adding the three base colors at different levels. For example: yellow = red + green. The IHS model is different; here, the intensity (sometimes also

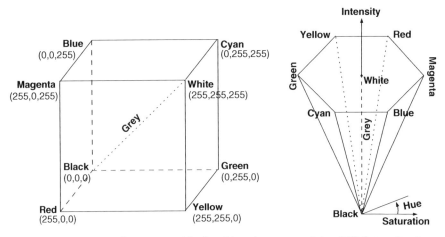

Fig. 8.8. Left: RGB (red, green, blue) cubic color space; right: IHS (intensity, hue, saturation) hexcone color space (adapted from Mather, 1999:99)

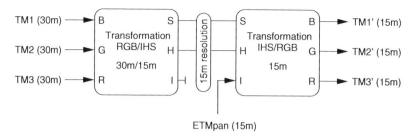

Fig. 8.9. Geometric resolution improvement of LANDSAT-TM7 data (IHS image fusion method). After conversion from RGB to IHS color model, the resolution is changed from 30m to 15m. The high resolution panchromatic channel replaces the original intensity channel before converting back to RGB color model

called "value") is a measure of color brightness, the hue corresponds with the dominant wavelength (which is related to color names), and the saturation describes degree of color purity.

Figure 8.8 shows both the RGB and the IHS color model. A pixel in the RGB color space has a specific position within the cube spanned by the coordinate axes, while the IHS color space forms a hexcone. The main advantage of RGB is that it is easy to understand; however, intensity changes are dependent on color settings. Thus, in the RGB model, a change in intensity always leads to a change in colors. The IHS color model preserves colors in case of intensity changes which is a major advantage of this model. Based on this feature, the IHS model can be used for image fusion, which we explain below. GRASS provides two color conversion modules, the i.rgb.his to convert an image from RGB to IHS and i.his.rgb to convert back from IHS to RGB.

8.5.2 Image fusion with the IHS transformation

For image fusion, two geometrically co-registered data sets are required. The acquisition time of these data sets should be very close to avoid possible modification of the result by land use changes. For IHS-fusion, the three RGB channels must first be transformed to the IHS color model. The general idea of IHS-fusion is to replace the intensity channel with a high resolution panchromatic channel for the back-transformation from the IHS to RGB color model. As a result, the color information in lower resolution is merged with the high spatial resolution of the panchromatic channel. In terms of GIS, a resolution change is required before back-transforming the images to achieve the higher spatial resolution in the output. A disadvantage of this fusion method is that this technique changes the spectral characteristics of the data.

As an example, we enhance the geometrical resolution of geocoded LANDSAT-TM7 color channels (each at 28.5m resolution) using the panchromatic ETMPAN channel (at 14.25m resolution) of the same satellite acquired

at the same time. Before starting the procedure outlined in Figure 8.9, the
contrast in the input channels should be enhanced with r.colors. The input
channels are then converted to the IHS color model with i.rgb.his at 28.5m
resolution. Then we set the region to the higher resolution defined by the
panchromatic channel. In case of LANDSAT-TM7, we change it from 28.5m
to 14.25m; for SPOT data, from 20m to 10m. To improve the geometric resolution, the original intensity image which resulted from the RGB to IHS transformation is replaced by the panchromatic channel for back-transformation
to the RGB color model. Finally, three new RGB channels at 14.25m resolution containing the multispectral information from the input channels are
generated. The GRASS procedure is as follows:

```
# if not done yet, apply a contrast stretch (histogram equal.)
g.region rast=lsat7_2002_10 -p
r.colors lsat7_2002_10 color=grey.eq
r.colors lsat7_2002_20 color=grey.eq
r.colors lsat7_2002_30 color=grey.eq

# RGB view of RGB channels
d.erase
d.rgb b=lsat7_2002_10 g=lsat7_2002_20 r=lsat7_2002_30

# RGB/IHS conversion
i.rgb.his blue=lsat7_2002_10 green=lsat7_2002_20 \
        red=lsat7_2002_30 hue=hue intensity=int saturation=sat

# IHS/RGB back conv. with ETMPAN replacing old intens. image
g.region rast=lsat7_2002_80 -p
i.his.rgb hue=hue intensity=lsat7_2002_80 \
      saturation=sat blue=etm.1_15 green=etm.2_15 red=etm.3_15

# color contrast enhancement
r.colors etm.1_15 color=grey.eq
r.colors etm.2_15 color=grey.eq
r.colors etm.3_15 color=grey.eq

# visualize higher resolution color composite
d.erase
d.rgb b=etm.1_15 g=etm.2_15 r=etm.3_15
```

The resulting color composite combines the three color channels with improved geometrical resolution. More complex merging procedures can also be
performed. For geological applications, the use of ratio calculations (generated
by r.mapcalc) is recommended, as they can be used as input into the fusion
instead of the common multispectral input channels.

Fig. 8.10. Left: Standard composite of LANDSAT-TM7 RGB channels (28.5m resolution); right: Image fusion of LANDSAT-TM7 RGB channels (28.5m) with ETMPAN (14.25m) with Brovey transform leading to resolution enhanced image (grey scale reproduction of color original)

8.5.3 Image fusion with Brovey transform

An alternate method for image fusion is the Brovey transform (described in Pohl and van Genderen, 1998, and among other methods in Zhou et al., 1998). The Brovey transform method is implemented in GRASS as i.fusion.brovey. The formula was originally developed for LANDSAT-TM5 and SPOT, but it also works well with LANDSAT-TM7. You need the panchromatic channel to be spatially co-registered to the multispectral channels. Image fusion based on Brovey transform for LANDSAT-TM7 data merges the channels 2, 4, and 5 (all at 28.5m resolution) with the panchromatic ETMPAN channel (at 14.25m resolution):

```
g.region rast=lsat7_2002_10 -p
d.erase
i.fusion.brovey -l ms1=lsat7_2002_10 ms2=lsat7_2002_20 \
    ms3=lsat7_2002_30 pan=lsat7_2002_80 out=brov
# it temporarily sets raster resolution to PAN resolution: 14.25

# original RGB
d.erase
d.rgb b=lsat7_2002_10 g=lsat7_2002_20 r=lsat7_2002_30

# Brovey transformed RGB
g.region -p rast=brov.red
d.erase
d.rgb b=brov.blue g=brov.green r=brov.red
d.vect streets_wake col=red
```

The result provides improved spatial resolution. You may consider modifying the color tables of the resulting channels to optimize the color quality to again achieve a near-natural color image:

```
i.landsat.rgb b=brov.blue g=brov.green r=brov.red
# note reversed B and G channel assignment
d.rgb g=brov.blue b=brov.green r=brov.red
```

Figure 8.10 shows an example for image fusion with LANDSAT-TM7 data.

8.6 Thematic classification of satellite data

One of the main goals of satellite remote sensing is to derive thematic maps describing the current land use/land cover of the earth's surface. In the GIS context, these maps are often used to update maps generated by conventional techniques. Common multispectral classification algorithms treat the multi-channel images as variables for a classification process. The resulting classes describe the dominating land use or land cover in a certain area, where the land use is considered locally homogeneous. Numerous classification methods have been developed; GRASS provides capabilities for a set of standard approaches. Due to its Open Source nature, additional methods can be directly implemented in C or other programming languages.

When reclassifying multispectral satellite data, the image data set is analyzed pixel-wise, with the values of all channels being taken into account for each pixel. The number of pixels covering the same geographic region depends on the number of channels. This group of pixel values describing the same small area is called the spectral vector. It describes its specific position in the feature space (compare Figure 8.5 earlier in this chapter) which contains all spectral vectors. Within the feature space, the classification algorithm tries to separate similar spectral vectors which vary depending on the observed object types as soil, vegetation, water bodies etc. Similar spectral vectors will be assigned to the same class. All classes are finally stored in a thematic map where each class describes the dominating land use type.

Local variations due to changes within and between the observed objects, terrain slope and aspect changes, and variations of the atmospheric conditions (haze, dust, clouds etc.) present a problem for the classification process. Depending on the observed area, data have to be radiometrically preprocessed as described in the previous section to minimize influences from slope/aspect and atmospheric effects.

In general, two strategies – *unsupervised* and *supervised* – are common for the classification of remote sensing data (Figure 8.11). For both methods, the classification process requires two major steps. First, the data have to be analyzed for similarities in their spectral responses and then, pixels have to be assigned to classes. The unsupervised method is fully automated based on image statistics, but it delivers only abstract class numbers. The main task then is to find a reasonable number of clusters/classes and assign ground truth information to these classes. The supervised classification requires user interaction, as training areas covering known land use have to be digitized. Image statistics are automatically derived from these training areas and used for the final classification.

Common classification methods (MLC - Maximum Likelihood classifier as described in Sections 8.6.1 and 8.6.2) are pixel-based. GRASS additionally provides a different method (SMAP - Sequential Maximum A Posteriori classifier, see Section 8.6.3) which also takes into account that neighborhood pixels may be similar. The fact that a neighborhood of similar pixels will lead to spatial autocorrelation is used to improve the result. Altogether, four different approaches for satellite analysis within two main groups of classification methods are available:

- Radiometric classification:
    - unsupervised classification (`i.cluster`, `i.maxlik` using the Maximum Likelihood classification (MLC) method),
    - supervised classification, and
    - partially supervised classification (`i.class`, `i.gensig`, `i.maxlik`),

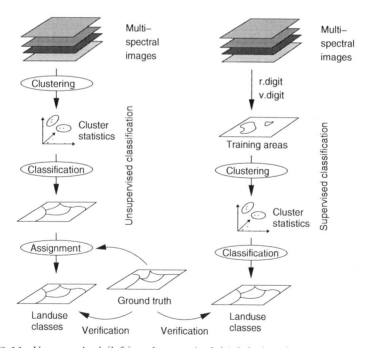

Fig. 8.11. Unsupervised (left) and supervised (right) classification procedures for multispectral data

- Combined radiometric/geometric supervised classification (i.gensigset, i.smap using the Sequential Maximum A Posteriori classification (SMAP) method)

A common problem in remote sensing of the environment are mixed pixels which cover various objects (field borders, urban areas, etc.). In this case, the mentioned methods will assign the pixel dominating object to a class, usually at a low confidence level. If appropriate, masking out settlements where lots of mixed pixels appear may be considered. Subpixel analysis methods such as "Spectral Mixture Analysis" are a way to overcome this problem (for a GRASS implementation, see Neteler, 1999).

Other classifiers such as Artificial Neural Networks (ANN), k-Nearest Neighbor Classification (kNN), Support Vector Machines (SVM), classification trees (multiclass), and other methods are implemented in i.pr (get from the GRASS AddOns SVN repository) and in the R statistical language. The latter can be linked to GRASS using the GRASS/R interface; for an introduction to R, see Section 10.2.

8.6.1 Unsupervised radiometric classification

Unsupervised classification is the automated assignment of raster pixels to different spectral classes. The assignment is based only on the image statistics. The unsupervised classification is a two-step approach. First, a clustering algorithm groups pixel values with similar statistical properties according to user definitions of minimum cluster size, separability, number of clusters, etc. This approach is similar to the creation of a map legend, where the number of signatures existing in a map is identified and visualized. The pixel clusters are image categories that can be related to land cover types on the ground. The iterative clustering algorithm computes the cluster mean values and covariance matrices (module i.cluster), adjusting these values while reading the image data set. The idea is to identify pixel clouds from the feature space which have similar reflectance values in the various channels. Each pixel cloud, grouped into clusters which represent land use classes, characterizes the spectral signature of a certain object which will be later assigned to a class.

This cluster information is used to perform the spatial assignment of the individual pixels to the derived clusters (module i.maxlik). The MLC determines which spectral class each cell in the image belongs to with the highest probability. Internally, a Chi-square test is run with changing thresholds until a predefined convergence is reached (stability of the pixel assignment during the iteration steps). The result is a new map containing the classes. The MLC also stores the confidence level for each pixel belonging to a certain class in a second map. This map is called "reject threshold map layer" or "rejection map" and contains one calculated confidence level for each reclassified cell in the reclass map. High values in the rejection map represent a high rejection probability for the assigned class. One of the possible uses for this map is as a

MASK, to identify cells in the reclassified image that have the lowest probability of being assigned to the correct class. It is important to know that MLC assumes that the spectral signatures for each class are normally distributed (i.e., Gaussian in nature) which is often unrealistic. For a detailed discussion, see various remote sensing books such as Mather (1999).

First step: Clustering of image data The unsupervised classification starts with collecting the image channels of interest (i.e., for optical data usually all reflective channels without thermal channel) into an image group using i.group (group parameter). It is important to also generate a subgroup containing the same channels because the classification modules will ask for the subgroup name.

The clustering process is performed with i.cluster. A set of parameters has to be specified to control the clustering. It is important to set the initial number of classes used for the first iteration ("number of initial classes"); for other parameters, you may use default values for the first try. They have the following meaning (class and cluster are used as synonyms, explanations are based on the U.S. Army CERL (1993) tutorial):

- *Minimum class size:* minimum number of pixels to define a cluster;
- *Class separation:* minimum separation below which clusters will be merged in the iteration process. It depends on the image data being reclassified and the number of final clusters that are statistically acceptable. Its determination requires experimentation, usual values range between 0.5 to 1.5. Note that as the minimum class separation is increased, the maximum number of iterations should also be increased to achieve this separation with a high percentage of convergence (see percent convergence);
- *Percent convergence:* point at which cluster means become stable during the iteration process. When clusters are being created, their means constantly change as pixels are assigned to them and the means are recalculated to include the new pixel. After all clusters have been created, i.cluster begins iterations that change cluster means by maximizing the distances between them. As these mean shift, a progressively higher convergence is approached. Because means will never become totally static, a percent convergence and a maximum number of iterations are supplied to stop the iterative process. The percent convergence should be reached before the maximum number of iterations. If the maximum number of iterations is reached, it is probable that the desired percent convergence was not reached. The number of iterations is reported in the cluster statistics in the report file;
- *Maximum number of iterations:* determines the maximum number of iterations which is greater than the number of iterations predicted to achieve the optimum percent convergence of the Chi-square test. If the number of iterations reaches the maximum designated by the user; the user may want to rerun i.cluster with a higher number of iterations;

- *Sampling intervals:* simplifies the calculations by grouping the pixels into blocks. If the system resources are about to be depleted due to a too small block size, i.cluster will warn the user. These numbers are optional with default values based on the size of the data set such that the total pixels to be processed is approximately 10,000 (consider round up). With appropriate hardware, the unrecommended sampling may become unnecessary.

For using i.cluster, define the group and subgroup names, and a sigfile name for the "result signatures" (which will store the cluster information for i.maxlik. Only if present (not in the first run), a seed name for a "Initial signature" may be specified. This allows you to use cluster information from a previous run or to use spectral signatures from another partial supervised classifications using i.class. Then, enter a reportfile name for the "Final report" file which will be written to the current directory. It contains statistical information about the clustering process. For a LANDSAT-TM5/7 scene, the classes ("Initial number of classes") should be initially set to 20. This is a test case – the number has to be changed depending on the results, especially when the convergence is not reached. After launching the command, the cluster analysis is running, generating the cluster statistics and the report file:

```
# store VIZ, NIR, MIR into group/subgroup
i.group group=lsat7_2002 sub=lsat7_2002 \
in=lsat7_2002_10,lsat7_2002_20,lsat7_2002_30,lsat7_2002_40,\
lsat7_2002_50,lsat7_2002_70

g.region rast=lsat7_2002_10 -p
i.cluster group=lsat7_2002 sub=lsat7_2002 sig=clst2002 \
        classes=10 report=rep_clst2002.txt
```

After checking the quality of the clustering process in the report file, an eventual modification of the parameters and one or more new runs of i.cluster are required.

Second step: Unsupervised classification of image data Now, the unsupervised classification based on the MLC algorithm can be done with i.maxlik. The module will assign all pixels in the satellite image to the spectral signatures (classes) derived by the previous clustering process. For starting i.maxlik, the image group and subgroup have to be defined as well as the signature file name which is the result of the clustering process performed with i.cluster. Additionally, a name for the new reclassified image and a name for the reject threshold map is needed. The latter stores pixel-wise the assignment confidence levels. As described above, this map represents the spatially localized errors which occurred when assigning each pixel to a class:

```
# MLC
i.maxlik group=lsat7_2002 sub=lsat7_2002 sig=clst2002 \
        class=lsat2002_maxlik rej=lsat2002_maxlik_rej
```

```
d.rast.leg lsat2002_maxlik
d.rast.leg lsat2002_maxlik_rej
```

With this command, GRASS computes the unsupervised classification. The resulting maps are displayed with **d.rast**. The reject threshold map contains one calculated confidence level for each classified cell in the classified image:

```
# select all areas with confidence level >= 70%
# of correct assignment
r.report lsat2002_maxlik_rej un=h
r.mapcalc "lsat2002_maxlik_rej_qual = \
          if(lsat2002_maxlik_rej >= 10, 1, null())"
r.report lsat2002_maxlik_rej_qual un=h
d.erase
d.histogram lsat2002_maxlik_rej
```

The filtered rejection map `lsat2002_maxlik_rej_qual` can be used as MASK to select the pixels with a high confidence level of assignment. In case that the quality of the classification process is not acceptable, the number of classes or other parameters need to be changed; subsequently, the clustering and MLC analysis must be repeated with the new values.

The classes in the classification map are then manually assigned to the appropriate land use types in the verification. The assignment of category labels is explained in Section 5.1.9. Alternatively, the map can be vectorized and attributes assigned with **v.db.update**. To change the map colors to more intuitive ones (water colored blue, etc.), the module **r.colors** is used.

8.6.2 Supervised radiometric classification

In a supervised classification, the classification process is supported by an interactive selection of known areas (for the general workflow, see Figure 8.11). Using visual inspection in the field or auxiliary training maps, areas with known land cover are selected and stored in a training map, which is used to identify the spectral signatures for the classification process. These known areas are also called "ground truth areas". It is important that the training areas are homogeneous samples. Since training areas cover several pixels, small local variations are included for the definition of the classes. For verification, the module **i.class** supports analyis of channel-wise histograms. A Gaussian distribution of the spectral responses is assumed and standard deviations are displayed in the histograms. These standard deviations can be modified to change the cluster statistics. The spectral signatures (grouped later into classes) are computed from the regional mean values of the training areas and their covariance matrices.

The training areas can either be digitized within the module **i.class** (covered in the first part of the following description) or prepared from auxiliary maps such as already available land use maps (second part of the following description).

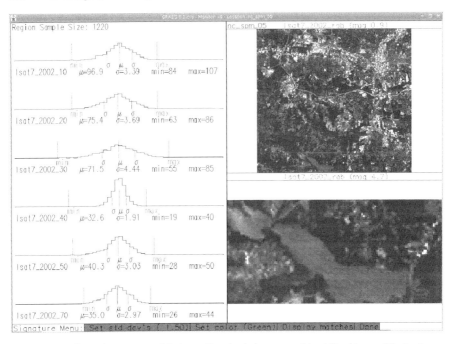

Fig. 8.12. Sample screen of interactive training area identification with i.class (LANDSAT-TM7 RGB composite image, North Carolina, Wake County)

Interactive selection of training areas The manual vectorization of training areas is accomplished with i.class. First, the satellite channels have to be joined into an image group and subgroup using i.group. Creating a natural or false color composite (see Section 8.1.3) which will be helpful for identification of training areas is recommended.

To start the module i.class, define an image group and subgroup. Also provide a name for the "result signature file". It will contain the spectral signatures for the later classification process. Optional is the "input signatures" which allows you to read in signatures from a previous run (e.g., in case you interrupted this procedure). We skip it for the first run. Then specify a "raster map to be displayed" which may be a previously generated natural or false color composite. This map, if not included in the image group, will not be considered for the image statistics:

```
g.region rast=lsat7_2002_rgb
i.class group=lsat7_2002 sub=lsat7_2002 map=lsat7_2002_rgb \
      outsig=outsig
```

The monitor display becomes divided into three parts. In the upper right corner, you see the image. In the lower right corner, zoomed map portions will be displayed when using the Zoom function. In the left section of the monitor, histograms for the selected training areas for all channels will be displayed. The training areas can be digitized by using the Define Region and the Draw Region buttons. When digitizing using the mouse, a vector line is drawn around the first training area. Keep in mind that the training area should cover a unique land use. To close the drawn polygon use Complete Region and leave the Draw Region menu with Done. An example screen is shown in Figure 8.12. To verify the cluster statistics, click on Analyze Region. Now the i.class module will search for spectral signatures based on the current training area within the image. The resulting histograms are shown in the left column of the monitor. The next step is to determine the class assignment to the image – the spatial distribution of the current spectral signatures can be overlayed as filled polygons with Display Matches (you can select a color for the area). If desired, the standard deviation can be set to a different value (Set Std Dev's). This way, you can try to improve the cluster statistics and display the matches again. After displaying the matching areas, you are asked whether to accept this signature. If yes, you can specify a Signature description in the terminal window for this spectral signature. You can then continue to digitize the next training area. With some experience, you will become familiar with the concept of this module. Please note that the vector lines of the digitized training areas are not stored. See the next paragraph for an alternate approach based on retrieving training areas from auxiliary maps.

To obtain good results, you should not digitize border pixels of any land use patch because such pixels often contain mixed spectral signatures. It is also important not to digitize very small areas since these will be ignored for statistical reasons (i.class will print a warning message, accordingly). Once you leave this module, the generated spectral signatures will be stored.

The spatial assignment of the data set pixels to the classes is done with i.maxlik. Specify the file generated by i.class as the "result signature" map. The other settings are the same as described in the previous section. Finally, the classification map and the "Reject threshold map" are created.

Generating training areas from auxiliary maps When additional maps with information about the current land use are available, they can be used to extract training areas. It is also useful to digitize training areas independently from i.class and store them in a separate map for later use/verification.

For raster maps, r.mapcalc (if-conditions) will be useful, for vector maps it will be the v.extract module. If training areas need to be digitized from a map, v.digit can be used (see Section 6.3.2). After digitizing, vector areas have to be converted to a raster map with v.to.rast. It is recommended to assign a vector label to each training area in v.digit. Otherwise, unlabeled areas will be hard to understand later. It is also possible to digitize from a raster image with r.digit.

The training map in the raster model is input for i.gensig. This module creates a signature file using the training area statistics similar to i.class. The spatial assignment of the pixels is subsequently handled by i.maxlik.

Partial supervised classification   The partial supervised classification is similar to the above-described unsupervised classification. The difference lies in the incorporation of training areas, which have to be defined prior to the application of i.cluster using i.class or i.gensig. After preparing spectral signatures from training areas, the clustering module i.cluster is started and the "result signature file" is generated with i.class or i.gensig is used as "seed signature" for i.cluster. Finally the i.maxlik is used to generate the classification map and "reject threshold map". Beyond this step the procedure is the same as for the unsupervised classification.

Also, in the reverse order, the hierarchical classification is a way to derive thematic maps from satellite data. Based on an unsupervised classification, potential training areas are identified and stored in a map. This map is a basis for a supervised classification as shown above. As signature files can be used across the modules, better results are eventually achieved through an iterative approach rather than a straight-forward classification.

8.6.3 Supervised combined geometric and radiometric classification

GRASS provides an additional sophisticated supervised classification tool. The algorithm is a combined radiometric/geometric classification method which is called *"SMAP – sequential maximum a posteriori – estimation"*. Unlike the pixel-based approach described above, this method uses an image pyramid approach which also takes neighborhood similarities into account (Schowengerdt, 1997:107, see also Ripley, 1996:167-168). This combination leads to a significant improvement of the classification results (Redslob, 1998). A second advantage is that the module also accepts a single channel data, so it can be used for image segmentation which we demonstrate later in Section 8.8. The SMAP implementation module is i.smap.

The steps for a SMAP-classification are as follows. First, the data set images are joined into a group with i.group. Training areas have to be digitized with v.digit, r.digit or generated from existing vector or raster maps. The training areas map has to be a raster map. Note that the number of training areas defines the number of classes. Similarly to the other classification methods, the training areas should cover several pixels; otherwise they will be ignored if the pixel number is too small. The spectral signatures are generated from the training map with i.gensigset. This module first queries the name of the training map, then the group and subgroup names. The module creates a "Subgroup signature file" which corresponds to the above mentioned "Result signature file".

	radiometric, unsupervised	radiometric, supervised		radio- and geometric, supervised
Preprocessing	i.cluster	i.class (monitor)	i.gensig (maps)	i.gensigset (maps)
Computation	i.maxlik	i.maxlik	i.maxlik	i.smap

Table 8.1. Classification methods in GRASS

The supervised classification can be performed from the spectral signatures with i.smap. Again, group and subgroup have to be entered, followed by the name of the recently generated "Subgroup signature file" and then the computation will start.

With a sufficient number of training areas, the results of this algorithm are superior to the MLC. A comparison of SMAP, MLC, and the ECHO classifier (the latter is not implemented in GRASS) can be found in McCauley and Engel (1995). Table 8.1 summarizes the main classification techniques available in GRASS.

8.7 Multitemporal analysis

The availability of multitemporal remote sensing data including long term satellite data is of great importance for environmental monitoring. Advanced analysis methodologies support monitoring of the earth's surface and atmosphere at different spatial and temporal scales. Many methods are described in the remote sensing literature; in this section, we will focus on a simple time series analysis with MODIS (Moderate Resolution Imaging Spectroradiometer) sensor data which is flown on board of the Terra and Aqua NASA satellites. Terra was launched in 12/1999, Aqua in 5/2002. The data analyzed here are Land Surface Temperatures (LST, MODIS level V004, pixel resolution nominally 1 km^2). To facilitate the use of these data which are originally delivered in HDF format and Sinusoidal projection, a spatial subset for parts of North Carolina was produced including application of MODIS quality assurance maps, outlier detection, reprojection to the nc_spm LOCATION and rescaling to degree Celsius. These processed data are available in a separate MAPSET modis2002lst. The data processing is outlined in Neteler (2005) and Rizzoli et al. (2007) The time series in MAPSET modis2002lst includes two daily maps from Terra/MODIS for 2002; from Aqua/MODIS, two daily maps are available from September 2002 onwards, July and August are incomplete. Due to this, four LST maps per day are available from 1st of September in the MAPSET.

The temporal order is as follows (given in approximate local NC time; the raster map metadata timestamps are given in UTC, as shown with r.info):

- Aqua overpass at 01:30 (e.g., aqua_lst_night20020921),
- Terra overpass at 10:30 (e.g., terra_lst_day20020921),
- Aqua overpass at 13:30 (e.g., aqua_lst_day20020921),
- Terra overpass at 22:30 (e.g., terra_lst_night20020921)

As a first step, we add the modis2002lst MAPSET to the map search path, list available maps and look at a selected day (21 Sept. 2002):

```
g.mapsets add=modis2002lst
g.list rast map=modis2002lst

# night overpass at 22:30h, temperatures in degree Celsius
d.rast.leg terra_lst_night20020921
d.vect us_natlas_hydrogp type=boundary col=blue
d.vect us_natlas_urban type=boundary
d.vect us_natlas_urban disp=attr attrcol=NAME type=centroid \
        where="AREA>=0.002" lsize=12

# day overpass at 10:30h
d.rast.leg terra_lst_day20020922
d.vect us_natlas_hydrogp type=boundary col=blue
d.vect us_natlas_urban type=boundary
d.vect us_natlas_urban disp=attr attrcol=NAME type=centroid \
        where="AREA>=0.002" lcol=black lsize=12
```

We can easily calculate the pixel-wise difference between the late evening and next morning LST using **r.mapcalc**,:

```
# difference night LST to next morning
r.mapcalc "diff_day_night=terra_lst_day20020922 - \
        terra_lst_night20020921"
d.rast.leg diff_day_night
d.vect us_natlas_hydrogp type=boundary col=black
d.vect us_natlas_urban type=boundary col=brown
d.vect us_natlas_urban disp=attr attrcol=NAME type=centroid \
        where="AREA>=0.002" lcol=black lsize=12
```

Not surprisingly, the lakes appear as very stable in temperature, while there are stronger differences for urban areas.

We can use these data to generate various time series indicators. A central problem to deal with in optical remote sensing are the clouds. Since cloud top temperatures are measured, they were filtered out in the data processing, leaving NULL data cells. For certain indices such as average temperature, we have to take care of this and perform NULL propagation. In the next example, we calculate daily LST averages for September 2002. In order to minimize the necessary steps, we write a simple UNIX shell script (a **for** loop with a sequence over all days, reformatting of the DAY number to match the map name style; for backticks usage, see Section 5.4.1). Using **r.series**, we can aggregate over time:

```
# bash shell script style
for d in `seq 1 30` ; do
 DAY=`echo $d | awk '{printf "%02d\n", $1}'`
 LIST=`g.mlist type=rast pat="*lst*200209${DAY}" sep=","`
```

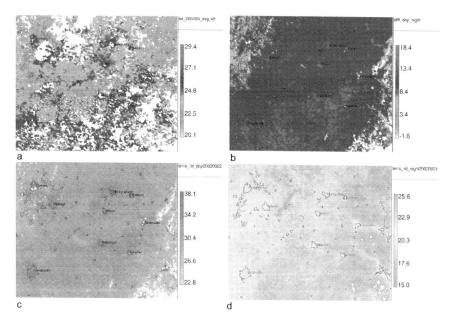

Fig. 8.13. LST maps derived from MODIS: a) Average Land Surface Temperature (LST) for September 2002 (in degree Celsius) showing urban heat islands while lake surroundings are relatively cool, b) difference between day and night LST on a given day, c) day LST, and d) night LST used to compute the differences

```
echo "$LIST"
# propagate NULL to avoid wrong average due to cloud passage
r.series -n $LIST out=lst_200209${DAY}_avg method=average
d.rast.leg lst_200209${DAY}_avg
d.vect us_natlas_urban type=boundary
done
```

We observe that due to the NULL propagation, often partially or completely empty daily average LST maps are generated. This cannot be avoided unless a complex time series reconstruction is performed which would use the perodicity of the diurnal and seasonal temperature profiles. However, for a monthly aggregation we may omit single days since the calculations are done on the pixel-wise time series.

We can aggregate the resulting daily average LST maps again to monthly average LST of September 2002:

```
LIST=`g.mlist type=rast pat="lst_200209*" sep=","`
echo "$LIST"
# here we don't propagate NULL
r.series $LIST out=lst_200209_avg method=average
r.colors lst_200209_avg col=gyr
```

```
# count number of "good" pixels in time series
r.series $LIST out=lst_200209_avg_count method=count
r.colors lst_200209_avg_count col=ryg

# show number of valid days
d.rast.leg lst_200209_avg_count
d.vect us_natlas_hydrogp type=boundary col=blue
d.vect us_natlas_urban type=boundary
d.vect us_natlas_urban disp=attr attrcol=NAME type=centroid \
       where="AREA>=0.002" lcol=black lsize=12

# only accept average LST map if sufficient pixels available
r.mapcalc "lst_200209_avg_filt=if(lst_200209_avg_count > 5, \
       lst_200209_avg, null())"

# show resulting filtered average LST map
d.rast.leg lst_200209_avg_filt
d.vect us_natlas_hydrogp type=boundary col=blue
d.vect us_natlas_urban type=boundary
d.vect us_natlas_urban disp=attr attrcol=NAME type=centroid \
       where="AREA>=0.002" lcol=black lsize=12
```

Despite some no data areas due to too many clouds, the resulting filtered average LST map clearly shows the high average temperature in urban areas in September 2002 (see Figure 8.13). Relative differences can be seen, for example between Raleigh and Durham with Durham being several degrees colder than Raleigh. The lakes and surroundings are significantly colder on average.

8.8 Segmentation and pattern recognition

Aerial photos can be used for land use/land cover classifications similar to satellite data. However, because often only a single channel is available, either in black-and-white, in visible colors or in the infrared spectral range, the classification of aerial photos is different from satellite images. Image segmentation is a method for semi-automated feature extraction, such as the land use/land cover classes or edges from remote sensing data. A segmentation algorithm is implemented in the SMAP module i.smap which we already introduced in Section 8.6.3 for multi-channel satellite data.

The SMAP algorithm exploits the fact that nearby pixels in an image are likely to belong to the same class. The module segments the image at various scales (resolutions) and uses the coarse scale segmentations to guide the finer scale segmentations (image pyramid). In addition to reducing the number of misclassifications, the SMAP algorithm generally produces results with larger connected regions of a fixed class which may be useful in numerous applications. The amount of smoothing that is performed in the segmentation is dependent on the behavior of the data in the image. If the data suggest that the nearby pixels often change class, then the algorithm will adaptively reduce the amount of smoothing. This ensures that excessively large regions are not formed. The SMAP segmentation algorithm attempts to improve segmentation accuracy by segmenting the image into regions rather than segmenting each pixel separately.

The size of the submatrix used for segmenting the image has a principle function of controlling memory usage; however, it can also have a subtle effect on the quality of the segmentation in the SMAP mode. The smoothing parameters for the SMAP segmentation are estimated separately for each submatrix. Therefore, if the image has regions with qualitatively different behavior, (e.g., natural woodlands and man-made agricultural fields) it may be useful to use a submatrix small enough so that different smoothing parameters may be used for each distinctive region of the image.

Generating a training map   The module i.smap runs with single images as well as multispectral images. The raster polygons resulting from the segmentation process may be vectorized later with r.to.vect. To classify, i.smap requires a training map containing spectral signature. This training map contains numbered raster polygons which cover selected areas with homogeneous land use/land cover. The pixels inside a training area are considered to be spectrally similar. The training map is analyzed by i.gensigset which generates statistical information from the input aerial image based on the training map. The training map can be digitized (r.digit or v.digit with v.to.rast subsequently). The image statistics derived by i.gensigset is input to i.smap. As opposed to the Maximum Likelihood Classifier algorithm which operates pixel-wise, the SMAP also considers spectral similarities of adjacent pixels. The module i.smap expects the aerial image listed in an image group and subgroup.

If the aerial photo is a color image, it may be analyzed in three channels. During import, a 24 bit aerial color image can be split into the red, green and blue channels. Also black-and-white images can be roughly split into three pseudo-color images with r.mapcalc (# operator, see Section 5.2). The # operator can be used to either convert map category values to their grey scale equivalents or to extract red, green, or blue components of a raster map into separate raster maps. The # operator has three forms: r#map, g#map and b#map. These extract the red, green, or blue pseudo-color components in the named raster map, respectively. For example,

```
g.region rast=ortho_2001_t792_1m -p
r.mapcalc "ortho_2001_t792_1m.r=r#ortho_2001_t792_1m"
r.mapcalc "ortho_2001_t792_1m.g=g#ortho_2001_t792_1m"
r.mapcalc "ortho_2001_t792_1m.b=b#ortho_2001_t792_1m"
d.erase
d.rast ortho_2001_t792_1m
```

```
d.rgb b=ortho_2001_t792_1m.b g=ortho_2001_t792_1m.g \
    r=ortho_2001_t792_1m.r
```

results in three pseudo-color channels which may be used for the segmentation. However, when working with 24bit images, the information contents is certainly higher.

Sometimes, it is useful to generate and use additional synthetic channels for the classification. For example, the module r.texture creates raster maps with textural features from image channels. The textural features are calculated from spatial dependence matrices at 0, 45, 90, and 135 degrees within a given moving window. For details, please refer to the related manual page.

A sample segmentation session   A sample segmentation procedure for the aerial image ortho_2001_t792_1m may be performed as follows:

```
i.group group=segment subgroup=segment in=ortho_2001_t792_1m

# digitizing training areas: fields, forest, roads, etc.
# save as map "training"
r.digit training
d.rast.leg training

# generate class statistics from training map
i.gensigset training gr=segment sub=segment sig=smapsig

# run segmentation
i.smap gr=segment sub=segment sig=smapsig out=ortho.smap
d.erase
d.rast ortho.smap
# overlay to original orthophoto
d.his i=ortho_2001_t792_1m h=ortho.smap
```

Depending on the training map, the result can become a vectorized land use/land cover map. However, this map would contain a lot of spurious areas. These can be filtered first with the script r.reclass.area. As an example, only areas greater than 0.0025 hectares will be preserved (note, this script only operates in georeferenced LOCATIONS):

```
# 5x5 pixel = 25m = 0.0025ha
r.reclass.area ortho.smap out=ortho.smap.filt greater=0.0025
d.rast ortho.smap.filt
```

The deleted areas are filled with NULL (no-data) values. To reassign values to these NULL cells, we can use a mode filter which replaces the NULL cells by the dominating land use value in the 5×5 neighborhood. The mode filter is implemented in r.neighbors:

```
# mode filter to replace NULL cells
r.neighbors ortho.smap.filt out=ortho.smap2 method=mode size=5
```

```
d.rast ortho.smap2
# vectorization and visual./report of land use/land cover map
r.to.vect -s ortho.smap2 out=ortho_smap2 feature=area
d.erase -f
d.rast ortho_2001_t792_1m
d.vect -c ortho_smap2 type=boundary col=yellow

v.info -c ortho_smap2
# now use v.db.update to assign text attributes in
# the "label" column of the table based on the "value" entries

# area report
v.report ortho_smap2 option=area units=h
```

We have now the land use/land cover map available as raster and vector maps.
Special care has to be taken for cast shadows resulting from sun which may
either be treated as a special class or reduced by image pre-processing.

9

Notes on GRASS programming

GRASS provides a unique opportunity to improve and extend GIS capabilities by a new code development. The GNU General Public License (GPL) keeps the code as Free Software, while protecting the rights of the individual authors as well of the users. Because the source code can be studied, modified and published again, there is an ongoing exchange of knowledge, methods and algorithms between GIS and software engineering experts. Free Software projects are gaining interest even in the proprietary GIS industry due to their stability and the transparent development process.

To make the development of GIS tools more efficient, GRASS provides a large GIS library with documented application programming interface (API). The code is portable on all common architectures and operating systems. The available programming documents enable potential developers to better estimate the workload of adding functionality to GRASS. The possibility of simply deriving a new tool or customized solution from existing source code greatly speeds up development. An important aspect of the GRASS development is the fact that the developers are at the same time advanced GRASS users who improve or extend the existing functionality, based on the needs of their daily GIS use in production.

9.1 GRASS programming environment

Important communication channel supporting GRASS development is the "GRASS developers mailing list".[1] Here advanced users exchange ideas and discuss problems related to code development or bugfixes. The core team members can submit changes at any time because the source code is managed in CVS (Concurrent Versioning System). Write access to CVS is granted by the GRASS Project Steering Committee (GRASS-PSC) to those developers who build a record of systematically contributing quality code. Documentation ("GRASS Programmer's Manual", GRASS Development Team, 2006 and

---

[1] "grass-dev" mailing list, http://grass.itc.it/mailman/listinfo/grass-dev

user manual) is automatically extracted from the source code (inline documentation, extracted with "doxygen" tool) into HTML and PDF formats and updated weekly. Access to the full source code, either as a released package, or as a weekly CVS snapshot, or directly extracted from CVS, allows the developers to study the code structure of a full featured GIS. We will describe the general code structure of GRASS 6 in greater detail below. The code structure is fairly standardized, following as much as possible ANSI/POSIX style. Figure 1.1 at the beginning of the book in Section 1.2 shows the current GRASS development model.

9.1.1 GRASS source code

The complete GRASS source code is available on the GRASS Web sites. It comprises around 550.000 lines of basically C, Makefiles, Tcl/Tk and Python code and shell scripts. Those who want to participate in the ongoing source code development of GRASS should learn more about the "GRASS-CVS". This electronic management tool (CVS: *Concurrent Versioning System*[2]) for the source code is used in GRASS development since December 1999. The idea of CVS is to enable the developers to have direct read/write access to the GRASS source code for independent development. CVS supports the centralized management of GRASS development, as the developers work on a single code base with restricted write, but public worldwide read access. You can find more information about this topic on the GRASS Web servers. Another advantage is that the CVS client minimizes data transfer after an initial download: during a subsequent synchronization of the local GRASS source code copy with the centralized CVS server ("cvs update" command) only new code changes are transferred through network. It is planned to migrate the repository to the more modern subversion (SVN) repository software which works in a similar way.

CVS snapshots which cover the latest development are generated on a weekly basis and made available in a package for download. This is useful for skilled users who do not want to learn CVS itself but who would like to follow the latest development. However, knowledge about how to compile source code packages is required.

To compile the source code it is recommended to first read the REQUIREMENTS.html file provided with the source code as well as the INSTALL file to make sure that you have all the necessary libraries installed on your system. After extracting the code (not needed when directly accessing the CVS server), the compilation is done with three steps:

```
# you may compile as normal user
./configure <parameters>
make
```

---

[2] CVS software, http://www.nongnu.org/cvs/

```
# installation usually requires 'root' permissions
su
make install
```

Note that the `configure` script may expect parameters depending on your local installation of required libraries. Please refer to the INSTALL file within the source code for further details and explanations.

If you want to compile code from the GRASS Addons SVN repository (see Section 3.1.1), you can compile GRASS modules (stored there in subdirectories) with:

```
make MODULE_TOPDIR=$HOME/grass63/
make install
```

The `MODULE_TOPDIR` path needs to be adapted to the directory where you installed GRASS.

9.1.2 Methods of GRASS programming

GRASS 6 is written in ANSI C programming language. Additionally, UNIX Shell scripts and a few PERL scripts are implemented. The current graphical user interface `gis.m` is based on Tcl/TK libraries, the `nviz` visualization tool additionally includes OpenGL function calls. A new Python/wxWidgets based graphical user interface is under development (**wxgrass**, available from the GRASS AddOns SVN repository).

The command line parser is part of the GIS library. It generates output in different programming languages. Several parameters are available to quickly prototype applications by simply launching a GRASS command:

- <grassmodule> `--script`: prints a boilerplate for a shell script with auto-generated graphical interface to the script;
- <grassmodule> `--tcltk`: prints out a Tcl/TK description of the module's parameters and flags;
- <grassmodule> `--interface-description`: prints out a XML description of the module's parameters and flags;
- <grassmodule> `--html-description`: prints out a HTML description of the module's parameters and flags.

A simple automatic Python/wxWidgets GUI builder based on the XML-based GRASS user interface descriptions along with a Document Type Definition (DTD) is available. Also a preliminary GRASS-SWIG interface has been implemented. SWIG is a software development tool that connects programs written in C and C++ with a variety of high-level programming languages; in GRASS, PERL and Python are currently supported.

The modular concept of GRASS provides huge potential for development. Apart from the standard use, two basic levels of programming are supported. Average users will use "script programming" to automate repeating processes and generate workflows, while advanced users can extend existing code or develop new modules based on the SWIG or C-API.

For script programming, there are no general limitations. After setting the specific environment variables as discussed below, more or less any UNIX compliant script language can be used, even MS-Windows batch scripts are supported with some limitations. Many UNIX shell scripts are already implemented, as well as PHP Hypertext Preprocessor[3] and others. To learn how to write customized scripts, take a look at the numerous scripts stored in $GISBASE/scripts/). The advantage of scripts is that they can utilize other UNIX tools such as awk, sed, or cut. If you are familiar with the command line usage of GRASS, it requires only a small step to start writing your own scripts. To write Web-based applications such as developing a "remotely controlled" GIS which dynamically generates Web map pages will certainly need more time: Besides functionality, issues like software ergonomics, speed, and security play an important role. In Section 10.4, we will demonstrate a simple example how to publish GRASS raster data with a fast Web mapping server.

We can only provide few introductory notes about GRASS C-programming here, as this GIS focused book does not intend to replace a general C-programming tutorial. Our objective is to depict the current source code structure to guide newcomers within the huge code base.

9.1.3 Level of integration

There are different levels of integrating new or external functionality into GRASS. We can distinguish between:

- links to external software:
  - loose coupling: linking other GIS to GRASS via data exchange through shared data formats (e.g., using GRASS in a heterogeneous network such as Linux/Samba/MS-Windows along with proprietary GIS software);
  - tight coupling: direct access to GRASS LOCATION through GDAL/OGR (e.g., to read maps from a LOCATION directly with UMN/MapServer);
- full integration: modification/extension of GRASS functionality (e.g., writing new code in C, Python, UNIX Shell, PERL, such as gstat and the GRASS/R interface) and/or using the SWIG interface.

In the following sections we provide several examples of scripts and programs illustrating the possibilities for extending GRASS capabilities. To improve legibility, we have typeset these scripts and programs in a language-sensitive formatting. You can download the larger scripts explained in this chapter from the Internet.[4]

---

[3] Sample application "Recent Earthquakes": map generated by GRASS on the fly with PHP, http://grass.itc.it/spearfish/php_grass_earthquakes.php

[4] GRASS Book Web site, http://www.grassbook.org

9.2 Script programming

UNIX Shell, PERL and Python scripts can be used to automate workflows which are repeatedly performed for different maps. The standard GRASS installation provides a set of scripts which, from the user's side, mostly behave like modules programmed in C language. The major drawback is that scripts usually run slower than compiled implementations (except for Python). Scripts are stored as ASCII files. Because the GRASS modules can be called within the scripts, complex applications can be developed. When writing such a script for the first time, it is useful to perform the task by running the commands step by step in command line mode. Then the UNIX `history` command allows you to view and save the previously entered commands. Another option is to use the UNIX `script` program which logs a session into a text file.

As an example, we write a simple script `d.rast.region` which first adjusts the current region settings to the map given as a parameter, and then displays the map:

```
#!/bin/sh
#
# This program is Free Software under the GNU GPL (>=V2).
# Adjust the current region settings to a raster map
# specified as parameter, then display the map

if test "$GISBASE" = ""; then
 echo "You must be in GRASS to run this program."
 exit
fi

#map name is first parameter:
MAP=$1

#zoom:
g.region rast=$MAP
#erase monitor and display map:
d.erase
d.rast $MAP
```

The first line is mandatory for UNIX Shell scripts. After performing a test to ensure that the script is executed in GRASS environment, the first parameter given on command line is stored into a script-internal variable. GRASS modules are then used to perform the desired task. The map name is stored in the new variable `$MAP`. It is a good programming practice to use easy to understand variable names and to document the functionality step-by-step. It is also important to set the proper file permissions after storing the script as a file `d.rast.region`:

```
chmod a+x d.rast.region
```

The script can be copied into the directory `$GISBASE/scripts/` or in a dedicated directory which has to be defined by the environmental variable `$GRASS_ADDON_PATH` before starting GRASS itself. This allows you to keep your scripts and modules physically separated from the standard GRASS. The path(s) defined in `$GRASS_ADDON_PATH` are added to the modules search path during startup.

The next example, demonstrating the potential of script programming, slightly extends the GRASS functionality by introduction of the `awk` tool. This script calculates general geostatistical information for raster images (adapted from Albrecht, 1992). You may save this script as `statistics.sh` and set the UNIX execute permissions:

```
#!/bin/sh
# This program is Free Software under the GNU GPL (>=V2).
# calculate univariate statistics for GRASS raster data

if test "$GISBASE" = ""; then
echo "You must be in GRASS to run this program."
exit
fi
MAP=$1
r.stats -1 $MAP | awk 'BEGIN {sum = 0.0 ; sum2 = 0.0}
NR == 1{min = $2 ; max = $1}
     {sum += $1 ; sum2 += $1 * $1 ; N++}
     {
      if ($1 > max) {max = $1}
      if ($1 < min) {min = $1}
     }
END{
print "Number of raster data samples N =",N
print "Minimum value  MIN =",min
print "Maximum value  MAX =",max
print "Variation      v   =",(max - ((min * -1) * -1))
print "Mean          MEAN =",sum / N
print "Variance       S2  =",(sum2 - sum * sum / N) / N
print "Standard deviation S =",sqrt((sum2 - sum * sum/N) / N)
print "Variation coeffic. V =",100*(sqrt((sum2 - sum*sum/N)/N)) \
     /(sum/N)
}'

exit 0
```

This script must be used from inside GRASS. The name of the input raster map is specified as a parameter. Within this script, the output of the GRASS module `r.stats` is piped to the program `awk`. The statistical calculations are

done within awk and the results are printed out. A modified version of the
above script, r.univar.sh, is provided with GRASS (note that there is also a
faster C implementation r.univar).

The next script example calculates the center of gravity of an area (area
centroid). You can use it to find the center of gravity of a region defined
by a watershed boundary. The script requires a watershed map generated
by r.watershed. Internally, all basins except the watershed of interest are
masked out with r.mask. You can download a r.centroid script from the
GRASS Book Web site. This script shows how UNIX commands and GRASS
modules can be combined. It also provides a parser (through g.parser), so the
script will run both in the command line mode, and, when started without
parameters, in the interactive mode, i.e. graphical user interface. The first
test checks if the module is started within GRASS, and exits if it is not. The
usage description is stored in a function to save space within the script and
to improve the legibility.

A simple parser is provided for command line processing with g.parser.
In our example, a raster map such as a watershed or other raster area map
are of interest. The module checks if the user requested the module help
description or otherwise accepts the first parameter as raster file name. Then
the ID number of the raster polygon is checked (which might be queried
with d.what.rast beforehand). This is followed by the centroid calculation
according to the centroid formula. The calculation requires the area of the
current raster polygon which is retrieved from r.report. Additionally, the
current resolution is needed. In case that MASK is present, it will be saved
and later restored. This is necessary, as the area of interest has to be selected
inside the input raster map with a new MASK. To see the r.centroid script
applied to the watersheds map basin_50K in our NC sample data set, you may
run:

```
# set region and display the input map
d.mon x0
g.region rast=basin_50K -p
d.erase
d.rast basin_50K

# calculate area and centroid coords for watershed no. 28
r.centroid map=basin_50K areanumber=28

# display centroid of watershed no. 28,
# coordinates taken from above r.centroid call
echo "636596.22 219735.74 28" | v.in.ascii out=grav_center \
                                fs=space
d.vect grav_center col=red icon=basic/circle
```

Generally you may face the problem that a script is not working as expected. To identify problem(s), you can add printing of variable contents with
echo $VARIABLE. An alternative recommended method to debug shell scripts
is to add the -x flag to the first line of the shell script:

```
#!/bin/sh -x
```

This switches the shell into an echo-mode, printing every line including variable content into the terminal window which simplifies error identification.

Further scripts are described in Albrecht (1992) and Shapiro and Westervelt (1992). Many scripts that you can learn from are included in GRASS (from inside GRASS change to $GISBASE/scripts/) and available from the GRASS Wiki. You can find further useful tips for GRASS script programming style as well as recommendations to avoid portability traps in the file SUBMITTING_SCRIPTS which is included in the source code.

9.3 Automated usage of GRASS

Due to the modular character of GRASS, a monolithic "GRASS program" does not exist. In fact, GRASS is a collection of modules which are run in a special environment. The structure allows GRASS to be completely controlled from outside through scripts.

9.3.1 Local mode: GRASS as GIS data processor

GRASS in batch mode   The usage of GRASS in batch mode requires setting of some environment variables, which can be also done manually or in scripts.[5] After setting these variables, the environment is defined and all GRASS modules can be used. The best approach is to write a working script, and then define it in a variable to be executed:

```
# bash shell style
GRASS_BATCH_JOB=my_grass6script.sh
export GRASS_BATCH_JOB
grass63 /path/to/LOCATION/MAPSET/

# or simply
GRASS_BATCH_JOB=my_grass6script.sh grass63 \
                /path/to/LOCATION/MAPSET/
```

To disable batch job processing again, type into the terminal:

```
unset GRASS_BATCH_JOB
```

Multiple sessions in the same LOCATION are best done when using different MAPSETs. The module-specific environment variables are further explained in the software documentation, check with:

```
g.manual variables
```

---

[5] GRASS in batch mode,
http://grass.gdf-hannover.de/wiki/GRASS_and_Shell

Once the minimum number of environment variables has been set, GRASS commands can be integrated into shell, Python, PERL, PHP and other scripts. When operating large batch jobs, it is recommended to remove temporary GRASS files from time to time by:

```
$GISBASE/etc/clean_temp
```

Automated creation of a LOCATION from external raster data   A nice application for running GRASS in batch mode is generate LOCATIONs from external GIS raster data. We use r.in.gdal for this purpose because the module supports many formats and it can read projections from metadata if provided:

```
#!/bin/sh
# This program is Free Software under the GNU GPL (>=V2).
# create a new LOCATION from a raster data set

#variables to customize:
# path to GRASS software main directory
GISBASE=/usr/local/grass63
# path to GRASS database
GISDBASE=$HOME/grassdata

#nothing to change below:
MAP=$1
LOCATION=$2

if [ $# -lt 2 ] ; then
echo "Script to create a new LOCATION from a raster data set"
echo "Usage:"
echo "  create_location.sh rasterfile location_name"
exit 1
fi

#generate temporal LOCATION:
TEMPDIR=$$.tmp
mkdir -p $GISDBASE/$TEMPDIR/temp

#save existing $HOME/.grassrc6
if test -e $HOME/.grassrc6 ; then
   mv $HOME/.grassrc6 /tmp/$TEMPDIR.grassrc6
fi

echo "LOCATION_NAME: $TEMPDIR" > $HOME/.grassrc6
echo "MAPSET: temp"           >> $HOME/.grassrc6
echo "DIGITIZER: none"        >> $HOME/.grassrc6
echo "GISDBASE: $GISDBASE"    >> $HOME/.grassrc6
export GISBASE=$GISBASE
```

```
export GISRC=$HOME/.grassrc6
export PATH=$PATH:$GISBASE/bin:$GISBASE/scripts

# import raster map into new location:
r.in.gdal -oe in=$MAP out=$MAP location=$LOCATION
if [ $? -eq 1 ] ; then
  echo "An error occured. Stop."
  exit 1
fi

#restore saved $HOME/.grassrc6
if test -f /tmp/$TEMPDIR.grassrc6 ; then
  mv /tmp/$TEMPDIR.grassrc6 $HOME/.grassrc6
fi

echo "Now launch GRASS with:"
echo " grass63 $GISDBASE/$LOCATION/PERMANENT"
```

In case that no projection information is found in the raster data set, you may use g.setproj to generate the projection information later. Note that g.setproj does not reproject any data. The above script accepts only raster data sets. For vector data, the v.in.ogr command would be used instead. Extending the script with data format as additional third parameter, it would become a universal solution for most spatial data formats.

You can launch GRASS directly without individually specifying the names of LOCATION, MAPSET and DATABASE as follows:

```
grass63 /usr/local/share/grassdata/nc_spm/user1
```

This is only successful if the LOCATION and MAPSET exist, otherwise an error message is shown.

9.3.2 Web based: PyWPS – Python Web Processing Service

GRASS can be used as GIS backbone for WebGIS applications by providing geoprocessing functionality over networks using Web Services. PyWPS (Python Web Processing Service[6]) implements the Web Processing Service standard from Open Geospatial Consortium, designed for Web based geoprocessing. The goal of PyWPS (Python Web Processing Service) is to provide GIS tools over a Web interface, with special focus on GRASS as reference implementation. PyWPS offers a user environment which can be easily extended for writing customized WPS services. It is written with native support for GRASS that makes GRASS modules directly accessible via Web. Demo applications covering vector network routing, line of sight analysis and more (using an AJAX based interface) are available on-line.[7]

---

[6] Python Web Processing Service, http://pywps.wald.intevation.org/
[7] PyWPS demo applications, http://pywps.ominiverdi.org

9.4 Notes on programming GRASS modules in C

This section briefly explains the GRASS code organization for a user with basic C language programming knowledge. While we cannot explain GRASS programming within this book, we provide a brief introduction to the huge code base. GRASS provides an ANSI C language API with several hundred GIS functions, from reading and writing maps to area and distance calculations for georeferenced data, as well as attribute handling and map visualization. All important aspects of GRASS programming are covered in the "GRASS Programmer's Manual" (available from the GRASS Web site). To understand the usage of the GRASS API, it is helpful to explore the existing modules. The general structure of the modules is similar, with each module stored in its own subdirectory of the GRASS source code. You can find further useful tips for GRASS programming style as well as recommendations to avoid portability traps in the file SUBMITTING. This file is included in the main directory of the source code.

It is important to know that a parser will launch the graphical user interface for a GRASS command if there are no command-line arguments entered by the user. Otherwise, it will run on command line. Command parameters and flags are defined within each module. They are used to ask user to define map names and other options.

GRASS source code structure The general GRASS 6 source code structure is as follows:

```
- db/              # database modules
- debian/          # debian control files
- demolocation/    # small location, used during compilation
- display/         # display modules
- doc/             # development documentation
- gem/             # GRASS extension manager
- general/         # general modules
- gui/             # graphical user interface (Tcl/Tk based)
- imagery/         # image processing
- include/         # include header files
- lib/             # set of GRASS libraries
- locale/          # message translations
- macosx/          # Mac OSX compilation files
- man/             # MAN manual pages (generated from HTML)
- misc/            # miscellaneous modules
- ps/              # postscript modules
- raster/          # raster modules
- raster3d/        # voxel modules
- rfc/             # RFC (request for comments) of GRASS-PSC
- rpm/             # RPM spec files
- scripts/         # shell scripts
- sites/           # legacy only
```

```
- swig/              # SWIG programming interface (PERL, Python)
- testsuite/         # test suite for regression testing
- tools/             # internal compilation tools
- vector/            # vector modules
- visualization/     # visualization modules
```

The GRASS libraries are stored in lib/. They provide the API in C (which is wrapped to other languages in swig/). This "GRASS programming library" (C API) is structured as follows (typical function name prefixes for related library functions are listed in squared brackets):

- GIS library: database routines (GRASS file management), memory management, parser (parameter identification on command line), projections, raster data management etc. [G_]
- vector library: management of area, line, and point vector data [Vect_]
- database (DBMI) library: attribute management [db_]
- image data library: image processing file management [I_]
- display library: graphical output to the monitor [D_]
- raster graphics library: display raster graphics on devices [R_]
- segment library: segmented data management [segment_]
- vask library: control of cursor keys etc. [V_]
- rowio library: for parallel row analysis of raster data [rowio_]

Modules consist of one or more C program files (∗.c), the local header files (∗.h) and a Makefile. The file Makefile contains instructions about files to be compiled and libraries to be used (GRASS and system libraries). The GRASS libraries are predefined as variables. The structure of Makefile follows special rules. A simple example illustrates a typical Makefile (it is important to generate indents with <TAB>, not with blanks!):

MODULE_TOPDIR = ../..

PGM = v.info

LIBES = $(VECTLIB) $(GISLIB)
DEPENDENCIES= $(VECTDEP) $(GISDEP)
EXTRA_INC = $(VECT_INC)
EXTRA_CFLAGS = $(VECT_CFLAGS)

include $(MODULE_TOPDIR)/include/Make/Module.make

default: cmd

The line default: cmd reads the compiler instructions from include/Make/∗.make files. Above this line several variables are set which contain the name of the GRASS module and needed library references. Numerous variables used here are pre-defined in:

`include/Make/Grass.make`

This meta-configuration file is created depending on the platform by `configure` script, which has to be run before a first compilation of GRASS. It contains information related to the compiler, paths to local libraries and include files etc. The other internal variables are defined in

`include/Make/Platform.make`

which is also generated by `configure`. These settings should be kept unchanged.

It is a good programming practice to subdivide a C program (here a GRASS module) into several files, organized by functionality. GRASS GIS library commands can be used in the source code when the code is linked against the related libraries. A short example of a raster module (file `main.c`):

```
/*
 * This program is Free Software under the GNU GPL (>=V2).
 * Conversion of LANDSAT-TM5 DNs to at-sensor radiances
 */

#include <stdio.h>
#include <string.h>
#include <math.h>
#include <grass/gis.h>
#include <grass/glocale.h>

int main(int argc, char *argv[])
{
    CELL *cellbuf;
    DCELL *result_cell;
    int nrows, ncols;
    int row, col;
    char *groupname;
    int fd;
    struct GModule *module;
    struct Option *group;
    struct Flag *quiet;

    module = G_define_module();
    module->description = _("Module to convert LANDSAT-TM5 "
        "digital numbers to at-sensor radiances");
    module->keywords = _("image processing, satellite");

    group = G_define_standard_option(G_OPT_I_GROUP);

    flag.quiet = G_define_flag();
    flag.quiet->key = 'q';
```

```
flag.quiet->description = _("Run quietly");

G_gisinit (argv[0]);

if (G_parser(argc,argv))
    exit(EXIT_FAILURE);
groupname = group->answer;

/* function defined in other file */
open_file(groupname, fd);

nrows = G_window_rows();
ncols = G_window_cols();
cellbuf = G_allocate_raster_buf(CELL_TYPE);
result_cell = G_allocate_raster_buf(DCELL_TYPE);

/* go row wise and col wise through image */
for (row = 0; row < nrows; row++) /* rows loop */
{
    /* read integer satellite channel */
    G_get_raster_row (fd, cellbuf, row, CELL_TYPE);
    for (col = 0; col < ncols; col++) /* cols loop */
    {
        /* the formula is defined in another file: */
        result_cell[col] = calc_new_pixel(cellbuf);
    } /* end cols loop */

    G_put_raster_row(fd, result_cell, DCELL_TYPE);
} /* end rows loop */

G_close_cell (fd);
exit(EXIT_SUCCESS);
}
```

The calculation is done row-wise and column-wise (see "for" loop). This draft program illustrates only the general structure of GRASS code, for copyright reasons it is not a real GRASS program. Please refer to the GRASS source code for the real world implementations.

Future of GRASS programming   At the time of writing this book new ideas for GRASS 7 are collected. In the near future, the development for GRASS 6 will be shifted to GRASS 7. Migration of the code to the new code repository will be accompanied by a code cleanup. Some code spread in various modules will be organized into new library functions. Existing library functions will be examined for consistency, and if needed, functions performing similar tasks will be merged. Also, at the module level, merging of modules with similar functionality will be done. In general, the goal is to provide a well defined, layered GRASS model with GRASS-Core, providing all library functions, GRASS-Base, providing basic modules for importing, exporting, displaying and basic manipulation of spatial data sets and extended GRASS packages including specialized add-on packages for image processing, hydrological modeling, volume data management and analysis, etc.

Important improvements will be a modernization of the raster processing library to implement tiling and caching for large data sets as well as adding support for calculations on computer grids and distributed systems. Also the image processing tools will benefit from these changes. For vector data processing, it is desired to add transactions which would protect the geometry in the event of an incomplete or otherwise unexpected termination of an editing session. Further optimizations concerning a better use of the spatial index are anticipated to reduce the computational overhead of building topology.

GRASS intends to be a general purpose GIS. The current GRASS 6 version is a major step to develop a reliable, intuitive to use, flexible GIS in terms of Free Software. Skilled users are invited to participate in its further development.

10

Using GRASS with other Open Source tools

GRASS is one of many Free Software projects in the GIS world; however, it is the only full featured free GIS at this time (see "Open Source GIS Survey" by Ramsey, 2006). A comprehensive list of more than 300 free GIS projects is available online at the FreeGIS Project Web site.[1] The use, development and support of Free GIS Software is promoted at this site, as well as the use and release of publicly available geographic data. Some free GIS projects can provide additional functionality for GRASS by addressing some of its unresolved issues or intentional constraints.

The software stack for Free Software GIS comprises system software, data processing tools, data serving tools, user interface tools, and end-user applications (Mitchell, 2005; Erle et al., 2005; Jolma et al., 2006, 2007). The OS-Geo foundation aims at promoting this software stack and encourages cross-pollination among the foundation projects.

Within this chapter, we first highlight selected procedures that extend the geostatistical analysis capabilities of GRASS. We focus on two statistics software packages, the gstat and the R project. This chapter does not try to cover the theory of geostatistics. Excellent other books on theory and applications are available, such as Cressie (1993); Bailey and Gatrell (1995), and Webster and Oliver (2001). In relation to the R program, it is useful to read Chambers and Hastie (1992); Venables and Ripley (2000), as well as Venables and Ripley (2002).

After a brief look at GPS related software tools, we close this chapter with a demonstration of fast Web mapping through UMN/MapServer and OpenLayers linked directly to GRASS for reading GIS data from a GRASS LOCATION.

Maas river bank soil pollution data   In the first section, we use the Maas river bank soil pollution data (Limburg, The Netherlands, Burrough and Mc-Donnell, 1998). These data are provided in the gstat package and are used in

---

[1] FreeGIS Project Web site, http://www.freegis.org

examples in its manual. The Maas river bank soil pollution data are sampled in a floodplain along the Dutch bank of the river Maas (Meuse) 3-5km north of Maastricht, not far from where the Maas enters the Netherlands (Borgharen, Itteren).[2] The river Maas is at the north-west border of the study site, traversing the area in north-east direction. Burrough and McDonnell (1998) use a subset of the same data in their book. A "maas" GRASS LOCATION was defined with following parameters: projection UTM, ellipsoid WGS84, zone 32, spatial extent 5650610N – 5652930N, 269870E – 272460E, resolution 10m, 232 rows and 259 columns. This LOCATION, including the pollution data stored as vector points, can be downloaded from the GRASS Web site.[3] The data set also includes two raster maps; they can be used to experiment with interpolation or other tasks. The point data contain the following columns (topsoil data were collected as bulk samples during fieldwork in 1990 within a radius of 5m according to Burrough and McDonnell, 1998:102, 309): East, north (UTM zone 32 coordinates in meters); x, y (local coordinates in meters); elev (elevation above local reference level in meters); d.river (distance from main river Maas channel in meters); Cd (cadmium in ppm); Cu (copper in ppm); Pb (lead in ppm); Zn (zinc in ppm); LOI (percentage organic matter loss on ignition); flfd (flood frequency class, 1: annual, 2: 2-5 years, 3: every 5 years); soil (3 unnamed soil types).

10.1 Geostatistics with GRASS and gstat

The gstat[4] package is Free Software for geostatistical modeling, prediction and simulation in one, two or three dimensions (Pebesma and Wesseling, 1998; Pebesma, 2001). It requires the gnuplot[5] graphical plotting software for the display of empirical variograms and variogram models.

A widely used geostatistical technique for interpolation and extrapolation which is not available in GRASS 6, is kriging. The theory and practice of kriging is described by a large volume of literature; here we briefly illustrate the principal usage with free software packages.

Using gstat, you can perform geostatistical analysis and modeling including computations of empirical (sample) variograms and cross variograms (or covariograms). The sample (co-)variograms can be generated from ordinary, weighted or generalized least squares residuals. Models can be fitted to these variograms to predict data distributions. Using weighted least squares, nested models are fitted to sample (co-)variograms. Restricted maximum likelihood estimation of partial sills is also implemented. Variograms are drawn using the

---

[2] Maas river bank soil pollution data descriptions: gstat package documentation; Burrough and McDonnell (1998):309-311 (subset)

[3] Maas river bank soil data GRASS LOCATION,
http://grass.itc.it/statsgrass/maas_grass6_location.tar.gz

[4] gstat software, http://www.gstat.org

[5] gnuplot software, http://gnuplot.sourceforge.net

plotting program gnuplot, when working in interactive variogram modeling user interface.

The gstat software provides prediction and estimation using a model that is the sum of a trend modeled as a linear function of polynomials of the coordinates or of user-defined base functions, and an independent or dependent, geostatistically modeled residual. This allows for simple, ordinary and universal kriging, simple, ordinary and universal cokriging, standardized cokriging, kriging with external drift, block kriging and "kriging the trend", as well as uncorrelated, ordinary or weighted least squares regression prediction. Simulation in gstat comprises uni- or multivariable conditional or unconditional multi-Gaussian sequential simulation of point values or block averages, or (multi-) indicator sequential simulation (features cited after Pebésma, 2001).

The gstat/GRASS interface allows the user to read point data from vector point and raster maps. This requires to have the GRASS support compiled into gstat. You can check your version with flag -v:

```
gstat -v
```

The line "with libraries" must list "grass" besides other supported formats (e.g., "grass gdal netcdf").

Output of gstat (prediction or simulation results) is written to raster and also to vector point maps. You need to run gstat from inside GRASS as the program requires the GRASS environment to internally set up the LOCATION definitions. When a subregion is set in GRASS, gstat will only interpolate or simulate the raster cells according to the current region. The variables of interest need to be floating point numbers (DOUBLE) or stored in a raster map. The instructions for gstat are stored in an ASCII file. The program gstat reads GRASS vector points data from the current MAPSET with the data() function. Variable positions are defined as:

```
x=1: coordinate column 1 contains the x-coordinate
y=2: coordinate column 2 contains the y-coordinate
z=3: coordinate column 3 contains the z-coordinate (optional)
v=1: data column 1 contains the first data (measurement)
      variable, when 0, a grid map is read
```

To illustrate how it works, we run a sample session based on the "Maas river bank" data set. First start GRASS with the Maas UTM LOCATION, then copy the Zn (zinc) concentrations vector points map to the current MAPSET:

```
grass63 /usr/local/share/grassdata/maas6/user1/
g.copy vect=Zn,zinc
v.info zinc
v.db.select zinc
```

The following example is based on the manual of gstat[6]. Store the following commands to the file gstat.maas1.zn in your home-directory:

---

[6] gstat manual, http://www.gstat.org/manual/

```
# two variables with (initial estimates of) variograms,
# start the variogram modeling user interface
data(zinc): 'Zn', x=1, y=2, v=1;
data(ln_zinc): 'Zn', x=1, y=2, v=1, log;
variogram(zinc): 10000 Nug() + 140000 Sph(800);
variogram(ln_zinc): 1 Nug() + 1 Sph(800);
```

The zinc concentrations are stored as first DOUBLE attribute (in ppm, reported by v.info -c zinc) and we select this data column through v=1. Run the analysis by:

```
gstat gstat.maas1.zn
```

The program starts to analyze the data and subsequently displays univariate statistics (the warning can be ignored):

```
gstat: Linux version 2.5.1 (02 April 2007)
using Marsaglia's random number generator
data(zinc): WARNING: Adapted sites library used for ...
GRASS site list zinc: 0 cat, 2 dim, 0 str, 1 dbl.
gstat/grass: 98 sites read successfully.
            zinc      (GRASS site list)
attribute:  col[1]     [x:] x_1    : [     269870,     272460]
n:              98     [y:] y_2    : [5.65061e+06,5.65293e+06]
sample mean:  481.031    sample std.:           398.808
data(ln_zinc): WARNING: Adapted sites library used for ...
GRASS site list zinc: 0 cat, 2 dim, 0 str, 1 dbl.
gstat/grass: 98 sites read successfully.
            zinc      (GRASS site list)
attribute:  log(col[1])  [x:] x_1   : [     269870,     272460]
n:              98       [y:] y_2    : [5.65061e+06,5.65293e+06]
sample mean:  5.87065   sample std.:          0.778309
[starting interactive mode]
press return to continue...
```

After pressing <ENTER>, we reach the main menu, which allows us to interactively analyze the loaded data set:

```
        gstat 2.5.1  (02 April 2007), gstat.maas1.zn

enter/modify data
choose variable    : zinc
calculate what     : semivariogram
cutoff, width      : 1159.03, 77.269
direction          : total
variogram model    : 10000 Nug(0) + 140000 Sph(800)
fit method         : no fit
>show plot <Tab>
[...]
Command: _
```

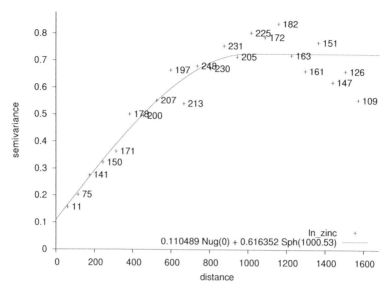

Fig. 10.1. gstat/GRASS: Semivariogram of zinc contaminations of the Maas river bank soil samples (variogram model: WLS, weights n(h))

After reaching the menu you can move around with cursor keys. Now choose the variable ln_zinc (logarithmic transformed zinc concentrations) with <ENTER>. Set the cutoff (lag distance) to 1600 and width to 70. Then select for fit methods "WLS, weights n(h)" (WLS is weighted least squares, other methods are also available). Now the variogram model will be fitted when hitting the <TAB> key or selecting show plot <Tab>. The resulting semivariogram is shown in Figure 10.1.

Zinc contamination kriging example In the next example (adapted from Pebesma, 2001:12), we will include a raster MASK and perform ordinary kriging prediction from the zinc data. This will result in a raster surface map with predicted distributed zinc contaminations and the predicted kriging error. The gstat instructions file looks as follows (store it as file gstat.maas2.zn):

```
# ordinary kriging prediction
#
data(zinc): 'zinc', x=1, y=2, v=1;
variogram(zinc): 0.0717 Nug(0) + 0.564 Sph(917.8);
mask: 'maasbank;
predictions(zinc): 'zinc_pr.map';
variances(zinc): 'zinc_var.map';
```

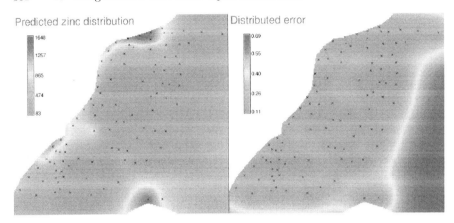

Fig. 10.2. gstat/GRASS: Ordinary kriging prediction of zinc contaminations on the Maas river bank. Left: predicted zinc contaminations [ppm], right: prediction error

In general, output grid maps are always written in the same format as the input mask map. Besides the GRASS format, gstat also supports other GIS formats. The Maas LOCATION contains a binary raster map maasbank which covers the river bank area. We can now run the kriging prediction:

```
d.mon x0
d.rast maasbank
d.vect zinc size=5
gstat gstat.maas2.zn
```

During calculations the program will report similar user messages:

```
gstat: Linux version 2.5.1 (02 April 2007)
Copyright (C) 1992, 2006 Edzer J. Pebesma
using Marsaglia's random number generator
data(zinc): WARNING: Adapted sites library used for ...
GRASS site list zinc: 0 cat, 2 dim, 0 str, 1 dbl.
gstat/grass: 98 sites read successfully.
            zinc    (GRASS site list)
attribute: col[1]      [x:] x_1   : [     269870,     272460]
n:              98     [y:] y_2   : [5.65061e+06,5.65293e+06]
sample mean:   481.031 sample std.:         398.808
[using ordinary kriging]
ncols 259
nrows 232
initializing maps ..
CREATING SUPPORT FILES FOR zinc_var.map
100% done
```

We have generated two new raster maps that represent the predicted zinc distribution (zinc_pr.map) and the distributed error (zinc_var.map). Using d.frame, we can display both maps side-by-side in the GRASS monitor:

```
r.colors zinc_pr.map col=gyr
r.colors zinc_var.map col=gyr
d.erase -f
d.frame -c at="0,100,0,50" frame=left
d.frame -c at="0,100,50,100" frame=right
d.frame -s left
d.rast zinc_pr.map
d.vect zinc col=black size=5
d.font FreeSans
echo "Predicted zinc distribution" | d.text
d.legend -s zinc_pr.map at=50,90,7,10
d.frame -s right
d.rast zinc_var.map
d.vect zinc size=5
echo "Distributed error" | d.text
d.legend -s zinc_var.map at=50,90,7,10
```

The result is shown in Figure 10.2. For further details and examples, please
refer to the **gstat** documents.

10.2 Spatial data analysis with GRASS and R

The "R data analysis programming language and environment" (Ihaka and
Gentleman, 1996, available from Internet[7]), a dialect of S (Becker et al., 1988),
is an extensible system which can be connected directly to GRASS. R consists
of a base package and extensions that can be downloaded from the project Web
site. A regular newsletter informs about recent changes and improvements.[8]
Besides classical methods, graphical and modern statistical techniques are
implemented in the base R library and supplementary packages. The latter
comprise packages for point pattern analysis, geostatistics, exploratory spatial data analysis and spatial econometrics. While R is a general data analysis
environment, it has been extensively used for modeling and simulation. The
R/GRASS interface substantially improves the geospatial analysis capabilities
of GRASS (Bivand, 2000; Bivand and Gebhardt, 2000; Bivand and Neteler,
2000; Furlanello et al., 2003; Bivand, 2007). The R/GRASS interface is integrated into the **sp** "spatial" classes as extension **spgrass6**[9]. Additional related
extensions such as **maptools** and **rgdal** are also avalable.

For the integration of R with GRASS, you need to run R from the GRASS
shell environment. The interface dynamically loads compiled GIS library functions into the R environment. The GRASS metadata about the LOCATION's
regional extent and raster resolution are transferred to R (internally calling

---

[7] R software, http://www.r-project.org
[8] R Newsletter, http://cran.r-project.org/doc/Rnews/
[9] Spatial data in R, http://r-spatial.sourceforge.net

g.region). The R/GRASS interface – like GRASS modules in general – assumes that the user needs to work at the current resolution, not the initial resolution of the map layer. The current interface is supporting raster and vector data.

Besides the base package of R, it is useful to install also the following contributed extensions: akima, fields, geoR, grid, gstat, lattice, MASS, scatterplot3d, spatial, and stepfun (available from the R Web site). Additional packages in terms of spatial-temporal analysis are focused on autocorrelation, spatial point patterns, time series, or wavelets.

Note that in this section, we omit the \ character for long lines as it is not allowed in R. Long, broken lines are indicated by the indent in the next line.

Installation of the R/GRASS interface The installation of the R/GRASS interface[10] is very easy and can be done by a single command (you probably have to be user "root" for this). If your computer is connected to the Internet, you can install packages within a R session. Start R and launch the command (we install "gstat" support at the same time):

```
# non-Mac users
install.packages("spgrass6", "gstat", dependencies = TRUE)

# Mac users
install.packages("spgrass6", "gstat", type="source",
                 dependencies = TRUE)
```

This will download the latest version of the package along with a set of dependency packages; it then unpacks, compiles and installs them. From time to time, the installed extensions should be checked for updates. The following command will download the R package list, compare to the local installation and upgrade installed packages if new versions are available:

```
update.packages()
```

Offline, you can install extra packages on command line:

```
# example
R CMD INSTALL spgrass6_0.3-7.tar.gz
```

As the interface is maturing, the version number is subject to change. Also the path to the R executable depends on your local installation. To start a quick exploration of standard R, simply run:

```
R
# list of demos (leave listings with <q>)
demo()
demo(graphics)
```

---

[10] R/GRASS interface, http://cran.r-project.org/src/contrib/

These demonstrations illustrate some of R's capabilities. As mentioned above, contributed packages extend immensely the functionality of GRASS in conjunction with the R/GRASS interface. To find out which packages are already available on your system, type within R:

```
library()
```

You can load an installed package by entering its name as a parameter for this function. Some packages also provide examples for their functions. We try an example for fitting a trend surface, the function is provided by library spatial:

```
library(spatial)
example(surf.ls)
?surf.ls
```

The ? command displays the function's help text (scroll down with <space>, leave with <q>). Help pages are also stored in HTML format, you can open them in a HTML browser with:

```
help.start()
?surf.ls
library(spgrass6)
?spgrass6
```

The pages provide package explanations and a local search engine.

To quit R, use the q() function. Before doing so, we look at the command history of the session:

```
history()
q()
```

The function history() works similar to the UNIX history command: It displays all commands used in the current R session. The function q() finishes the R session. When leaving R with q(), you will be asked: "Save workspace image? [y/n/c]:". If answering <y>, the session objects are stored within the local directory into the hidden file .RData and the command history into file .Rhistory. When launching R next time in this directory, objects and history will be read into the system and you can continue with your work.

When using R in batch mode, you can develop GRASS scripts which direct utilize R functionality in a GRASS user environment. In Section 10.2.3, we show an example.

10.2.1 Reading GRASS data into R

To illustrate how to apply R to your data, we present examples based on the North Carolina data set. As an exercise, we will analyze precipitation normals. After starting GRASS with North Carolina nc_spm LOCATION and setting the region to the precipitation map (along with 1000m resolution), we launch R within the GRASS terminal:

```
grass63 $HOME/grassdata/nc_spm/user1
g.region vect=precip_30ynormals res=1000 -ap
R
```

If you need to change the region settings later within R, you can use the `system()` function to call `g.region`. Within R the R/GRASS interface is loaded as follows:

```
library(spgrass6)
# show metadata
str(gmeta6())
```

By this, we have loaded the library of interface functions and loaded GRASS' metadata about the location into the R environment. Function `str(gmeta6())` prints more detailed information about the metadata.

Now we are ready to perform geospatial analysis of GRASS raster and vector data. As a prerequisite, to process these spatial data in R, we import some maps into the R environment. To start, we load the North Carolina 30 year monthly and annual precipitation normals `precip_30ynormals` and the vector map `nc_state` representing the state political boundaries:

```
precip30n <- readVECT6("precip_30ynormals", ignore.stderr=TRUE)
# verify that it is a SpatialPointsDataFrame
class(precip30n)

summary(precip30n)
 [...]
Number of points: 136
Data attributes:
      cat              station           lat               long
 Min.   :  1.00   Min.   :310090   Min.   :33.99   Min.   :-84.02
 1st Qu.: 34.75   1st Qu.:312511   1st Qu.:35.20   1st Qu.:-81.68
 Median : 68.50   Median :314950   Median :35.64   Median :-80.02
 Mean   : 68.50   Mean   :314952   Mean   :35.57   Mean   :-80.04
 3rd Qu.:102.25   3rd Qu.:317396   3rd Qu.:35.91   3rd Qu.:-78.34
 Max.   :136.00   Max.   :319675   Max.   :36.50   Max.   :-75.62
      elev              jan              feb              mar
 Min.   :   2.438 Min.   : 77.98 Min.   : 77.98 Min.   : 97.79
 1st Qu.:  38.252 1st Qu.:103.38 1st Qu.: 87.63 1st Qu.:109.47
 Median : 214.884 Median :109.22 Median : 91.95 Median :114.30
 Mean   : 300.203 Mean   :114.56 Mean   : 98.07 Mean   :122.22
 3rd Qu.: 396.697 3rd Qu.:117.16 3rd Qu.: 99.89 3rd Qu.:124.02
 Max.   :1615.440 Max.   :207.26 Max.   :181.86 Max.   :235.46
 [...]

nc_state <- readVECT6("nc_state", ignore.stderr=TRUE)
```

The `summary()` function computes the summary characteristics for the precipitation data set including univariate statistics (minimum, maximum, 1st and 3rd quartile, median, mean and the number of NA's).

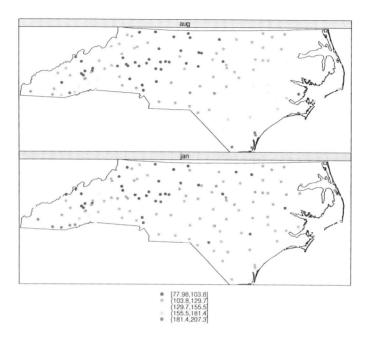

Fig. 10.3. R/GRASS: meteorological stations with 30 year precipitation normals in North Carolina, monthly values for January and August

Since we are working with spatial data, we can plot the R object `precip30n` as a simple point map with the NC state map as background:

```
plot(nc_state, axes=TRUE)
plot(precip30n, add=TRUE, lwd=2, col="brown")
```

This map only shows the locations of the meteorological stations in North Carolina available in our data set. To see all data objects which are currently loaded into R, use the `ls()` function. Next we list the variables in the R object:

```
ls()
  [1] "nc_state" "precip30n"

names(precip30n)
  [1] "cat"  "station" "lat"   "long"  "elev"  "jan"   "feb"
  [8] "mar"  "apr"     "may"   "jun"   "jul"   "aug"   "sep"
 [15] "oct"  "nov"     "dec"   "annual"

# spatial plot of selected months, load color library first
library(RColorBrewer)
spl <- list("sp.polygons", nc_state)

# spatial plot of map in two monthly lattice panels
```

```
spplot(precip30n, c("jan", "aug"),
  col.regions=brewer.pal(5,"Spectral"),sp.layout=list(spl))
# annual precipitation
spplot(precip30n, "annual",
  col.regions=brewer.pal(5,"Spectral"),sp.layout=list(spl))
```

The two plotted maps show different patterns in winter and summer precipitation (see Figure 10.3). The annual precipitation shows average precipitation along the costline, higher precipitation in the South-West NC and lower values in the Piedmont areas (map not shown here). You can find more examples of spatial plotting in R in the "gallery" at the "Spatial data in R" Web site[11].

10.2.2 Kriging in R

The R software also supports kriging, based on the gstat embedded as an extension of R. Here, we show how to transfer data from GRASS to R, perform geostatistical analysis and transfer the resulting maps back to GRASS. We again use the precipitation normals for this exercise, continuing the previous R session.

Interpolation from point data requires an existing spatial grid frame that will be filled with the interpolated values. The best way to generate the grid frame is to resample an available GRASS raster map. Since we later want to include the elevation of the meteorological station in the analysis (given the mountainous areas in the western part of NC), we use the map elev_state_500m from GRASS and cookie-cut it to the boundaries of NC state using MASK (MASK is respected by readRAST6()). We will use this masked elevation map later as newdata parameter in predict.gstat() for the kriging. Since we are still in the R session and don't want to leave it, we call the GRASS commands via the system() function:

```
# prepare data in GRASS, set region to full state first
system("g.region vect=nc_state res=1000 -ap")
system("v.to.rast nc_state out=nc_state_mask use=val val=1")
system("r.mask nc_state_mask")
elev <- readRAST6("elev_state_500m", ignore.stderr=TRUE)
system("r.mask -r nc_state_mask")

# plot elevation map with stations point list on top
pts <- list("sp.points", precip30n, lwd=2, col="brown")
spplot(elev,col.regions=terrain.colors(20),sp.layout=list(pts))
```

The resulting map is shown in Figure 10.4.

From Figure 10.3, we observe a spatial trend from West to East in January (which corresponds to variable "long" – longitude may be relevant besides the

---

[11] "Spatial data in R" Web site, http://r-spatial.sourceforge.net

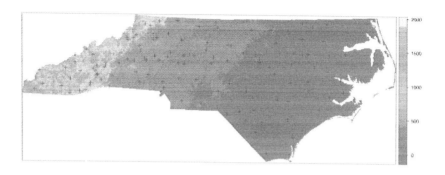

Fig. 10.4. R/GRASS: meteorological stations used for the 30 year precipitation normals in North Carolina drawn over elevation model

elevation). Also the annual precipitation shows a trend. Such trends are important for kriging and variogram estimation. We perform a series of attempts to fit an exponential model variogram while minimizing the prediction error:

```
library(gstat)

# fix projection string which should be identical
proj4string(elev) <- CRS(proj4string(precip30n))

# first attempt, using longitude and station elevation
t1 <- gstat(formula=annual ~ long + elev, data=precip30n)
annual_vgm1 <- variogram(t1)
efitted1 <- fit.variogram(annual_vgm1, vgm(psill=200,
        model="Exp", range=50000, nugget=100))
plot(annual_vgm1, efitted1, main="Exponential model variogram 1")
# we use the fitted variogram as model
t2 <- gstat(formula=annual ~ long + elev, data=precip30n,
        model=efitted1)
# calculate prediction error
pe1 <- gstat.cv(t2, debug.level=0, random=FALSE)$residual
sqrt(mean(pe1^2))
 [1] 115.7851

# second attempt, using second order of longitude instead
t3 <- gstat(formula=annual ~ long + I(long^2) + elev,
        data=precip30n)
```

Exponential model variogram 3

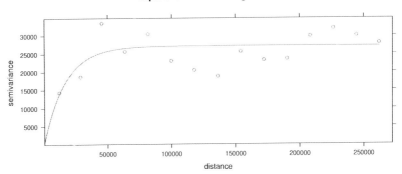

Fig. 10.5. R/GRASS: Exponential variogram model for annual precipitations in NC using elevation as variable

```
annual_vgm2 <- variogram(t3)
efitted2 <- fit.variogram(annual_vgm2, vgm(psill=200,
        model="Exp", range=50000, nugget=100))
plot(annual_vgm2, efitted2, main="Exponential model variogram 2")
t4 <- gstat(formula=annual ~ long + I(long^2) + elev,
        data=precip30n, model=efitted2)
pe2 <- gstat.cv(t4, debug.level=0, random=FALSE)$residual
sqrt(mean(pe2^2))
[1] 115.7715
```

The prediction error is almost identical for the second order model. We can now try without longitude and rely only on elevation as variable:

```
# third attempt, using only elevation
t5 <- gstat(formula=annual ~ elev, data=precip30n)
annual_vgm3 <- variogram(t5)
efitted3 <- fit.variogram(annual_vgm3, vgm(psill=200,
        model="Exp", range=50000, nugget=100))
plot(annual_vgm3, efitted3, main="Exponential model variogram 3")
t6 <- gstat(formula=annual ~ elev, data=precip30n,
        model=efitted3)
pe3 <- gstat.cv(t6, debug.level=0, random=FALSE)$residual
sqrt(mean(pe3^2))
[1] 115.9228
```

The second order model shows the (relatively) lowest prediction error. Figure 10.5 shows the exponential model variogram of the third attempt which we will use for the prediction. We can verify the 3rd model also in a different way, creating a simple plot of elevation versus annual precipitation:

```
plot(annual ~ elev, data=precip30n)
```

While precipitation is highly variable at sea level and higher altitudes, it is less variable below 350m elevation (low elevations are dominant in NC). We will use the last model to perform spatial prediction of annual rainfall in North Carolina:

```
# prepare SpatialPixelsDataFrame for predition, matrix will
# be filled then
names(elev) <- "elev"
elev_SP <- as(elev, "SpatialPixelsDataFrame")
proj4string(elev_SP) <- CRS(proj4string(precip30n))

# the annual ~ elev model, using universal kriging
annual_precip <- predict(t6, newdata=elev_SP)
annual_precip$var1.pe <- sqrt(annual_precip$var1.var)
summary(annual_precip)
 [...]
 Data attributes:
    var1.pred            var1.var            var1.pe
  Min.    : 948.4   Min.    : 1169    Min.    : 34.19
  1st Qu.:1213.6    1st Qu.:17213     1st Qu.:131.20
  Median :1239.6    Median :20814     Median :144.27
  Mean    :1277.4   Mean    :19918    Mean    :139.83
  3rd Qu.:1296.4    3rd Qu.:23476     3rd Qu.:153.22
  Max.    :2305.9   Max.    :31798    Max.    :178.32


spl <- list("sp.points", precip30n, cex=0.7, col="brown")
pal <- colorRampPalette(brewer.pal(4, "Blues"))

# plot predictive map of precipitation
plot1 <- spplot(annual_precip,"var1.pred",sp.layout=list(spl),
        col.regions=pal(20),
        main="Universal kriging of annual precipitation in NC")
# plot variance map
plot2 <- spplot(annual_precip, "var1.pe",sp.layout=list(spl),
        col.regions=pal(20), main="Prediction error map")

print(plot1, split = c(1,1,1,2), more = TRUE)
print(plot2, split = c(1,2,1,2), more = FALSE)

# or, alternatively
image(annual_precip, "var1.pred", col=pal(20))
points(precip30n, cex=0.7, pch=3, col="brown")
```

The first map shows the annual rainfall. Additionally, a second map is generated which represents the spatial distribution of predictive error (see Figure 10.6). Finally, we can write these two maps back into GRASS:

```
writeRAST6(annual_precip , "var1.pred")
system("g.rename rast=var1.pred,annual_precip_ukrig")
```

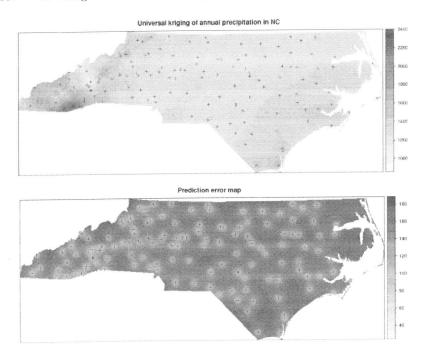

Fig. 10.6. R/GRASS: Universal kriging of annual precipitations in NC using elevation as variable

```
system("r.info -r annual_precip_ukrig")
 min=950.364014
 max=2313.140381
writeRAST6(annual_precip , "var1.pe")
system("g.rename rast=var1.pe,annual_precip_pe")
q()
```

The GRASS maps can be visualized as usual with **d.rast** or **d.rast.leg**, or in **nviz**. You can compare the result to the splines interpolation performed in Section 6.10.1:

```
# apply same color table
r.colors annual_precip_ukrig rast=precip_anntopo_500m
d.rast.leg precip_anntopo_500m
d.rast.leg annual_precip_ukrig

# calculate differences
r.mapcalc "diff_krig_rst=annual_precip_ukrig - \
          precip_anntopo_500m"
r.univar diff_krig_rst
r.colors diff_krig_rst col=gyr
```

```
d.rast.leg diff_krig_rst
nviz diff_krig_rst
```

In this exercise, we did not calculate the precipitation at 500m resolution which influences somewhat the difference map, but the results from both methods are very close. You may try to perform the kriging procedure at full resolution. The highest differences are observed in the mountainous areas; additionally, the map generated with kriging method shows some local maxima along the coast line which suggests further tuning of the parameters (compare also the variance map in Figure 10.6).

10.2.3 Using R in batch mode

R supports batch mode processing for a fully scripted usage. Within GRASS (maybe also scripted) geospatial data analysis can be automated. The desired analysis methods have to be stored in a text file. As an example, we want to create a batch job which calculates the empirical cumulative distribution function (ECDF) of a given map. We use environment variables to pass parameters so that we can write the script as a generic script for any filename. Save the following command sequence as script R_ecdf.batch:

```
# usage:
# export R_INMAP=elev_state_500m
# g.region rast=elev_state_500m res=1000 -pa
# R BATCH R_ecdf.batch
# -> will write 'ecdfplot.pdf'

# read input map from environment variable $R_INMAP
mapname <- Sys.getenv("R_INMAP")
library(spgrass6)
map <- readRAST6(mapname)
# summary(map)

# write graph to PDF
pdf("ecdfplot.pdf")

tstr <- c("ECDF (hypsometric integral):", mapname)
# plot ECDF
mapecdf <- ecdf(map$elev_state_500m)
plot(mapecdf,verticals=TRUE,do.points=F, \
    xlab="elevation", main=tstr)
for (i in seq(0,2000,50)) lines(c(i,i),c(0,1),lty=3,col="red2")
dev.off()

# cleanup workspace
rm(list = ls(all = TRUE))
```

The last command is needed to avoid that the loaded data are stored in the R workspace file .RData. Alternatively, the flag --no-save can be used when running the script. To run it, you have to define the GRASS raster map name to be analyzed in the environment variable $R_INMAP. We define it before starting the batch job (here bash shell syntax):

```
grass63 /usr/local/share/grassdata/nc_spm/user1
g.region rast=elev_state_500m res=1000 -pa
R CMD BATCH R_ecdf.batch
```

The function (and potential error) messages are echoed in the file R_ecdf.batch.Rout for batch process verification. In the example above, a plot of the ECDF function in PDF format is included. You should find this file in the current directory if no error occurs. The resulting PDF file ecdfplot.pdf contains the graph showing the hypsometric integral of the input map (here: elev_state_500m). When using the above approach with environment variables, complex (pseudo) GRASS scripts can be written to extend functionality of GRASS through R.

10.3 GPS data handling

The FreeGIS Web portal lists a set of programs freely available to handle GPS data including the data transfer from a GPS device to GIS and eventual datum transformations. The solutions are highly dependent on the GPS device. We only refer to few software packages:

- GPS Manager (GPSMan[12]) is a graphical manager of GPS data that supports the preparation, inspection, and editing of GPS data in a user friendly environment. GPSMan supports communication and real-time logging with both Garmin and Lowrance receivers and accepts real-time logging information in NMEA 0183 from any GPS receiver;
- GPStrans[13] is a program which transfers track, route, and waypoint data to and from various Garmin GPS;
- GPSBabel[14] reads and writes GPS waypoints in a variety of formats. Backends include GPX, Magellan and Garmin serial protocols, Geocaching.com, GPSMan, Garmin Mapsource, Magellan Mapsend, and many others. It runs on various operating systems.

GRASS itself provides two scripts: v.in.garmin as well as v.in.gpsbabel (the first requires gpstrans, the second gpsbabel). We have already shown GPS import into GRASS in Section 4.2.2.

---

[12] GPS Manager, http://www.ncc.up.pt/gpsman/
[13] GPStrans, http://gpstrans.sourceforge.net
[14] GPSBabel, http://www.gpsbabel.org

10.4 WebGIS applications with UMN/MapServer and OpenLayers

An excellent, fast and flexible Open Source mapping Web software is UMN/MapServer, one of the OSGeo projects. On a basic level, the program is run through CGI (Common Gateway Interface). In this case, it only requires a definition file (so-called map file) and a HTML template (and GIS data of course) to respond to a variety of spatial requests like making maps, scale-bars, and point, area and feature queries. It also supports Web Services (WMS, WFS, WCS, etc.), both as a server as well as a client to integrate Web Services into an UMN/MapServer application. The installation is quite convenient as the configuration of the Web mapping interface can be done without high level programming. A draft map file can be even generated with QGIS which allows you to save a project as map file. However, some editing is needed before deploying your own UMN/MapServer.

UMN/MapServer reads common GIS raster and vector formats. When GDAL has been compiled with GRASS plugin support, UMN/MapServer directly reads raster and vector data from a GRASS LOCATION through GDAL. Figure 10.7 provides a screenshot of the simple demonstrational GRASS-UMN/MapServer which is implemented at GRASS Web site; also an OpenLayers Spearfish demo is available.[15]

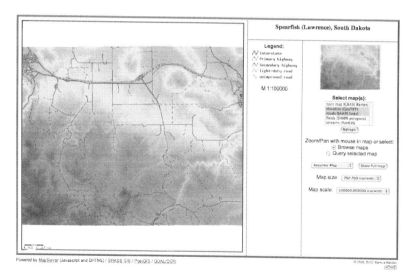

Fig. 10.7. Screenshot of GRASS / UMN/MapServer demonstrational Web site as implemented at FBKITC-irst

---

[15] Simple demonstrational GRASS UMN/MapServer (Spearfish data),
`http://grass.itc.it/start.html`
OpenLayers Spearfish application, `http://grass.itc.it/spearfish/lite.html`

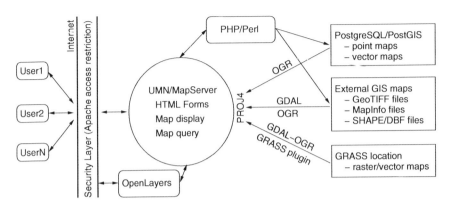

Fig. 10.8. Sample UMN/MapServer implementation model

For more complex applications, UMN/MapServer can be enhanced using Java, JavaScript, PHP or other Internet technologies. The PHP/MapScript extension provides access to the underlying UMN/MapServer C API. The UMN/MapServer software is freely available from the UMN/MapServer Web site.[16] It can be extended with several extensions such as "ka-Map", "p.mapper", "Chameleon" and others. A recent alternative with innovative interface is OpenLayers[17] which supports data integration from Web Services, local files and more. It also reads UMN/MapServer map files directly.

A more complex implementation using several Free Software tools is shown in Figure 10.8. Requirements for this implementation are: Web server such as Apache Server (with PHP support), UMN/MapServer, GDAL/OGR, PROJ4, GDAL/OGR-GRASS plugin, GRASS and PostgreSQL/PostGIS. OpenLayers can work on top of an UMN/MapServer backend, as alternative interface, optionally aggregating additional (external) Web Services.

As mentioned, to deploy a customized UMN/MapServer or OpenLayers application, essentially two files are required:

- UMN/MapServer definition file: to be stored in a `map-script/` directory in parallel to the `htdocs/` directory;
- a HTML template file which goes into the HTML space (into `htdocs/` directory).

At the GRASS Book Web site (see Chapter 1), samples of these files for the North Carolina data set can be downloaded along with installation instructions. Numerous public WebGIS applications are accessible on the Internet to get further inspiration.

---

[16] UMN/MapServer project, `http://mapserver.gis.umn.edu`
[17] OpenLayers Web site, `http://www.openlayers.org`

A

Appendix

A.1 Selected equations used in GRASS modules

This appendix section includes equations for selected GRASS modules to provide theoretical background of the methods used in these modules and give users opportunity to assess advantages and limitations of their functionality. The equations are also helpful for those who would like to improve or extend the modules.

Basic Statistics

Several GRASS modules, such as r.univar, v.univar, r.stats, r.series, r.neighbors compute basic statistical measures. We list the relevant equations here for reference, but check the source code to see the exact implementation. In the following equations, we are using $\bar{x}$ for mean estimated from sample of the population; replace it by μ when you are working with the entire population.

Arithmetic mean:

$$\bar{x} = \frac{1}{n}(x_1 + x_2 + \ldots + x_n) = \frac{1}{n}\sum_{i=1}^{n} x_i \tag{A.1}$$

Arithmetic mean $\bar{x}$ has the same units as x_i.

Median:
The median is the value below which 50% of the sample lie. To find the median, the data have to be ordered from the smallest to the highest. In case of an odd number of samples, it is the middle value; in case of an even number of samples, it is as half way between the two middle samples. Median has the same units as as the samples.

Mode:
The mode is defined as the most frequently occuring measurement in a data set. Continuous data need to be discretized into intervals to compute the mode. For data with normal distribution the mean, median and mode lead to the same value (in the limit of a large data sample).

Mean absolute deviation:

$$MD = \frac{1}{n} \sum_{i=1}^{n} |x_i - \bar{x}| \qquad (A.2)$$

The mean absolute deviation is an expression of dispersion about the mean. If the given values are deviations (e.g., provided by v.surf.rst), then we need arithmetic mean of absolute values:

$$\bar{x}_D = \frac{1}{n}(|x_1| + |x_2| + \ldots + |x_n|) = \frac{1}{n} \sum_{i=1}^{n} |x_i| \qquad (A.3)$$

Arithmetic mean of absolute values has the same units as x_i.

Variance for a population and for a sample:

$$\sigma^2 = \frac{1}{n} \sum_{i=1}^{n}(x_i - \bar{x})^2 \qquad \sigma^2 = \frac{1}{n-1} \sum_{i=1}^{n}(x_i - \bar{x})^2 \qquad (A.4)$$

The variance is another measure of dispersion, it is mean squared deviation and has the squared units of x_i.

Standard deviation for a population and for a sample:

$$\sigma = \sqrt{\frac{1}{n} \sum_{i=1}^{n}(x_i - \bar{x})^2} \qquad \sigma = \sqrt{\frac{1}{n-1} \sum_{i=1}^{n}(x_i - \bar{x})^2} \qquad (A.5)$$

The standard deviation is the positive square root of the variance and has the same units as x_p.

Coefficient of variation:

$$v = \frac{\sigma}{|\bar{x}|} * 100 \qquad (A.6)$$

The coefficient of variation is the ratio of the standard deviation to the mean and is dimensionless. It is expressed as a percentage when multiplied by 100 as in the above equation.

Skewness:

$$skewness = \frac{1}{n} \sum_{i=1}^{n} \left(\frac{x_i - \bar{x}}{\sigma} \right)^3 \tag{A.7}$$

Skewness is zero for any symmetric distribution. A distribution with a long tail towards larger values has a positive skewness (left skewed, typical for remote sensing images, Schowengerdt, 1997:118). Skewness is dimensionless and sensitive to outliers.

Kurtosis:

$$kurtosis = \left[\frac{1}{n} \sum_{i=1}^{n} \left(\frac{x_i - \bar{x}}{\sigma} \right)^4 \right] - 3 \tag{A.8}$$

Kurtosis is zero for a normal distribution. If a distribution has a positive kurtosis, then the peak is sharper than of a Gaussian distribution. Kurtosis is dimensionless and sensitive to outliers.

Covariance:

$$covariance = \frac{1}{n-1} \sum_{i=1}^{n} (x_{im} - \bar{x}_m)(x_{in} - \bar{x}_n) \tag{A.9}$$

Interpolation

Bilinear and bicubic interpolation Bilinear interpolation uses 4 neighboring cells to compute the unknown value z by performing linear interpolation in east-west and then north-east direction, leading to the following function:

$$z = a_0 + a_1 * x + a_2 * y + a_3 * x * y \tag{A.10}$$

Note that the above function is not linear, in spite of its name. Using the notation from the figure below the function can written as follows:

$$z = z12*(1-u)*(1-v) + z22*u*(1-v) + z11*(1-u)*v + z21*u*v \tag{A.11}$$

where $u = dx/px, v = dy/py$.

Bicubic interpolation uses 16 cells leading to the following function:

$$z = a_0 + a_1 x + a_2 y + a_3 x^2 + a_4 xy + a_5 y^2 + a_6 x^2 y + a_7 xy^2 + a_8 x^2 y^2$$
$$+a_9 x^3 + a_{10} y^3 + a_{11} x^3 y + a_{12} xy^3 + a_{13} x^3 y^2 + a_{14} x^2 y^3 + a_{15} 3x^3 y^3 \quad \text{(A.12)}$$

The coefficients are derived from the height at the four vertices, together with three partial derivatives at each vertex estimated using the neighboring vertices.

Inverse distance weighted interpolation (IDW)   The method is based on an assumption that the value at an unsampled point can be approximated as a weighted average of values at points within a certain cut-off distance, or from a given number m of the closest points (typically 10 to 30). Weights are usually inversely proportional to a power of distance (Watson, 1992; Burrough, 1986) which, at an unsampled location $\mathbf{r} = (x, y)$, leads to an estimator:

$$F(\mathbf{r}) = \sum_{i=1}^{m} w_i z(\mathbf{r}_i) = \frac{\sum_{i=1}^{m} z(\mathbf{r}_i) / |\mathbf{r} - \mathbf{r}_i|^p}{\sum_{j=1}^{m} 1 / |\mathbf{r} - \mathbf{r}_j|^p} \quad \text{(A.13)}$$

where p is a parameter (typically $p = 2$, for more details on the influence of this parameter, see Watson, 1992). GRASS modules use $p = 2$ and $m = 12$ as default values.

Regularized Spline with Tension   The function is a sum of a *trend function* and a *radial basis function* with an explicit form which depends on the choice of the measure of smoothness, for more details see Mitasova and Mitas (1993) and Mitasova et al. (1995):

$$z(\mathbf{r}) = T(\mathbf{r}) + \sum_{j=1}^{N} \lambda_j R(\mathbf{r}, \mathbf{r}^{[j]}) . \quad \text{(A.14)}$$

The trend function $T(\mathbf{r})$ is given by

$$T(\mathbf{r}) = \sum_{l=1}^{M} a_l f_l(\mathbf{r}) \quad \text{(A.15)}$$

where $\{f_l(\mathbf{r})\}$ is a set of linearly independent functions (monomials) which have zero smooth seminorm. $R(\mathbf{r}, \mathbf{r}^{[j]})$ is a radial basis function with an explicit form which depends on the choice of weights for partial derivatives in the smooth seminorm. See Mitasova and Mitas (1993); Mitasova et al. (1995) for the RST smoothness seminorm, which includes derivatives of all orders with their weights decreasing with the increasing derivative order.

RST can be generalized to an arbitrary dimension and the corresponding d-variate formula for the radial basis function is given by

$$R_d(\mathbf{r}, \mathbf{r}_j) = R_d(|\mathbf{r} - \mathbf{r}_j|) = R_d(r) = \varrho^{-\delta}\,\gamma\,(\delta, \varrho) - \frac{1}{\delta} \qquad \text{(A.16)}$$

where $r = |\mathbf{r} - \mathbf{r}_j|$, $\delta = (d-2)/2$, and $\varrho = (\varphi r/2)^2$. Further, φ is a generalized tension parameter, and $\gamma(\delta, \varrho)$ is the incomplete gamma function, not to be confused with semivariogram (Abramowitz and Stegun, 1964). For the special cases $d = 2, 3, 4$ (s.surf.rst, s.vol.rst, s.volt.rst, respectively), the equation A.16 can be rewritten as:

$$R_2(r) = -\left[\mathrm{E}_1(\varrho) + \ln\varrho + C_E\right] \qquad \text{(A.17)}$$

$$R_3(r) = \sqrt{\frac{\pi}{\varrho}}\,\mathrm{erf}\left(\sqrt{\varrho}\right) - 2 \qquad \text{(A.18)}$$

$$R_4(r) = \frac{1 - e^{-\varrho}}{\varrho} - 1 \qquad \text{(A.19)}$$

where $C_E = 0.577215\ldots$ is the Euler constant, $\mathrm{E}_1(\varrho)$ is the exponential integral function and $\mathrm{erf}(\sqrt{\varrho})$ is the error function (Abramowitz and Stegun, 1964), while the trend function is a constant ($M = 1$):

$$T(\mathbf{x}) = a_1\,, \quad d = 2, 3, 4 \qquad \text{(A.20)}$$

The coefficients $a_1, \{\lambda_j\}$ are obtained by solving the following system of linear equations:

$$a_1 + \sum_{j=1}^{N} \lambda_j \left[R(\mathbf{r}^{[i]}, \mathbf{r}^{[j]}) + \delta_{ji} w_0/w_j\right] = z^{[i]}\,, \quad i = 1, \ldots, N \qquad \text{(A.21)}$$

$$\sum_{j=1}^{N} \lambda_j = 0\,. \qquad \text{(A.22)}$$

where w_0/w_j are positive smoothing weights.

Topographic analysis

Topographic parameters slope, aspect and curvatures are computed using the principles of differential geometry and derived, for example, by Krcho (1973, 1991), and Mitasova and Hofierka (1993). First we introduce the following simplifying notations:

$$f_x = \frac{\partial z}{\partial x}, \qquad f_y = \frac{\partial z}{\partial y}, \qquad f_{xx} = \frac{\partial^2 z}{\partial x^2}, \qquad f_{yy} = \frac{\partial^2 z}{\partial y^2}, \qquad f_{xy} = \frac{\partial^2 z}{\partial x \partial y} \qquad \text{(A.23)}$$

and

$$p = f_x^2 + f_y^2 , \quad q = p + 1 \quad . \tag{A.24}$$

The steepest slope angle γ in degrees or percent, and aspect angle α in degrees are computed from gradient $\nabla f = (f_x, f_y)$ (its direction is upslope) as follows

$$\gamma = \arctan\sqrt{p} \qquad \gamma[\%] = 100.\sqrt{p} \tag{A.25}$$

$$\alpha = \arctan\frac{f_y}{f_x} \qquad (\alpha = 0 \text{ in west direction}) \quad . \tag{A.26}$$

Sometimes we need to compute change of the surface in a direction given by an angle α. The directional derivative of the surface $z = g(x, y)$ can be computed as

$$E = \frac{\partial g}{\partial s} = \frac{\partial g}{\partial x}\cos\alpha + \frac{\partial g}{\partial y}\sin\alpha \tag{A.27}$$

where (x, y) are the georeferenced coordinates, and α is aspect (given direction).

Curvatures In general, a surface has different curvatures in different directions. For applications in geosciences, the curvature in gradient direction (profile curvature) is important because it reflects the change in slope angle and thus controls the change of velocity of mass flowing downwards along the slope curve. The curvature in a direction perpendicular to the gradient (tangential curvature) reflects the change in aspect angle and influences the divergence/convergence of water flow. Both curvatures are measured in the normal plane. Equations for these curvatures can be derived using the general equation for curvature κ of a plane section through a point on a surface (Rektorys, 1969; Mitasova and Mitas, 1993).

The equation for the profile curvature $\kappa_s[m^{-1}]$ is

$$\kappa_s = \frac{f_{xx}f_x^2 + 2f_{xy}f_xf_y + f_{yy}f_y^2}{p\sqrt{q^3}} \quad . \tag{A.28}$$

The equation for tangential curvature $\kappa_t[m^{-1}]$ at a given point is derived as the curvature of normal plane section in a direction perpendicular to gradient (direction of tangent to the contour line)

$$\kappa_t = \frac{f_{xx}f_y^2 - 2f_{xy}f_xf_y + f_{yy}f_x^2}{p\sqrt{q}} \quad . \tag{A.29}$$

The positive and negative values of profile and tangential curvature can be combined to define the basic geometric relief forms (Krcho, 1973, 1991; Dikau, 1989). Each form has a different type of flow. Convex and concave forms in gradient direction have accelerated and deccelerated flow, respectively, and convex and concave forms in tangential direction exhibit converging and diverging flow, respectively.

Other types of curvatures, such as the principle, mean, or Gauss curvatures as well as curvatures in an arbitrary direction can be computed directly from the interpolation function.

Gradient and curvatures for volumes Volumes can be modeled by a trivariate interpolation function in the general form of $w = f(x, y, z)$. When this function is differentiable at least up to the 2nd order, the topographic parameters for volumes (3D) can be computed directly from its partial derivatives (Mitasova et al., 1995). First, we introduce simplifying notations for partial derivatives of this function:

$$f_x = \frac{\partial f}{\partial x}, \quad f_y = \frac{\partial f}{\partial y}, \quad f_z = \frac{\partial f}{\partial z},$$

$$f_{xx} = \frac{\partial^2 f}{\partial x^2}, f_{xy} = \frac{\partial^2 f}{\partial x \partial y}, f_{xz} = \frac{\partial^2 f}{\partial x \partial z},$$

$$f_{yy} = \frac{\partial^2 f}{\partial y^2}, f_{yz} = \frac{\partial^2 f}{\partial y \partial z}, f_{zz} = \frac{\partial^2 f}{\partial z^2}.$$

(A.30)

Volume geometry parameters are also derived from differential geometry, using additional independent spatial coordinate (z). Theoretically, such topographic parameters can be derived up to N-dimensional space (see Hofierka, 1997b). For a three-dimensional cartesian space these parameters have the following form:
Size of gradient:

$$|\nabla f| = \sqrt{f_x{}^2 + f_y{}^2 + f_z^2}$$

(A.31)

Direction of gradient can be defined by two angles.
Horizontal angle A_n:

$$A_n = \arctan\left(\frac{f_y}{f_x}\right)$$

(A.32)

and vertical angle B_n:

$$B_n = \arctan\left(\frac{\sqrt{f_x^2 + f_y^2}}{f_z}\right)$$

(A.33)

The change of gradient size in its direction has the following form:

$$\frac{\partial |\nabla f|}{\partial n} = \frac{f_x^2 f_{xx} + 2f_{xz}f_x f_z + 2f_{xy}f_x f_y + f_y^2 f_{yy} + 2f_{yz}f_y f_z + f_z^2 f_{zz}}{f_x^2 + f_y^2 + f_z^2}$$

(A.34)

When we note principal curvatures in 3D cartesian space as k_1, k_2, k_3, then the Gauss-Kronecker curvature K can by expressed as:

$$K = k_1 * k_2 * k_3 \tag{A.35}$$

The mean curvature M is:

$$M = \frac{k_1 + k_2 + k_3}{3} \tag{A.36}$$

In cartesian system these equations can be expressed as follows:

$$K = \frac{f_{xz}^2 f_{yy} + f_{yz}^2 f_{xx} + f_{xy}^2 f_{zz} - f_{xx} f_{yy} f_{zz} - 2 f_{xy} f_{yz} f_{xz}}{\left(\sqrt{1 + f_x^2 + f_y^2 + f_z^2} \right)^5} \tag{A.37}$$

$$H = \frac{\begin{vmatrix} h_{11} & h_{12} & h_{13} \\ h_{21} & h_{22} & h_{23} \\ h_{31} & h_{32} & h_{33} \end{vmatrix}}{3 \left(1 + f_x^2 + f_y^2 + f_z^2 \right)} \tag{A.38}$$

where:

$$h_{11} = \frac{-f_{xx}}{\sqrt{1 + f_x^2 + f_y^2 + f_z^2}} + 2(1 + f_x^2) \tag{A.39}$$

$$h_{12} = h_{21} = \frac{-f_{xy}}{\sqrt{1 + f_x^2 + f_y^2 + f_z^2}} + 2 f_x f_y \tag{A.40}$$

$$h_{22} = \frac{-f_{yy}}{\sqrt{1 + f_x^2 + f_y^2 + f_z^2}} + 2(1 + f_y^2) \tag{A.41}$$

$$h_{13} = h_{31} = \frac{-f_{xz}}{\sqrt{1 + f_x^2 + f_y^2 + f_z^2}} + 2 f_x f_z \tag{A.42}$$

$$h_{23} = h_{32} = \frac{-f_{yz}}{\sqrt{1 + f_x^2 + f_y^2 + f_z^2}} + 2 f_y f_z \tag{A.43}$$

$$h_{33} = \frac{-f_{zz}}{\sqrt{1 + f_x^2 + f_y^2 + f_z^2}} + 2(1 + f_z^2) \tag{A.44}$$

Estimation of partial derivatives To compute the above described equations for gradients and curvatures, we need to estimate first and second order partial derivatives.

In the RST-based modules, partial derivatives of RST functions are used. First, several definitions are introduced:

$$\eta = \frac{\varphi}{2} \tag{A.45}$$

$$R'(r_j) = 2 \frac{1 - e^{-(\eta r_j)^2}}{r_j} \tag{A.46}$$

$$R''(r_j) = 2 \frac{\left(2(\eta r_j)^2 + 1\right) e^{-(\eta r_j)^2} - 1}{r_j^2} \tag{A.47}$$

Partial derivatives for the bivariate RST basis function can then be expressed as follows:

$$\frac{\partial R(r_j)}{\partial x} = R'(r_j) \frac{(x - x^{[j]})}{r_j} , \qquad l = 1, 2 \tag{A.48}$$

$$\frac{\partial^2 R(r_j)}{\partial x_1^2} = R''(r_j) \frac{(x - x^{[j]})^2}{r_j^2} + R'(r_j) \frac{(y - y^{[j]})^2}{r_j^3} \tag{A.49}$$

whereas the derivatives, according to y, are found easily from Equation A.49 by exchange of x to y. The mixed derivative is given by

$$\frac{\partial^2 R(r_j)}{\partial x \partial x} = \left[R''(r_j) - \frac{R'(r_j)}{r_j} \right] \frac{(x - x^{[j]})(y - y^{[j]})}{r_j^2} \tag{A.50}$$

These expressions for first and second order derivatives are used for the computation of slope, aspect and curvatures in the modules s.surf.rst, v.surf.rst and r.resamp.rst. Optionally, the values of these partial derivatives are output by the module s.surf.rst when using the flag -d.

Partial derivatives for trivariate RST :

$$R'(r_j) = \frac{1}{r_j \sqrt{\pi}} \exp\left[-\left(\frac{\varphi r_j}{2} \right)^2 \right] - \frac{1}{\varphi r_j^2} erf\left(\frac{\varphi r_j}{2} \right) \tag{A.51}$$

$$R''(r_j) = \frac{2}{\varphi r_j^3} erf\left(\frac{\varphi r_j}{2} \right) - \sqrt{\pi} \left(\frac{2}{r_j^2} + \frac{\varphi^2}{2} \right) \exp\left[-\left(\frac{\varphi r_j}{2} \right)^2 \right] \tag{A.52}$$

In r.slope.aspect, second order polynomial approximation of a surface defined by given point and its 3×3 neighborhood is used leading to the following equations for the partial derivatives (as used in Horn's formula, see Horn, 1981):

$$z(x, y) = a_0 + a_1 x + a_2 y + a_3 xy + a_4 x^2 + a_5 y^2 \tag{A.53}$$

By fitting this polynomial to the 9 grid points (the given point $z_{i,j}$ and its 3 x 3 neighborhood, as shown below), using weighted least squares, we can derive the coefficients of this polynomial as well as its partial derivatives ($f_x = a_1$, $f_y = a_2$, $f_{xx} = 2a_4$, $f_{yy} = 2a_5$, $f_{xy} = a_3$):

```
i-1,j+1 ----- i,j+1 ----- i+1,j+1
   |           |           |
   |           |           |
   |           |           |
i-1,j ------- i,j --------- i|1,j
   |           |           |
   |           |           |
   |           |           |
i-1,j-1 ----- i,j-1 ----- i+1,j-1
```

$$f_x = \frac{(z_{i-1,j-1} - z_{i+1,j-1}) + (2z_{i-1,j} - 2z_{i+1,j}) + (z_{i-1,j+1} - z_{i+1,j+1})}{8\Delta x}$$

(A.54)

$$f_y = \frac{(z_{i-1,j-1} - z_{i-1,j+1}) + (2z_{i,j-1} - 2z_{i,j+1}) + (z_{i+1,j-1} - z_{i+1,j+1})}{8\Delta y}$$

(A.55)

where Δx and Δy is the resolution (grid spacing) in the east-west and north-south direction respectively.

Let us denote $D(i,\delta) = z_{i,j+1} + z_{i,j-1} - 2z_{i,j}$ and $D(\delta,j) = z_{i+1,j} + z_{i-1,j} - 2z_{i,j}$. Then we can write

$$f_{xx} = \frac{D(\delta, j+1) + (4z_{i-1,j} + 4z_{i+1,j} - 8z_{i,j}) + D(\delta, j-1)}{6(\Delta x)^2}$$

(A.56)

$$f_{yy} = \frac{D(i-1, \delta) + (4z_{i,j+1} + 4z_{i,j-1} - 8z_{i,j}) + D(i+1, \delta)}{6(\Delta y)^2}$$

(A.57)

$$f_{xy} = \frac{(z_{j-1,i-1} - z_{j-1,i+1}) - (z_{j+1,x-1} - z_{j+1,i+1})}{4\Delta x \Delta y}$$

(A.58)

where $z_{i,j}$ is the elevation value at row j column i, Δx is the east-west grid spacing and Δy is the north-south grid spacing (resolution).

Insolation

Equations for computation of solar energy related parameters used in r.sun (Hofierka, 1997a; Hofierka and Súri, 2002, further citations in the manual page of r.sun). The clear-sky solar radiation model applied in this module is based on the work undertaken for development of European Solar Radiation Atlas (Scharmer and Greif, 2000; Page et al., 2001; Rigollier et al., 2000).

Solar geometry

Declination d [rad]:

$$\delta = \arcsin(0.3978\sin(j - 1.4 + 0.0355\sin(j - 0.0489))) \qquad (A.59)$$

where:

$$j = 2\pi\,day/365.25 \quad \text{[rad]}$$

Position of the sun in respect to a horizontal plane:

$$sinh_0 = C_{31}\cos T + C_{33} \qquad (A.60)$$

$$cos A_0 = \frac{C_{11}\cos T + C_{13}}{\sqrt{(C_{22}\sin T)^2 + (C_{11}\cos T + C_{13})^2}} \qquad (A.61)$$

where:

$C_{11} = \sin\varphi\cos\delta$
$C_{13} = -\cos\varphi\sin\delta$
$C_{22} = \cos\delta$
$C_{31} = \cos\varphi\cos\delta$
$C_{33} = \sin\varphi\sin\delta$

Position of the sun in respect to an inclined plane:

$$\sin\delta_{exp} = C'_{31}\cos(T - \lambda') + C'_{33} \qquad (A.62)$$

where:

$C'_{31} = \cos\varphi'\cos\delta$
$C'_{33} = \sin\varphi'\sin\delta$
$\sin\varphi' = -\cos\varphi\sin\gamma_N\cos A_N + \sin\varphi\cos\gamma_N$
$\tan\lambda' = -\frac{\sin\gamma_N\sin A_N}{\sin\varphi\sin\gamma_N\cos A_N + \cos\varphi\cos\gamma_N}$

Sunrise/sunset over a horizontal plane:

$$\cos(Th_{r,s}) = -\frac{C_{33}}{C_{31}} \qquad (A.63)$$

Sunrise/sunset over an inclined plane:

$$\cos(Tp_{r,s} - \lambda') = -\frac{C'_{33}}{C'_{31}} \qquad (A.64)$$

Extraterrestrial irradiance on a plane perpendicular to the solar beam G_0 $[W/m^2]$

$$G_0 = I_0 \epsilon \qquad (A.65)$$

where:
$$\epsilon = 1 + 0.03344 \cos(j - 0.048869)$$
$$\text{values j and 0.048869 are in radians.}$$

Extraterrestrial irradiance on a horizontal plane G_{0h} $[W/m^2]$

$$G_{0h} = G_0 \sin h_0 \qquad (A.66)$$

Beam irradiance on a horizontal plane B_h $[W/m^2]$

$$B_h = G_0 e^{(-0.8662\ T_{LK}\ m\ \delta_R(m))} \sin h_0 \qquad (A.67)$$

where:
$$p/p_0 = e^{(-z/8434.5)}$$
$$\Delta h_0^{ref} = 0.061359(0.1594 + 1.123 h_0 + 0.065656 h_0^2)/(1 + 28.9344 h_0 + 277.3971 h_0^2)$$
$$h_0^{ref} = h_0 + \Delta h_0^{ref}$$
$$m = (p/p_0)/(\sin h_0^{ref} + 0.50572(h_0^{ref} + 6.07995)^{-1.6364})$$
$$\text{where values } h_0^{ref} \text{ and 6.07995 are in degree}$$
$$\delta_R(m) = 1/(6.6296 + 1.7513m - 0.1202m^2 + 0.0065m^3 - 0.00013m^4)$$
$$\text{if m} \leq 20$$
$$\delta_R(m) = 1/(10.4 + 0.718m) \qquad\qquad\qquad\qquad \text{if m} > 20$$

Beam irradiance on an inclined plane B_i $[W/m^2]$

$$B_i = G_0 e^{(-0.8662\ T_{LK}\ m\ \delta_R(m))} \sin \delta_{exp} \qquad (A.68)$$

Diffuse irradiance on a horizontal plane B_h $[W/m^2]$

$$D_h = G_0 F_d(h_0) Tn(T_{LK}) \qquad (A.69)$$

where:
$$Tn(T_{LK}) = -0.015843 + 0.030543\ T_{LK} + 0.0003797\ T_{LK}^2$$
$$F_d(h_0) = A_1 + A_2 \sin h_0 + A_3 \sin^2 h_0$$
$$A_1' = 0.26463 - 0.061581 T_{LK} + 0.0031408 T_{LK}^2$$
$$A_1 = 0.0022/Tn(T_{LK})$$

$$A_1 = A_1'$$

$$\text{if } A_1' \; Tn(T_{LK}) < 0.0022$$

$$\text{if } A_1' \; Tn(T_{LK}) \geq 0.0022$$

$$A_2 = 2.04020 + 0.018945 \; T_{LK} + 0.011161 \; T_{LK}^2$$
$$A_3 = -1.3025 + 0.039231 \; T_{LK} + 0.0085079 \; T_{LK}^2$$

Diffuse irradiance on an inclined plane D_i [W/m^2]

$$D_i = D_h F_x \tag{A.70}$$

where:

if plane is in shade (e.g. $\delta_{exp} < 0$ and $h_0 \geq 0$):
$$F_x = F(\gamma_N)$$
$$F(\gamma_N) = r_i(\gamma_N) + \left(\sin\gamma_N - \gamma_N \cos\gamma_N - \pi \sin^2\left(\tfrac{\gamma_N}{2}\right)\right) 0.252271$$

if plane is sunlit under clear sky:
$$\text{if } h_0 \geq 0.1 rad:$$
$$F_x = F(\gamma_N)(1 - K_b) + K_b \sin\delta_{exp}/\sin h_0$$
$$\text{if } h_0 < 0.1 rad:$$
$$F_x = F(\gamma_N)(1 - K_b) + K_b \sin\gamma_N \cos A_{LN}/(0.1 - 0.008 h_0)$$
$$A_{LN}^* = A_0 - A_N$$
$$A_{LN} = A_{LN}^*$$

$$A_{LN} = A_{LN}^* - 2\pi$$

$$A_{LN} = A_{LN}^* + 2\pi$$

$$\text{if } -\pi \leq A_{LN}^* \leq \pi$$

$$\text{if } A_{LN}^* > \pi$$

$$\text{if } A_{LN}^* < -\pi$$

$$F(\gamma_N) = r_i(\gamma_N)$$
$$+ \left(\sin\gamma_N - \gamma_N \cos\gamma_N - \pi\sin^2\left(\tfrac{\gamma_N}{2}\right)\right)(0.00263 - 0.712 K_b - 0.6883 K_b^2)$$
$$K_b = B_h/G_{0h}$$
$$r_i(\gamma_N) = (1 + \cos\gamma_N)/2$$

Diffuse ground reflected irradiance on an inclined plane R_i [W/m^2]

$$R_i = \rho_y G_h r_g(\cdot\gamma_N) \tag{A.71}$$

where:
$$r_g(\gamma_N) = (1 - \cos\gamma_N)/2$$
$$G_h = B_h + D_h$$

$$\text{with } G_h \text{ in } [W/m^2]$$

Symbols

- Position of the grid cell (solar plane):
  φ geographical latitude [rad],
  z elevation above sea level [m],
  γ_N slope angle [rad],
  A_N aspect (orientation, azimuth) – angle between the projection of the
  normal on the horizontal plane and east [rad],
  φ' relative geographical latitude of an inclined plane [rad],
  λ' relative geographical longitude [rad].
- Parameters of the surface (plane):
  ρ_g mean ground albedo.
- Date-related parameters:
  day day number 1-365 (366),
  j Julian day number expressed as a day angle [rad],
  T time of computation [decimal hours/rad],
  $Th_{r,s}$ time of sunrise and sunset over the local horizon,
  $Tp_{r,s}$ time of sunrise and sunset over the inclined grid cell (plane),
  δ solar declination [rad],
  ϵ correction of the variation of sun-earth distance from its mean value.
- Solar position:
  h_0 solar altitude – an angle between sun and horizon [rad],
  A_0 solar azimuth – an angle between sun and meridian measured from
  east [rad],
  A_{LN} angle between the vertical plane containing the normal to the surface
  and vertical plane passing through the center of the solar disc [rad],
  δ_{exp} solar incidence angle - an angle between sun and the (inclined) plane
  [rad].
- Solar radiation:
  I_0 solar_const $= 1367$ W/m^2,
  G_0 extraterrestrial irradiance on a plane perpendicular to the solar beam
  [W/m^2],
  G_h $G_h = B_h + D_h$ – global solar irradiance on a horizontal plane [W/m^2],
  G_i $G_i - B_i + D_i + R_i$ – global solar irradiance on an inclined plane [W/m^2],
  B_h beam irradiance on a horizontal plane [W/m^2],
  B_i beam irradiance on an inclined plane [W/m^2],
  D_h diffuse irradiance on a horizontal plane [W/m^2],
  D_i diffuse irradiance on an inclined plane [W/m^2],
  R_i diffuse ground reflected irradiance on an inclined plane [W/m^2].
- Parameters of the atmosphere:
  p/p_0 correction of station elevation [-],
  T_{LK} Linke turbidity factor [-],
  T_L corrected Linke turbidity factor ($T_L = 0.8662\,T_{LK}$), see Kasten (1996),
  m relative optical air mass [-],
  $\delta_R(m)$ Rayleigh optical thickness [-].

- Parameters of the radiation transmission:
  $F_d(h_0)$ diffuse solar elevation function,
  $Tn(T_{LK})$ diffuse transmission function,
  $F(\gamma_N)$ function accounting for the diffuse sky irradiance distribution,
  K_b proportion between beam irradiance and extraterrestrial solar irradiance on a horizontal plane,
  $r_i(\gamma_N)$ fraction of the sky dome viewed by an inclined plane [-],
  $r_g(\gamma_N)$ fraction of the ground viewed by an inclined plane [-].

Walking person

Anisotropic movement of a person between different geographic locations:

$$T = a.\Delta S + b.\Delta H_u + c.\Delta H_d + d.\Delta H_s \qquad (A.72)$$

where T is time of movement in seconds, ΔS is the distance covered in meters, $\Delta H_u, H_d, H_s$ is the altitude difference in meters when going uphill, downhill, and steep downhill respectively. The a, b, c, d parameters represent speed under different conditions and are linked to a underfoot condition (a=1/walking_speed), b underfoot condition and cost associated to movement uphill, c underfoot condition and cost associated to movement moderate downhill, d underfoot condition and cost associated to movement steep downhill. Moving downhill is beneficial up to a specific slope value threshold, after that it becomes unfavourable. The default slope value threshold (slope factor) is -0.2125, corresponding to $\tan(12)$ downslope. The default values for a, b, c, d are those proposed by Langmuir (0.72, 6.0, 1.9998, -1.9998), based on man walking effort in standard conditions. Total cost is estimated as a linear equation combining movement and friction costs using the λ parameter:

$$TC = MC + \lambda * FC \qquad (A.73)$$

where TC is total cost, MC is movement time cost and FC is friction costs, (see Aitken, 1977; Fontanari, 2001).

A.2 Landscape process modeling

Wetness index

$$W = \ln \frac{A}{\tan \beta} \qquad (A.74)$$

where A is the upslope area per unit contour width [m] and β is the slope angle in [deg].

Universal Soil Loss Equation (USLE, RUSLE)

$$A = RKLSCP \tag{A.75}$$

where A is average annual soil loss in ton/(acre.year)=0.2242kg/(m^2.year), R is rainfall factor in (hundreds of ft-tonf.in)/(acre.hr.year)=17.02 (MJ.mm)/(ha.hr.year), K is soil erodibility factor in (ton acre.hr)/(hundreds of acre ft-tonf.in)=0.1317(ton.ha.hr)/(ha.MJ.mm), LS is a dimensionless topographic (length-slope) factor, C is a dimensionless land cover factor, and P is a dimensionless prevention measures factor. The modified factor, representing topographic potential for erosion at a point on the hillslope, is a function of the upslope area per unit width and the slope angle:

$$LS = (m + 1) \left(\frac{U}{22.1} \right)^m \left(\frac{\sin \beta}{0.09} \right)^n \tag{A.76}$$

where U is the upslope area per unit width (measure of water flow) in meters (m^2/m), β is the slope angle in degree, 22.1 is the length of the standard USLE plot in meters and $0.09 = 9\% = 5.15°$ is the slope of the standard USLE plot. The values of exponents range for $m = 0.2 - 0.6$ and $n = 1.0 - 1.3$, where the lower values are used for prevailing sheet flow and higher values for prevailing rill flow. When nothing is known about the type of flow, $m = 0.4$ and $n = 1.3$ are usually used (see RUSLE for ArcView[1]).

Unit Stream Power Based Erosion/Deposition model (USPED) The *Unit Stream Power Based Erosion/Deposition model (USPED*, Mitasova and Mitas, 2001) estimates a simplified case of erosion/deposition using the idea originally proposed by Moore and Burch (1986). It combines the *RUSLE* parameters and upslope contributing area per unit width A to estimate the sediment flow T:

$$T \approx RKCPA^m(sin\beta)^n. \tag{A.77}$$

The upslope area and slope are not normalized, because T is an estimate of sediment flow $[kg/(ms)]$ (rather than soil detachment $[kg/(m^2s)]$). The net erosion/deposition $D[kg/(m^2s)]$ is then computed as a divergence of sediment flow:

$$D = \nabla \cdot (T\mathbf{s_0}) = \frac{d(T \cos \alpha)}{dx} + \frac{d(T \sin \alpha)}{dy}, \tag{A.78}$$

where α in degrees is the aspect of the terrain surface (direction of flow). The exponents m, n control the relative influence of water and slope terms and reflect the impact of different types of flow. The typical range of values is $m = 1.0 - 1.6, n = 1.0 - 1.3$, with the higher values reflecting the pattern for

---

[1] RUSLE for ArcView,
http://abe.www.ecn.purdue.edu/~engelb/agen526/gisrusle/gisrusle.html

prevailing rill erosion with more turbulent flow when erosion sharply increases with the amount of water. Lower exponent values close to $m = n = 1$ better reflect the pattern of compounded, long term impact of both rill and sheet erosion and averaging over a long term sequence of large and small events.

A.3 Definition of SQLite-ODBC connection

At time of this writing, there is only OpenOffice.org Base which can be used as powerful graphical user interface to SQL databases. While GRASS can be directly connected to SQLite, OpenOffice.org Base needs an ODBC based driver, the "sqliteodbc" library[2] and of course ODBC. In this example, we define a sample SQLite-ODBC connection.

After installation of the "sqliteodbc" library, on UNIX-based systems, add the driver to /etc/odbcinst.ini (either add next lines or create as new file if the file does not yet exist):

```
[SQLite]
Description=SQLite ODBC Driver
Driver=/usr/local/lib/libsqlite3odbc.so
Setup=/usr/local/lib/libsqlite3odbc.so
```

Now the SQLite driver is available for ODBC. The next step is to add the definition(s) of the database you want to connect to. Add an ODBC Data Source Name (DSN) to your definition file at $HOME/.odbc.ini (replace "user" and "mymapset" with the correct entries):

```
[nc_sqlite]
Description=North Carolina SQLite DB
Driver=SQLite
Database=/home/user/grassdata/nc_spm/mymapset/sqlite.db
# optional lock timeout in milliseconds
Timeout=2000
```

See Section 6.2.1 for GRASS related usage notes of the SQLite driver. For installation under MS-Windows see the "sqliteodbc" Web page.

---

[2] Web site of "sqliteodbc" library, http://www.ch-werner.de/sqliteodbc/

References

Abramowitz, M. and Stegun, I. 1964. *Handbook of Mathematical Functions*. New York: Dover.

Aitken, R. 1977. *Wilderness areas in Scotland*. Ph.D. thesis, University of Aberdeen, U.K.

Albrecht, J. 1992. GTZ-handbook GRASS. Technical report, Univ. of Vechta. Electr. Doc. http://grass.itc.it/gdp/.

Alexandrov, A., Kolmogorov, A., and Lavrent'ev, M. 1989. *Mathematics: Its content, methods and meanings*, volume 2. Cambridge (MA): MIT Press, 6th edition.

Arge, L., Chase, J., Halpin, P., Toma, L., Vitter, J., Urban, D., and Wickremesinghe, R. 2003. Efficient flow computation on massive grid terrain datasets. *GeoInformatica* **7** (4): 283–313.

Bailey, T. and Gatrell, A. 1995. *Interactive Spatial Data Analysis*. Essex: Pearson.

Baker, W. 2001. The r.le programs for multiscale analysis of landscape structure using the GRASS geographical information system. Technical report, Department of Geography and Recreation, University of Wyoming.

Baker, W. and Cai, Y. 1992. The r.le programs for multiscale analysis of landscape structure using the GRASS geographical information system. *Landscape Ecology* **7** (4): 291–302.

Barsi, J., Schott, J., Palluconi, F., Helder, D., Hook, S., Markham, B., Chander, G., and O'Donnell, E. 2003. Landsat TM and ETM+ thermal band calibration. *Canadian Journal of Remote Sensing* **29** (2): 141–153.

Bartelme, N. 1995. *Geoinformatik: Modelle, Strukturen, Funktionen*. Heidelberg: Springer.

Becker, R., Chambers, J., and Wilks, A. 1988. *The New S Language*. London: Chapman & Hall.

Bivand, R. 2000. Using the R statistical data analysis language on GRASS 5.0 GIS database files. *Computers & Geosciences* **26** (9-10): 1043–1052.

Bivand, R. 2007. Using the R–GRASS interface. *OSGeo Journal* **1**: 36–38. Electr. Doc. http://www.osgeo.org/journal/volume1.

Bivand, R. and Gebhardt, A. 2000. Implementing functions for spatial statistical analysis using the R language. *Journal of Geographical Systems* **2** (3): 307–317.

Bivand, R. and Neteler, M. 2000. Open Source Geocomputation: Using the R Data Analysis Language Integrated with GRASS GIS and PostgreSQL Data Base Systems. In *Proc., 5th conference on GeoComputation, University of Greenwich, U.K.*, pp. 23–25. Electr. Doc.
http://reclus.nhh.no/gc00/gc009.htm.

Blazek, R., Neteler, M., and Micarelli, R. 2002. The new GRASS 5.1 vector architecture. In *Proc., Open Source Free Software GIS – GRASS users conference, Trento, Italy, Sept. 2002* (eds. Ciolli, M. and Zatelli, P.), pp. 11–13. Electr. Doc.
http://www.ing.unitn.it/~grass/conferences/GRASS2002/
proceedings/proceedings/pdfs/Blazek_Radim.pdf.

Brandon, R., Kludt, T., and Neteler, M. 1999. Archaeology and GIS – the Linux way: Using GRASS and Linux to analyze archaeological data. *Linux Journal* **7**: 50–54. Feature article.

Brovelli, M., Cannata, M., and Longoni, U. 2004. LIDAR data filtering and DTM interpolation within GRASS. *Transactions in GIS* **8** (2): 155–174.

Brown, W., Astley, M., Baker, T., and Mitasova, H. 1995. GRASS as an integrated GIS and visualization environment for spatio-temporal modeling. In *Proc., Auto-carto XII, ACSM/ASPRS, Charlotte, NC*, pp. 89–99.

Bugayevskiy, L. and Snyder, J. 2000. *Map Projections: A reference manual*. London, Philadelphia: Taylor & Francis.

Burrough, P. 1986. *Principles of GIS for Land Resources Assessment*. Oxford: Clarendon Press.

Burrough, P. and McDonnell, R. 1998. *Principles of Geographical Information Systems*. New York: Oxford University Press.

Cannata, M. 2006. *A GIS embedded approach for Free & Open Source Hydrological Modelling*. Ph.D. thesis, Department of Geodesy and Geomatics, Polytechnic of Milan, Italy.

Chambers, J. and Hastie, T. 1992. *Statistical Models in S*. London: Chapman & Hall.

Chavez, P., Guptill, S., and Bowell, J. 1984. Image processing techniques for thematic mapper data. *Proc., Am. Soc. Photogr.* **2**: 728–743.

Chavez Jr., P. 1996. Image-based atmospheric corrections – revisited and improved. *Photogr. Eng. & Rem. Sens.* **62** (9): 1025–1036.

Clarke, K. 2002. *Getting Started with Geographic Information Systems*. New Jersey: Prentice Hall, 4th edition.

Cressie, N. 1993. *Statistics for Spatial Data*. New York: Wiley.

Curtin, K. M. 2007. Network Analysis in Geographic Information Science: Review, Assessment, and Projections. *Cartography and Geographic Information Science* **34** (2): 103–111.

Desmet, P. and Govers, G. 1996. A GIS procedure for automatically calculating the USLE LS factor on topographically complex landscape units. *J. Soil and Water Conservation* **51** (5): 427–433.

Dikau, R. 1989. The application of a digital relief model to landform analysis in geomorphology. In *Three dimensional applications in Geographic Information Systems* (ed. Raper, J.), pp. 51–77. Taylor & Francis, London.

Ehlschlaeger, C. and Goodchild, M. 1994. Dealing with uncertainty in categorical coverage maps: Defining, visualizing, and managing errors. *Proc., Workshop on Geographic Information Systems at the Conf. on Information and Knowledge Management, Gaithersburg, MD* pp. 86–91.

Erle, S., Gibson, R., and Walsh, J. 2005. *Mapping Hacks*. Cambridge: O'Reilly & Associates.

Evenden, G. 1995. Cartographic projection procedures for the Unix environment – a user's manual. Technical report, U.S. Geological Survey Open File Report. Electr. Doc. http://proj.maptools.org.

Fontanari, S. 2001. Sviluppo di metodologie GIS per la determinazione dell'accessibilità territoriale come supporto alle decisioni nella gestione ambientale, M.Sc. thesis, University of Trento, Italy.

Fortune, S. 1987. A sweepline algorithm for Voronoi diagrams. *Algorithmica* **2**: 153–174.

Furlanello, C., Neteler, M., Merler, S., Menegon, S., Fontanari, S., Donini, A., Rizzoli, A., and Chemini, C. 2003. GIS and the RandomForest predictor: Integration in R for tick-borne disease risk assessment. In *Proc., Distributed Statistical Computing, Vienna, Austria, March 2003* (eds. Hornik, K. and Leisch, F.).

Gebbert, S. 2007. Groundwater modeling. Dipl. Ing. thesis, Technical University Berlin, Germany.

GRASS Development Team, ed. 2006. *GRASS 6.2 Programmer's Manual. Geographic Resources Analysis Support System*. FBKITC-irst, Trento, Italy. Electr. Doc. http://grass.itc.it/devel/.

GSFC/NASA 2001. Landsat-7 Science Data User's Handbook. Technical report, NASA. Electr. Doc. http://ltpwww.gsfc.nasa.gov/IAS/handbook/handbook_htmls/chapter11/chapter11.html.

Haan, C., Barfield, B., and Hayes, J. 1994. *Design Hydrology and Sedimentology for Small Catchments*. New York: Academic Press.

Hake, G. and Grünreich, D. 1994. *Kartographie*. Berlin: de Gruyter, 7th edition.

Hofierka, J. 1997a. Direct solar radiation modelling within an Open GIS environment. *Proc., Geographical Information '97: Third Joint European Conference & Exhibition on Geographical Information., Vienna, Austria, April 1997* pp. 575–584.

Hofierka, J. 1997b. *Modeling Natural Phenomena in a GIS Environment*. Ph.D. thesis, Comenius University, Bratislava.

Hofierka, J., Mitasova, H., and Neteler, M. 2007. Terrain parameterization in GRASS. In *Geomorphometry: concepts, software, applications* (eds. Hengl,

T. and Reuter, H.), pp. 301–318. Office for Official Publications of the European Communities, Luxembourg, EUR 22670 EN. Electr. Doc. http://www.geomorphometry.org.

Hofierka, J., Parajka, J., Mitasova, H., and Mitas, L. 2002. Multivariate interpolation of precipitation using regularized spline with tension. *Transactions in GIS* **6** (2): 135–150.

Hofierka, J. and Súri, M. 2002. The solar radiation model for Open Source GIS: Implementation and applications. In *Proc., Open Source Free Software GIS – GRASS users conference, Trento, Italy, Sept. 2002* (eds. Ciolli, M. and Zatelli, P.), pp. 11–13.

Horn, B. 1981. Hill shading and the reflectance map. *Proc. of the IEEE* **69** (1): 14–47.

Hutchinson, M. and Bischof, R. 1983. A new method for estimating the spatial distribution of mean seasonal and annual rainfall applied to the hunter valley, new south wales. *Australian Meteorological Magazine* **31**: 179–184.

Ihaka, R. and Gentleman, R. 1996. R: A language for data analysis and graphics. *J. of Comp. and Graph. Stat.* **5** (3): 299–314.

Jolma, A., Ames, D., Horning, N., Mitasova, H., Neteler, M., Racicot, A., and Sutton, T. 2007. Environmental Modeling and Management using Free and Open Source Geospatial Tools. In *Encyclopedia of GIS* (eds. Shekhar, S. and Xiong, H.). Springer.

Jolma, A., Ames, D., Horning, N., Neteler, M., Racicot, A., and Sutton, T. 2006. Free and Open Source geospatial tools for environmental modeling and management. In *Proc., iEMSs 2006, 3rd Biennial meeting of the Intl. Env. Mod. and Softw. Soc. July 9-13, 2006, Burlington, Vermont, USA, W13: Open geospatial tools and methods in environmental modeling and management* (ed. Voinov, A.). Electr. Doc. http://www.iemss.org/iemss2006/papers/w13/pp.pdf.

Journel, A. 1996. Modelling uncertainty and spatial dependence: Stochastic imaging. *Int. J. of Geogr. Inf. Sys.* **10** (5): 517–522.

Kasten, F. 1996. The linke turbidity factor based on improved values of the integral Rayleigh optical thickness. *Solar Energy* **56** (3): 239–244.

Kinner, D., Mitasova, H., and Stallard, R. 2005. GIS Database and Stream Network Analysis for the Upper Río Chagres Basin, Panama In *The Río Chagres: A Multidisciplinary Perspective of a Tropical River Basin* (ed. Harmon, R.), pp. 83–95. New York: Kluwer Academic/Plenum Publishers.

Krcho, J. 1973. Morphometric analysis of relief on the basis of geometric aspect of field theory. *Acta Geographica Universitae Comenianae* Geographica Physica 1, Bratislava, SPN.

Krcho, J. 1991. Georelief as a subsystem of landscape and the influence of morphometric parameters of georelief on spatial differentiation of landscape-ecological processes. *Ecology /CSFR/* **10**: 115–157.

Longley, P., Goodchild, M., Maguire, D. J., and Rhind, D. 2005. *Geographical Information Systems: Principles, Techniques, Management and Applications*. London: Wiley.

Maling, D. 1992. *Coordinate Systems and Map Projections*. Elmsford, New York: Pergamon Press, 2nd edition.

Mandelbrot, B. 1983. *The Fractal Geometry of Nature*. New York: Freeman.

Mather, P. 1999. *Computer Processing of Remotely-sensed Images*. Chichester: Wiley.

McCauley, J. and Engel, B. 1995. Comparison of Scene Segmentations: SMAP, ECHO and Maximum Likelihood. *IEEE Trans. on Geosc. & Rem. Sens.* **33** (6): 1313–1316.

Mitas, L., Brown, W., and Mitasova, H. 1997. Role of dynamic cartography in simulations of landscape processes based on multi-variate fields. *Comp. & Geosc.* **23**: 437–446. Electr. Doc. `http://skagit.meas.ncsu.edu/~helena/gmslab/lcgfin/cg-mitas.html`.

Mitas, L. and Mitasova, H. 1998. Distributed soil erosion simulation for effective erosion prevention. *Water Resources Research* **34** (3): 505–516.

Mitas, L. and Mitasova, H. 1999. Spatial interpolation. In *Geographical Information Systems: Principles, Techniques, Management and Applications*. (eds. Longley, P., Goodchild, M., Maguire, D., and Rhind, D.), pp. 481–492. New York: Wiley.

Mitasova, H. and Hofierka, J. 1993. Interpolation by Regularized Spline with Tension: II. Application to Terrain Modeling and Surface Geometry Analysis. *Math. Geol.* **25**: 657–667. Electr. Doc. `http://skagit.meas.ncsu.edu/~helena/gmslab/papers/listsj.html`.

Mitasova, H., Hofierka, J., Zlocha, M., and Iverson, L. 1996. Modeling topographic potential for erosion and deposition using GIS. *Int. J. of Geogr. Inf. Sci.* **10** (5): 629–641.

Mitasova, H. and Mitas, L. 1993. Interpolation by Regularized Spline with Tension: I. Theory and Implementation. *Math. Geol.* **25**: 641–655. Electr. Doc. `http://skagit.meas.ncsu.edu/~helena/gmslab/papers/listsj.html`.

Mitasova, H. and Mitas, L. 2001. Multiscale soil erosion simulations for land use management. In *Landscape erosion and landscape evolution modeling*. (eds. Harmon, R. and Doe, W.), pp. 321–347. New York: Kluwer Academic/Plenum Publishers.

Mitasova, H., Mitas, L., Brown, W., Gerdes, D., Kosinovsky, I., and Baker, T. 1995. Modeling spatially and temporally distributed phenomena: New methods and tools for GRASS GIS. *Int. J. of GIS, Special issue on Integrating GIS and Environmental modeling* **9** (4): 433–446.

Mitasova, H., Mitas, L., and Harmon, R. 2005a. Simultaneous Spline Interpolation and Topographic Analysis for LiDAR Elevation Data: Methods for Open Source GIS. *IEEE GRSL* **2** (4): 375–379. Electr. Doc. `http://skagit.meas.ncsu.edu/~helena/gmslab/papers/listsj.html`.

Mitasova, H., Overton, M., and Harmon, R. 2005b. Geospatial analysis of a coastal sand dune field evolution: Jockey's Ridge, North Carolina. *Geomorphology* **72**: 204–221. Electr. Doc. http://skagit.meas.ncsu.edu/~helena/gmslab/papers/listsj.html.

Mitasova, H., Thaxton, C., Hofierka, J., McLaughlin, R., Moore, A., and Mitas, L. 2005c. Path sampling method for modeling overland water flow, sediment transport, and short term terrain evolution in Open Source GIS. In *Proc., XVth International Conf. on Computational Methods in Water Resources (CMWR XV)* (eds. Miller, C., Farthing, M., Gray, V., and Pinder, G.), pp. 1479–1490. Elsevier.

Mitchell, T. 2005. *Web Mapping Illustrated*. Cambridge: O'Reilly & Associates.

Moore, I. and Burch, G. 1986. Modeling erosion and deposition: Topographic effects. *Transactions ASAE* **29**: 1624–1640.

Moore, I. and Wilson, J. 1992. Length-slope factors for the revised universal soil loss equation: Simplified method of estimation. *Journal of Soil and Water Conservation* **47**: 423–428.

Moran, M., Jackson, R., Slater, P., and Teillet, P. 1992. Evaluation of simplified procedures for retrieval of land surface reflectance factors from satellite sensor output. *Rem. Sens. Env.* **41**: 169–184.

Neteler, M. 1999. Spectral Mixture Analysis von Satellitendaten zur Bestimmung von Bodenbedeckungsgraden im Hinblick auf die Erosionsmodellierung. M.Sc. thesis, Univ. of Hannover, Germany.

Neteler, M. 2000. *GRASS-Handbuch*. Geosynthesis 11. Univ. of Hannover. Der praktische Leitfaden zum Geographischen Informationssystem GRASS.

Neteler, M. 2001. Towards a stable Open Source GIS: Status and Future Directions in GRASS Development. In *The Geomatics Workbook* (ed. Brovelli, M.). Polytec. di Milano, Italy, 2nd edition. Electr. Doc. http://geomatica.ing.unico.it/workbooks2/.

Neteler, M. 2005. Time series processing of MODIS satellite data for landscape epidemiological applications. *International Journal of Geoinformatics* **1** (1): 133–138.

Oliver, C. and Quegan, S. 1998. *Understanding Synthetic Aperture Radar Image*. London: Artech House.

O'Rourke, J. 1998. *Computational Geometry in C*. Cambridge: Cambridge University Press, 2nd edition.

Page, J., Albuisson, M., and Wald, L. 2001. The European Solar Radiation Atlas: A Valuable Digital Tool. *Solar energy* **71**: 81–83.

Pebesma, E. 2001. Gstat user's manual. Technical report, Dept. of Physical Geography, Faculty of Geosciences, Utrecht University. Gstat 2.3.3., Electr. Doc. http://www.gstat.org.

Pebesma, E. J. and Wesseling, C. G. 1998. Gstat: a Program for Geostatistical Modelling, Prediction and Simulation. *Comp. & Geosc.* **24** (1): 17–31.

Poeter, E., Hill, M., Banta, E., Mehl, S., and Christensen, S. 2005. UCODE_2005 and six other computer codes for Universal Sensitivity Analysis, calibration, and uncertainty evaluation. Technical report, USGS.

Pohl, C. and van Genderen, J. 1998. Multisensor image fusion in remote sensing: Concepts, methods and application. *Int. J. of Rem. Sens.* **19**: 823–854.

Powell, M. 1992. Tabulation of thin plate splines on a very fine two-dimensional grid. In *Numerical Methods of Approximation Theory* (eds. Braess, D. and Schumaker, L.), pp. 221–244. Univ. of Cambridge, U.K.

Ramsey, P. 2006. The State of Open Source GIS. Technical report, Refractions Research Inc. Electr. Doc. http://www.refractions.net/white_papers/oss_briefing/.

Rase, W. 1998. Visualisierung von Planungsinformationen: Modellierung und Darstellung immaterieller Oberflächen. Technical report, Bundesamt für Bauwesen und Raumordnung, Forschungen H. 89, Bonn.

Raymond, E. 1987. The cathedral and the bazaar. Technical report. Electr. Doc. http://www.ccil.org/~esr/writings/cathedral-paper.html.

Raymond, E. 1999. *The Cathedral and the Bazaar. Musings on Linux and Open Source by an accidental revolutionary.* Cambridge: O'Reilly & Associates.

Redslob, M. 1998. *Radarfernerkundung in Niedersächsischen Hochmooren.* Ph.D. thesis, Inst. f. Landschaftspfl. u. Natursch., Univ. of Hannover.

Rektorys, K. 1969. *Survey of Applicable Mathematics.* Cambridge, MA: MIT Press and London: Iliffe Books Ltd.

Richards, J. and Xiuping, J. 1999. *Remote Sensing Digital Image Analysis: An Introduction.* Heidelberg: Springer, 3rd edition.

Rigollier, C., Bauer, O., and Wald, L. 2000. On the clear sky model of the ESRA – European Solar Radiation Atlas – with respect to the Heliosat method. *Solar energy* **68**: 33–48.

Rigon, R., Ghesla, E., Tiso, C., and Cozzini, A. 2006. *The Horton machine: a system for DEM analysis.* Università degli studi di Trento. Dipartimento di Ingegneria Civile e Ambientale, Trento.

Ripley, B. 1996. *Pattern Recognition and Neural Networks.* Cambridge: Cambridge University Press.

Rizzoli, A., Neteler, M., Rosà, R., Versini, W., Cristofolini, A., Bregoli, M., Buckley, A., and Gould, E. 2007. Early detection of TBEv spatial distribution and activity in the Province of Trento assessed using serological and remotely-sensed climatic data. *Geospatial Health* **1** (2): 169–176.

Robinson, A., Morrison, J., Muehricke, P., Kimerling, A., and Guptill, S. 1995. *Elements of Cartography.* New York: Wiley, 6th edition.

Scharmer, K. and Greif, J. 2000. *The European Solar Radiation Atlas.* Paris: Les Presses de l'Ecole des Mines. Vol. 2: Database and exploitation software.

Schowengerdt, R. 1997. *Remote Sensing: Models And Methods for Image Processing.* San Diego: Academic Press, 2nd edition.

Shapiro, M. and Westervelt, J. 1992. r.mapcalc. An algebra for GIS and image processing. Technical report, US-Army CERL, Champaign, Illinois. Electr. Doc. http://grass.itc.it/gdp/.

Snyder, J. 1987. Map projections, a working manual. Technical report, U.S. Geological Survey Professional Paper 1395. Department of the Interior. Washington, D.C.

Súri, M. and Hofierka, J. 2004. A new GIS-based Solar Radiation Model and its application to photovoltaic assessments. *Transactions in GIS* **8** (2): 175–190.

Talmi, A. and Gilat, G. 1977. Method for smooth approximation of data. *J. of Computational Physics* **23**: 93–123.

Ullah, I., Barton, M., and Mitasova, H. 2007. Mediterranean Landscape Dynamics Project. Technical report, Arizona State University.

U.S. Army CERL 1993. GRASS 4.1 Reference Manual. Technical report, U.S. Army Corps of Engineers, Construction Engineering Research Laboratories, Champaign, Illinois.

Venables, W. and Ripley, B. 2000. *S Programming*. New York: Springer.

Venables, W. and Ripley, B. 2002. *Modern Applied Statistics with S*. New York: Springer, 4th edition. Supplements: http://www.stats.ox.ac.uk/pub/MASS4/.

Vermote, E., Tanré, D., Deuzé, J., Herman, M., and Morcrette, J. 1997. Second Simulation of the Satellite Signal in the Solar Spectrum, 6S: An Overview. *IEEE Trans. Geosc. Rem. Sens.* **35** (3): 675–686.

Wahba, G. 1990. Spline models for observational data. Technical report, CNMS-NSF Regional Conference series in applied mathematics, 59, SIAM, Philadelphia, Pennsylvania.

Watson, D. 1992. *Contouring: A Guide to the Analysis and Display of Spatial Data*. Oxford: Pergamon.

Webster, R. and Oliver, M. 2001. *Geostatistics for Environmental Scientists*. Chichester: Wiley.

Wheeler, D. 2003. Why open source: Look at the numbers. Technical report. Electr. Doc. http://www.dwheeler.com/oss_fs_why.html.

Wood, J. 1996. *The Geomorphological Characterisation of Digital Elevation Models*. Ph.D. thesis, Dep. of Geogr., Univ. of Leicester, U.K. Electr. Doc. http://www.geog.le.ac.uk/jwo/research/dem_char/thesis/.

Zhou, J., Civico, D., and Silander, J. 1998. A wavelet transform method to merge LANDSAT-TM and SPOT panchromatic data. *Int. J. of Rem. Sens.* **19** (4): 743–757.

Index

Made in the USA
Lexington, KY
13 July 2011